SIMPLIFIED BUILDING DESIGN FOR WIND AND EARTHQUAKE FORCES

Other titles in the
PARKER–AMBROSE SERIES OF SIMPLIFIED DESIGN GUIDES

Harry Parker and James Ambrose
Simplified Design of Concrete Structures, 7th Edition

Harry Parker, John W. MacGuire and James Ambrose
Simplified Site Engineering, 2nd Edition

James Ambrose
Simplified Design of Building Foundations, 2nd Edition

Harry Parker and James Ambrose
Simplified Design of Steel Structures, 6th Edition

James Ambrose
Simplified Design of Masonry Structures

James Ambrose and Peter D. Brandow
Simplified Site Design

Harry Parker and James Ambrose
Simplified Mechanics and Strength of Materials, 5th Edition

Marc Schiler
Simplified Design of Building Lighting

James Patterson
Simplified Design for Building Fire Safety

James Ambrose
Simplified Engineering for Architects and Builders, 8th Edition

James Ambrose
Simplified Design of Wood Structures, 5th Edition

William Bobenhausen
Simplified Design of HVAC Systems

James Ambrose
Simplified Design of Building Structures, 3rd Edition

James Ambrose and Jeffrey E. Ollswang
Simplified Design for Building Sound Control

SIMPLIFIED BUILDING DESIGN FOR WIND AND EARTHQUAKE FORCES

Third Edition

JAMES AMBROSE

**Formerly Professor of Architecture
University of Southern California
Los Angeles, California**

DIMITRY VERGUN

**University of Southern California
Los Angeles, California**

JOHN WILEY & SONS, INC.

New York · Chichester · Weinheim · Brisbane · Singapore · Toronto

Copyright © 1995 by John Wiley & Sons, Inc.

Library of Congress Cataloging in Publication Data:

Ambrose, James E.
    Simplified building design for wind and earthquake forces / James
Ambrose, Dimitry Vergun. — 3rd ed.
        p. cm. — (Parker/Ambrose series of simplified design guides)
    "A Wiley-Interscience Publication."
    Includes bibliographical references and index.
    ISBN 0-471-30958-3 (cloth)
    ISBN 0-471-19211-2 (paper)
    1. Wind resistant design.   2. Earthquake resistant design.
I. Vergun, Dimitry, 1933–   . II. Title. III. Series.
TA658.48.A48   1995
693′.852—dc20   94-30569

Printed in the United States of America

10 9 8 7 6 5 4 3 2 1

CONTENTS

v

PREFACE

This book is prepared for use by persons desiring comprehensive treatment of the practical considerations of design for the effects of wind and earthquakes on buildings. It is developed for a general readership that is not highly trained or experienced in engineering work, including most architects, builders, landscape architects, and others who are seriously involved in the work of building design and construction. While written to serve as a basic text or self-study resource, the book also presents a lengthy explanation of practical design and construction for working professionals.

This is the third outing for this material since the publication of the first edition in 1980. If anything, the need and urgency for attention to these topics is greater now. Keeping up with changes in codes, available technology, and design and construction practices is an enduring task. This book retains the fundamental treatments of the earlier editions and presents much up-dated material and some slight expansion of the coverage of the topics.

Quite late in the development of the manuscript for this book an event occurred that had a major effect on the work. This was the Northridge earthquake of January 17, 1994, in the northern portion of the Los Angeles area. Being residents of the area, both authors were quite profoundly affected—both personally and professionally—by this earthquake and its disastrous effects. Although some lessons from this major seismic event were learned quickly from observations, the major impact on current design practices will be slower in developing. The work in this book is based primarily on the current edition of the *Uniform Building Code*, published in the late spring of 1994. Although that code edition followed the January earthquake, it was not possible to respond to the lessons from the event, but surely the next edition (1997?) will do so. This is, of course, the nature of the hopscotch game of keeping up with the latest codes and evolving practices.

We are grateful to many people and organizations for support of the work for this book. As usual, the publishers of the *Uniform Building Code*, the International Conference of Building Officials, has granted permission to reproduce extensive materials from their latest publication, the 1994 edition. Sources of other reproduced materials are cited with the materials in the book. A major source of refinement for this book is the continuous feedback (friendly and otherwise) from our students, design clients, academic and professional colleagues, and others who trouble themselves to give us their comments.

It must be acknowledged that this work would not exist without the support and major effort put forth by the many people at John Wiley & Sons, our publishers. For this project we wish to thank in particular our publisher, Peggy Burns; our editors, Everett Smethurst and Amanda Miller; and the hard-working people of the Wiley production department, especially Robert J. Fletcher IV and Milagros Torres.

As always, we are also grateful to our families, friends, and co-workers for their support and their tolerance of our occasional absorption with this task.

JAMES AMBROSE
DIMITRY VERGUN

Los Angeles, California

INTRODUCTION

The purpose of this book is to provide a source of study and reference for the topics of wind and earthquake effects as they pertain to the design of building structures. The treatment of these subjects is aimed at persons not trained in structural engineering but who have some background experience in the analysis and design of simple structures. Material presented includes the development of background topics, such as basic aspects of wind and earthquake effects and fundamentals of dynamic behavior, as well as the pragmatic considerations of design of structures for real situations.

As implied by the title, the scope of the work is limited. This limitation is manifested in the level of complexity of the problems dealt with and in the techniques used, principally with regard to the degree of difficulty in mathematical analysis and the sophistication of design methods. In order to set these limits we have assumed some specific minimal preparation by the reader, and individual readers should orient themselves with regard to these assumptions. For those with some lack of preparation, the list of references in the Bibliography may be useful for supplementary study. For the reader with a higher capability in mathematics or a more intensive background in applied mechanics and structural analysis, this work may serve as a springboard to more rigorous study of the topics.

The majority of the mathematical work in the applied design examples is limited to relatively simple algebra and geometry. In the treatment of the fundamentals of dynamics and in the explanation of some of the formulas used in analysis and design it is occasionally necessary to use relationships from trigonometry, vector analysis, and calculus. The reader with this level of mathematical background will more fully appreciate the rational basis for the formulas, although their practical application will usually involve only simple algebra and arithmetic. Persons expecting to pursue the study of these topics beyond the scope of this book are advised to prepare themselves with work in mathematics that proceeds to the level of advanced calculus, partial differential equations, and matrix methods of analysis.

A minimal preparation in the topics of applied mechanics and structural analysis and design is assumed. This includes the topics of statics, elementary strength of materials, and the design of simple elements of wood, steel, and concrete structures for buildings. When some of the examples involve the analysis of indeterminate structures, the work presented is done with simplified, approximate methods that should be reasonably well understood by the reader with minimal background. For a more rigorous and exact analysis of such problems, or for the study of more complex problems, the reader is advised to pursue a general study of the analysis of indeterminate structures.

A third area of assumed background knowledge is that of the ordinary materials and methods of building construction as practiced in the United States. It is assumed that the reader has a general familiarity with the ordinary processes of building construction and with the codes, standards, and sources of general data for structures of wood, steel, masonry, and concrete.

A major reference used for this work is the 1994 edition of the *Uniform Building Code* (Ref. 1), hereinafter called the *UBC*. The design examples in this book use the general requirements, the analytical procedures, and some of the specific data from this reference. Much of the material from the *UBC* that relates directly to problems of wind and earthquakes is reprinted in Appendix B of this book. It is recommended, however, that the reader have a copy of the entire code available because it contains considerable additional material pertinent to the use of specific materials, to structural design requirements in general, and to various problems of building planning and construction.

In real design situations individual buildings generally fall under the jurisdiction of a particular local code. Most large cities, many counties, and some states have their own individual codes. In many cases these codes are based primarily on one of the so-called "model" codes, such as the *UBC*, with some adjustments and additions for specific local conditions and practices. The reader who expects to work in a particular area is advised to obtain a copy of the code with jurisdiction in that area and to compare its provisions with those of the *UBC* as they are used in this work.

Building codes, including the *UBC*, are occasionally updated to keep them abreast of current developments in research, building practices, analytical and design techniques, and so on. The publishers of the *UBC* have generally followed a practice of issuing a new edition every three years. For reference in any real design work the reader is advised to be sure that the code he is using is the one with proper jurisdiction and is the edition currently in force. This precaution regarding use of dated materials applies also to other reference sources, such as handbooks, industry brochures, detailing manuals, and so on.

Use of the word *simplified* does not mean to imply that all design for wind and earthquakes can be reduced to simple methods. On the contrary, many problems in this area represent highly complex, and as yet far from fully understood, situations in structural design, situations that demand considerable seriousness, competency, and effort by professional engineers and researchers. We have deliberately limited the material in this book to that which we believe can be relatively easily understood and mastered by persons in the beginning stages of study of the design of structures. For those whose work will be limited to the relatively simple situations presented by the examples in this book, mastery of this material will provide useful working skills. For those who expect to continue their studies into more advanced levels of analysis and design, this material will provide a useful introduction.

Computations

In professional design firms, structural computations are now commonly done with computers, particularly when the work is complex or repetitive. Anyone aspiring to participation in professional design work is advised to acquire the background and experience necessary to the application of computer-aided techniques. The computational work in this book is simple and can be performed easily with a pocket calculator. The reader who has not already done so is advised to obtain one. The "scientific" type with eight-digit capacity is quite sufficient.

For the most part, structural computations can be rounded off. Accuracy beyond the third place is seldom significant, and this is the level used in this work. In some examples more accuracy is carried in early stages of the computation to ensure the desired degree in the final answer. Most of the work in this book, however, was performed on an eight-digit pocket calculator.

Symbols

The following "shorthand" symbols are frequently used:

Symbol	Reading
$>$	is greater than
$<$	is less than
$\geqslant$	greater than or equal to
$\leqslant$	less than or equal to
$6'$	six feet
$6''$	six inches
Σ	the sum of
ΔL	change in L

Notation

Use of standard notation in the general development of work in mechanics and strength of materials is complicated by the fact that there is some lack of consistency in the notation currently used in the field of structural design. Some of the standards used in the field are developed by individual groups (notably those relating to a single basic material, wood, steel, concrete, masonry, etc.) which each have their own particular notation. Thus the same type of stress (e.g., shear stress in a beam) or the same symbol (f_c) may have various representations in structural computations. To keep some form of consistency in this book, we use the following notation, most of which is in general agreement with that used in structural design work at present.

a	(1) Moment arm; (2) acceleration; (3) increment of an area
A	Gross (total) area of a surface or a cross section
b	Width of a beam cross section
B	Bending coefficient
c	Distance from neutral axis to edge of a beam cross section
d	Depth of a beam cross section or overall depth (height) of a truss
D	(1) Diameter; (2) deflection
e	(1) Eccentricity (dimension of the mislocation of a load resultant from the neutral axis, centroid, or simple center of the loaded object); (2) elongation

E   Modulus of elasticity (ratio of unit stress to the accompanying unit strain)

f   Computed unit stress

F   (1) Force; (2) allowable unit stress

g   Acceleration due to gravity

G   Shear modulus of elasticity

h   Height

H   Horizontal component of a force

I   Moment of inertia (second moment of an area about an axis in the plane of the area)

J   Torsional (polar) moment of inertia

K   Effective length factor for slenderness (of a column: *KL/r*)

M   Moment

n   Modular ratio (of the moduli of elasticity of two different materials)

N   Number of

p   (1) Percent; (2) unit pressure

P   Concentrated load (force at a point)

r   Radius of gyration of a cross section

R   Radius (of a circle, etc.)

s   (1) Center-to-center spacing of a set of objects: (2) distance of travel (displacement) of a moving object; (3) strain or unit deformation

t   (1) Thickness; (2) time

T   (1) Temperature; (2) torsional moment; (3) fundamental period of vibration of building

V   (1) Gross (total) shear force; (2) vertical component of a force

w   (1) Width; (2) unit of a uniformly distributed load on a beam

W   (1) Gross (total) value of a uniformly distributed load on a beam; (2) gross (total) weight of an object

Δ (delta)   Change of

θ (theta)   Angle

μ (mu)   Coefficient of friction

Σ (sigma)   Sum of

ϕ (phi)   Angle

Special notation is used in every field of engineering. The special notation used in the areas of wind and seismic design is generally reflected in the building codes. We use here the notation from the *UBC* in these areas, which is explained in the chapters that deal with the topics of design for wind and seismic effects.

PART I

GENERAL CONSIDERATIONS

1

WIND EFFECTS ON BUILDINGS

Wind is moving air. The air has a particular mass (density or weight) and moves in a particular direction at a particular velocity. It thus has kinetic energy of the form expressed as

$$E = \tfrac{1}{2}mv^2$$

When the moving fluid air encounters a stationary object, there are several effects that combine to exert a force on the object. The nature of this force, the many variables that affect it, and the translation of the effects into criteria for structural design are dealt with in this chapter.

1.1 WIND CONDITIONS

The wind condition of concern for building design is primarily that of a wind storm, specifically high-velocity, ground-level winds. These winds are generally associated with one of the following situations.

Tornadoes

Tornadoes occur with some frequency in the midwest and occasionally in other parts of the United States. In coastal areas they are usually the result of ocean storms that wander ashore. Although the most violent effects are at the center of the storm, high-velocity winds in a large surrounding area often accompany these storms. In any given location the violent winds are usually short in duration as the tornado dissipates or passes through the area.

Hurricanes

Whereas tornadoes tend to be relatively short-lived (a few hours at most), hurricanes can sustain storm wind conditions for several days. Hurricanes occur with some frequency in the Atlantic and Gulf coastal areas of the United States. Although they originate and develop their greatest fury over the water, they often stray ashore and can move some distance inland before dissipating. As with tornadoes, the winds of highest velocity occur at the eye of the hurricane, but major winds can develop in large surrounding areas, often affecting coastal areas some distance inland even when the hurricane stays at sea.

Local Peculiar Wind Conditions

An example of wind conditions peculiar to one locality are the Santa Ana winds of southern California. These winds are recurrent conditions caused by the peculiar geographic and climatological conditions of an area. They can sometimes result in local wind velocities of the level of those at the periphery of tornadoes and hurricanes and can be sustained for long periods.

Sustained Local Wind Conditions

Winds that occur at great elevations above sea level are an example of sustained local wind conditions. Such winds may possibly never reach the extremes of velocity of storm conditions, but they can require special consideration because of their enduring nature.

Local and regional meteorological histories are used to predict the degree of concern for or likelihood of critical wind conditions in a particular location. Building codes establish minimum design requirements for wind based on this experience and the statistical likelihood it implies. The map in the *UBC* Fig. 16-1 (Appendix B) shows the variation of critical wind conditions in the United States.

Of primary concern in wind evaluation is the maximum velocity that is achieved by the wind. Maximum velocity usually refers to a sustained velocity and not to gust effects. A gust is essentially a pocket of higher-velocity wind within the general moving fluid air mass. The resulting effect of a gust is that of a brief increase, or surge, in the wind velocity, usually of not more than 15% of the sustained velocity and for only a fraction of a second in duration. Because of both its higher velocity and its slamming effect, the gust actually represents the most critical effect of the wind in most cases.

Winds are measured regularly at a large number of locations. The standard measurement is at 10 meters (approximately 33 ft) above the surrounding terrain, which provides a fixed reference with regard to the drag effects of the ground surface. The graph in Fig. 1.1 shows the correlation between wind velocity and various wind conditions. The curve on the graph is a plot of a general equation used to relate wind velocity to equivalent static pressure on buildings, as discussed in Sec. 1.3.

Although wind conditions are usually generalized for a given geographic area, they can vary considerably for specific sites because of the nature of the surrounding terrain, of landscaping, or of nearby structures. Each individual building design should consider the possibilities of these localized site conditions.

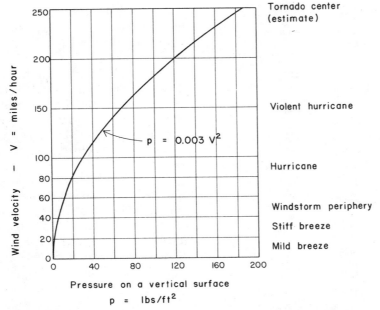

FIGURE 1.1 Relation of wind velocity to pressure on a stationary object.

1.2 GENERAL WIND EFFECTS

The effects of wind on stationary objects in its path can be generalized as in the following discussions (see Fig. 1.2).

Direct Positive Pressure

Surfaces facing the wind and perpendicular to its path receive a direct impact effect from the moving mass of air, which generally produces the major portion of force on the object unless it is highly streamlined in form.

Aerodynamic Drag

Because the wind does not stop upon striking the object but flows around it like a liquid, there is a drag effect on surfaces that are parallel to the direction of the wind. These surfaces may also have inward or outward pressures exerted on them, but it is the drag effect that adds to the general force on the object in the direction of the wind path.

Negative Pressure

On the leeward side of the object (opposite from the wind direction) there is usually a suction effect, consisting of pressure outward on the surface of the object. By comparison to the direction of pressure on the windward side, this is called *negative pressure*.

These three effects combine to produce a net force on the object in the direction of the wind that tends to move the object along with the wind. In addition to these there are other possible effects on the object that can occur due to the turbulence of the air or to the nature of the object. Some of them are as follows.

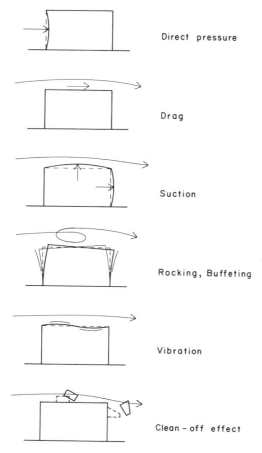

FIGURE 1.2 General effects of wind.

Rocking Effects. During wind storms, the wind velocity and its direction are seldom constant. Gusts and swirling winds are ordinary, so that an object in the wind path tends to be buffeted, rocked, flapped, and so on. Objects with loose parts, or with connections having some slack, or with highly flexible surfaces (such as fabric surfaces that are not taut) are most susceptible to these effects.

Harmonic Effects. Wind can produce vibration, whistling, flutter, and so on. These effects can occur at low velocities as well as with wind storm conditions. This is a matter of some match between the velocity of the wind and the natural period of vibration of the object or of its parts.

Clean-Off Effect. The friction effect of the flowing air mass tends to smooth off the objects in its path. This fact is of particular concern to objects that protrude from the general mass of the building, such as canopies, parapets, chimneys, and signs.

The critical condition of individual parts or surfaces of an object may be caused by any one, or some combination, of the above effects. Damage can occur locally or be total with regard to the object. If the object is resting on the ground, it may be collapsed or may be slid, rolled over, or lifted from its position. Various aspects of the wind, of the object in the path of the wind, or of the surrounding environment determine the critical wind effects. With regard to the wind itself some considerations are the following:

The magnitude of sustained velocities

The duration of high-level velocities

The presence of gust effects, swirling, and so on

The prevailing direction of the wind (if any)

With regard to objects in the path of the wind some considerations are the following:

The size of the object (relates to the relative effect of gusts, to variations of pressure above ground level, etc.)

The aerodynamic shape of the object (determines the critical nature of drag, suction, uplift, etc.)

The fundamental period of vibration of the object or of its parts

The relative stiffness of surfaces, tightness of connections, and so on

With regard to the environment, possible effects may result from the sheltering or funneling caused by ground forms, landscaping, or adjacent structures. These effects may result in an increase or reduction of the general wind effects or in turbulence to produce a very unsteady wind condition.

The actual behavior of an object during wind storm conditions can be found only by subjecting it to a real wind situation. Wind tunnel tests in the laboratory are also useful, and because we can create the tests more practically on demand, they have provided much of the background for data and procedures used in design.

1.3 CRITICAL WIND EFFECTS ON BUILDINGS

The major effects of wind on buildings can be generalized to some degree because we know a bracketed range of characteristics that cover the most common conditions. Some of the general assumptions made are as follows:

Most buildings are boxy or bulky in shape, resulting in typical aerodynamic response.

Most buildings present closed, fairly smooth surfaces to the wind.

Most buildings are fit snugly to the ground, presenting a particular situation for the drag effects of the ground surface.

Most buildings have relatively stiff structures, resulting in a fairly limited range of variation of the natural period of vibration of the structure.

These and other considerations allow for the simplification of wind investigation by permitting a number of variables to be eliminated or to be lumped into a few modifying constants. For unusual situations, such as elevated buildings, open structures, highly flexible structures, and unusual aerodynamic shapes, it may be advisable to do more thorough investigation, including the possible use of wind tunnel tests.

The primary effect of wind is visualized in the form of pressures normal to the building's exterior surfaces. The basis for this pressure begins with a conversion of the kinetic energy of the moving air mass into an equivalent static pressure using the basic formula

$$p = Cv^2$$

in which C is a constant accounting for the air mass, the units used, and a number of the assumptions previously described. With the wind in miles per hour (mph) and the pressure in pounds per square foot (psf), the C value for the total wind effect on a simple box-shaped building is approximately 0.003, which is the value used in deriving the graph in Fig. 1.1. It should be noted that this pressure does not represent the actual effect on a single building surface, but rather the *entire* effect of all surface pressures visualized as a single pressure on the windward side of the building.

Building codes provide data for establishing the critical wind velocity and for determining the design wind pressures for the investigation of wind effects on a particular building. Considerations involve the variables of the building size, shape, and degree of openness, of the sheltering effect of the surrounding terrain, and numerous concerns for special situations. A discussion of code criteria is given in Sec. 1.4.

The general effects of wind on stationary objects were described in Sec. 1.2. These effects are translated into building design criteria as explained in the following discussions.

Inward Pressure on Exterior Walls

Surfaces directly facing the wind are usually required to be designed for the full base pressure, although this is somewhat conservative, because the windward force usually accounts for only about 60% of the total force on the building. Designing for only part of the total force is, however, partly compensated for by the fact that the base pressures are not generally related to gust effects which tend to have less effect on the building as a whole and more effect on parts of the building.

Suction on Exterior Walls

Most codes also require suction on exterior walls to be the full base pressure, although the preceding comments about inward pressure apply here as well.

Pressure on Roof Surfaces

Depending on their actual form, as well as that of the building as a whole, nonvertical surfaces may be subjected to either inward or suction pressures because of wind. Actually, such surfaces may experience both types of pressure as the wind shifts direction. Most codes require an uplift (suction) pressure equal to the full design pressure at the elevation of the roof level. Inward pressure is usually related to the actual angle of the surface as an inclination from the horizontal.

Overall Force on the Building

Overall horizontal force is calculated as a horizontal pressure on the building silhouette, as described previously, with adjustments made for height above the ground. The lateral resistive structural system of the building is designed for this force.

Horizontal Sliding of the Building

In addition to the possible collapse of the lateral resistive system, there is the chance that the total horizontal force may slide the building off its foundations. For a tall building with fairly shallow foundations, this may also be a problem for the force transfer between the foundation and the ground. In both cases, the dead weight of the building generates a friction that helps to resist this force.

Overturn Effect

As with horizontal sliding, the dead weight tends to resist the overturn, or toppling, effect. In practice, the overturn effect is usually analyzed in terms of the overturn of individual vertical elements of the lateral resistive system rather than for the building as a whole.

Wind on Building Parts

The clean-off effect discussed previously is critical for elements that project from the general mass of the building. In some cases codes require for such elements a design pressure higher than the base pressure, so that gust effects as well as the clean-off problem are allowed for in the design.

Harmonic Effects

Design for vibration, flutter, whipping, multinodal swaying, and so on, requires a dynamic analysis and cannot be accounted for when using the equivalent static load method. Stiffening, bracing, and tightening of elements in general may minimize the possibilities for such effects, but only a true dynamic analysis or a wind tunnel test can assure the adequacy of the structure to withstand these harmonic effects.

Effect of Openings

If the surface of a building is closed and reasonably smooth, the wind will slip around it in a fluid flow. Openings or building forms that tend to cup the wind can greatly affect the total wind force on the building. It is difficult to account for these effects in a mathematical analysis, except in a very empirical manner. Cupping of the wind can be a major effect when the entire side of a building is open, for example. Garages, hangars, band shells, and other buildings of similar form must be designed for an increased force that can only be estimated unless a wind tunnel test is performed.

Torsional Effect

If a building is not symmetrical in terms of its wind silhouette, or if the lateral resistive system is not symmetrical within the building, the wind force may produce a twisting effect. This effect is the result of a misalignment of the centroid of the wind force and the centroid (called *center of stiffness*) of the lateral resistive system and will produce an added force on some of the elements of the structure.

Although there may be typical prevailing directions of wind in an area, the wind must be considered to be capable of blowing in any direction. Depending on the building shape and the arrangement of its structure, an analysis for wind from several possible directions may be required.

1.4 BUILDING CODE REQUIREMENTS FOR WIND

Model building codes such as the *UBC* (Ref. 1) are not legally binding unless they are adopted by ordinances by some state, country, or city. Although smaller communities usually adopt one of the model codes, states, counties, and cities with large populations usually develop their own codes using one of the model codes as a basic reference. In the continental United States the *UBC* is generally used in the west, the Southern Standard Building Code in the southeast, and the *BOCA Code* in the rest of the country.

Where wind is a major local problem, local codes are usually more extensive with regard to design requirements for wind. However, many codes still contain relatively simple criteria for wind design. One of the most up-to-date and complex standards for wind design is contained in the *Minimum Design Loads for Buildings and Other Structures*, published by the American Society of Civil Engineers (ASCE) (Ref. 2).

Complete design for wind effects on buildings includes a large number of both architectural and structural concerns. Of primary concern for the work in this book are those requirements that directly affect the design of the lateral bracing system. The following is a discussion of some of the requirements for wind as taken from the 1994 edition of the *UBC* (Ref. 1), which is in general conformance with the material presented in the ASCE standard just mentioned. Reprints of some of the materials referred to are given in Appendix B.

Basic Wind Speed

This is the maximum wind speed (or velocity) to be used for specific locations. It is based on recorded wind histories and adjusted for some statistical likelihood of occurrence. For the continental United States the wind speeds are taken from *UBC*, Fig. 16.1 (see Appendix B). As a reference point, the speeds are those recorded at the standard measuring position of 10 meters (approximately 33 ft) above the ground surface.

Exposure

This refers to the conditions of the terrain surrounding the building site. The ASCE standard (Ref. 2) describes four conditions (A, B, C, and D), although the *UBC* uses only three (B, C, and D). Condition C refers to sites surrounded for a distance of one half mile or more by flat, open terrain. Condition B has buildings, forests, or ground surface irregularities 20 ft or more in height covering at least 20% of the area for a distance of 1 mile or more around the site. Condition D refers to exceptionally open situations with flat terrain and a location at the shoreline of a large body of water.

Wind Stagnation Pressure (q_s)

This is the basic reference equivalent static pressure based on the critical local wind speed. It is given in *UBC* Table 16-F (see Appendix B) and is based on the following formula as given in the ASCE standard

$$q_s = 0.00256V^2$$

Example: For a wind speed of 100 mph,

$$q_s = 0.00256V^2 = (0.00256)(100)^2$$
$$= 25.6 \text{ psf}$$

which is rounded off to 26 psf in the *UBC* table.

Design Wind Pressure

This is the equivalent static pressure to be applied normal to the exterior surfaces of the building and is determined from the formula

$$p = C_e C_q q_s I$$
[*UBC* Formula (18-1), Sec. 1618]

where

p = design wind pressure, psf

C_e = combined height, exposure, and gust factor coefficient as given in *UBC* Table 16-G (see Appendix B)

C_q = pressure coefficient for the structure or portion of structure under consideration as given in *UBC* Table 16-H (see Appendix B)

q_s = wind stagnation pressure at 30 ft given in *UBC* Table 16-F (see Appendix B)

I = importance factor as given in *UBC* Table 16-L (see Appendix B)

The design wind pressure may be positive (inward) or negative (outward, suction) on any given surface. Both the sign and the value for the pressure are given in the *UBC* table. Individual building surfaces, or parts thereof, must be designed for these pressures.

Design Methods

Two methods are described in the code for the application of the design wind pressures in the design of structures. For design of individual elements particular values are given in *UBC* Table 16-H for the C_q coefficient to be used in determining p. For the primary bracing system the C_q values and their use is to be as follows:

1. *Normal Force Method.* In this method wind pressures are assumed to act simultaneously normal to all exterior surfaces. This method is required to be used for gabled rigid frames and may be used for any structure.

2. *Projected Area Method.* In this method the total wind effect on the

building is considered to be a combination of a single inward (positive) horizontal pressure acting on a vertical surface consisting of the projected building profile and an outward (negative, upward) pressure acting on the full projected area of the building in plan. This method may be used for any structure less than 200 ft in height, except for gabled rigid frames. This is the method generally employed by building codes in the past.

Uplift

Uplift may occur as a general effect, involving the entire roof or even the whole building. It may also occur as a local phenomenon such as that generated by the overturning moment on a single shear wall. In general, use of either design method will account for uplift concerns.

Overturning Moment

Most codes require that the ratio of the dead load resisting moment (called the restoring moment, stabilizing moment, etc.) to the overturning moment be 1.5 or greater. When this is not the case, uplift effects must be resisted by anchorage capable of developing the excess overturning moment. Overturning may be a critical problem for the whole building, as in the case of relatively tall and slender tower structures. For buildings braced by individual shear walls, trussed bents, and rigid frame bents, overturning is investigated for the individual bracing units. Method 2 is usually used for this investigation, except for very tall buildings and gabled rigid frames.

Drift

Drift refers to the horizontal deflection of the structure due to lateral loads. Code criteria for drift are usually limited to requirements for the drift of a single story (horizontal movement of one level with respect to the next above or below). The *UBC* does not provide limits for wind drift. Other standards give various recommendations, a common one being a limit of story drift to 0.005 times the story height. For masonry structures wind drift is sometimes limited to 0.0025 times the story height. As in other situations involving structural deformations, effects on the building construction must be considered; thus the detailing of curtain walls or interior partitions may affect limits on drift.

Combined Loads

Although wind effects are investigated as isolated phenomena, the actions of the structure must be considered simultaneously with other phenomena. The requirements for load combinations are given by most codes, although common sense will indicate the critical combinations in most cases. With the increasing use of load factors the combinations are further modified by applying different factors for the various types of loading, thus permitting individual control based on the reliability of data and investigation procedures and the relative significance to safety of the different load sources and effects. Required load combinations are described in Sec. 1603.6 of the *UBC*.

Special Problems

The general design criteria given in most codes are applicable to ordinary buildings. More thorough investigation is recommended (and sometimes required) for special circumstances such as the following:

Tall Buildings. These are critical with regard to their height dimension as well as the overall size and number of occupants inferred. Local wind speeds and unusual wind phenomena at upper elevations must be considered.

Flexible Structures. These may be affected in a variety of ways, including vibration or flutter as well as the simple magnitude of movements.

Unusual Shapes. Open structures, structures with large overhangs or other projections, and any building with a complex shape should be carefully studied for the special wind effects that may occur. Wind tunnel testing may be advised or even required by some codes.

Use of code criteria for various ordinary buildings is illustrated in the design examples in Part III.

1.5 GENERAL DESIGN CONSIDERATIONS FOR WIND

The relative importance of design for wind as an influence on the general building design varies greatly among buildings. The location of the building is a major consideration, the basic design pressure varying by a factor of 2.4 from the lowest wind speed area to the highest on the *UBC* map. Other important variations include the dead weight of the construction, the height of the building, the type of structural system (especially for lateral load resistance), the aerodynamic shape of the building and its exposed parts, and the existence of large openings, recessed portions of the surface, and so on.

The following is a discussion of some general considerations of design of buildings for wind effects. Any of these factors may be more or less critical in specific situations.

Influence of Dead Load

Dead load of the building construction is generally an advantage in wind design, because it is a stabilizing factor in resisting uplift, overturn, and sliding and tends to reduce the incidence of vibration and flutter. However, the stresses that result from various load combinations, all of which include dead load, may offset these gains when the dead load is excessive.

Anchorage for Uplift, Sliding, and Overturn

Ordinary connections between parts of the building may provide adequately for various transfers of wind force. In some cases, such as with lightweight elements, wind anchorage may be a major consideration. In most design cases the adequacy of ordinary construction details is considered first and extraordinary measures are used only when required. Various situations of anchorage are illustrated in the examples in Part III.

Critical Shape Considerations

Various aspects of the building form can cause increase or reduction in wind effects. Although it is seldom as critical in building design as it is for racing cars or aircraft, streamlining can improve the relative efficiency of the building in wind resistance. Some potential critical situations, as shown in Fig. 1.3, are as follows:

1. Flat versus curved forms. Buildings with rounded forms, rather than rectangular forms with flat surfaces, offer less wind resistance.

2. Tall buildings that are short in horizontal dimension are more critical for overturn and possibly for the total horizontal deflection at their tops.

3. Open-sided buildings or buildings with forms that cup the wind tend to catch the wind, resulting in more wind force than that assumed for the general design pressures. Open structures must also be investigated for

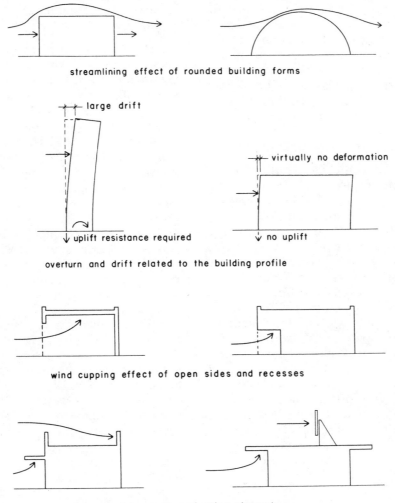

streamlining effect of rounded building forms

overturn and drift related to the building profile

wind cupping effect of open sides and recesses

increased force on projecting elements

FIGURE 1.3 Wind effects related to building form.

major outward force on internal surfaces.

4. Projections from the building. Tall parapets, solid railings, cantilevered balconies and canopies, wide overhangs, and free standing exterior walls catch considerable wind and add to the overall drag effect on the building. Signs, chimneys, antennae, penthouses, and equipment on the roof of a building are also critical for

the clean-off effect discussed previously.

Relative Stiffness of Structural Elements

In most buildings the lateral resistive structure consists of two basic elements: the horizontal distributing elements and the vertical cantilevered or braced frame elements. The manner in which the horizontal elements distribute forces and the manner in which the vertical elements share forces are critical considerations in wind analysis. The relative stiffness of individual elements is the major property that affects these relationships. The various situations that occur are discussed and illustrated in the examples in Part III.

Stiffness of Nonstructural Elements

When the vertical elements of the lateral resistive system are relatively flexible, as with rigid frames and wood shear walls that are short in plan length, there may be considerable lateral force transferred to nonstructural elements of the building construction. Wall finishes of masonry veneer, plaster, or drywall can produce relatively rigid planes whose stiffnesses exceed those of the structures over which they are placed. If this is the case, the finish material may take the load initially, with the structure going to work only when the finish fails. This result is not entirely a matter of relative stiffness, however, because the load propagation through the building also depends on the attachments between elements of the construction. This problem should be considered carefully when developing the details of the building construction.

Allowance for Movement of the Structure

All structures deform when loaded. The actual dimension of movement may be insignificant, as in the case of a concrete shear wall, or it may be considerable, as in the case of a slender steel rigid frame. The effect of these movements on other elements of the building construction must be considered. The case of transfer of load to nonstructural finish elements, as just discussed, is one example of this problem. Another critical example is that of windows and doors. Glazing must be installed so as to allow for some movement of the glass with respect to the frame. The frame must be installed so as to allow for some movement of the structure of the building without load being transferred to the window frame.

All these considerations should be kept in mind in developing the general design of the building. If the building form and detail are determined and the choice of materials made before any thought is given to structural problems, it is not likely that an intelligent design will result. This is not to suggest that structural concerns are the most important concerns in building design but merely that they should not be relegated to afterthoughts.

2

EARTHQUAKE EFFECTS
ON BUILDINGS

Earthquakes are essentially vibrations of the earth's crust caused by subterranean ground faults. They occur several times a day in various parts of the world, although only a few each year are of sufficient magnitude to cause significant damage to buildings. Major earthquakes occur most frequently in particular areas of the earth's surface that are called *zones of high probability*. However, it is theoretically possible to have a major earthquake anywhere on the earth at some time.

During an earthquake the ground surface moves in all directions. The most damaging effects on structures are generally the movements in a direction parallel to the ground surface (that is, horizontally) because of the fact that structures are routinely designed for vertical gravity loads. Thus, for design purposes the major effect of an earthquake is usually considered in terms of horizontal force similar to the effect of wind.

A general study of earthquakes includes consideration of the nature of ground faults, the propagation of shock waves through the earth mass, the specific nature of recorded major quakes, and so on. We do not present here a general discussion of earthquakes but concentrate on their influence on the design of structures for buildings. Some of the references in the Bibliography may be used for a study of the general nature of earthquakes, and the reader is urged to study these if lacking such a background of knowledge.

2.1 CHARACTERISTICS OF EARTHQUAKES

Following a major earthquake, it is usually possible to retrace its complete history through the recorded seismic shocks over an extended period. This period may cover several weeks, or even years, and the record will usually show several shocks preceding and following the major one. Some of the minor shocks may be of significant magnitude themselves, as well as being the foreshocks and aftershocks of the major quake.

A major earthquake is usually rather short in duration, often lasting only a few seconds and seldom more than a minute or so. During the general quake, there are usually one or more major peaks of magnitude of motion. These peaks represent the maximum effect of the quake. Although the intensity of the quake is measured in terms of the energy release at the location of the ground fault, the critical effect on a given structure is determined by the ground movements at the location of the structure. The extent of these movements is affected mostly by the distance of the structure from the epicenter, but they are also influenced by the geological conditions directly beneath the structure and by the nature of the entire earth mass between the epicenter and the structure.

Modern recording equipment and practices provide us with representations of the ground movements at various locations, thus allowing us to simulate the effects of major earthquakes. Figure 2.1 shows the typical form of the graphic representation of one particular aspect of motion of the ground as recorded or as interpreted from the recordings for an earthquake. In this example the graph is plotted in terms of the acceleration of the ground in one horizontal direction as a function of elapsed time. For use in physical tests in laboratories or in computer modeling, records of actual quakes may be "played back" on structures in order to analyze their responses.

These playbacks are used in research and in the design of some major structures to develop criteria for design of lateral resistive systems. Most building design work, however, is done with criteria and procedures that have been evolved through a combination of practical experience, theoretical studies, and some empirical relationships derived from research and testing. The results of the current collective knowledge are put forth in the form of recommended design procedures and criteria that are incorporated into the building codes.

Although it may seem like a gruesome way to achieve it, we advance our level of competency in design every time there is a severe earthquake that results in some ma-

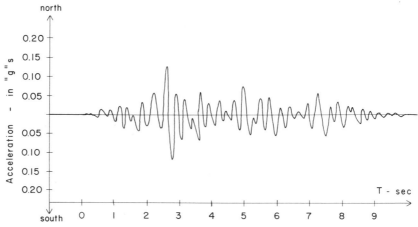

FIGURE 2.1   Characteristic form of ground acceleration graph for an earthquake.

jor structural damage to buildings. Engineering societies and other groups routinely send investigating teams to the sites of major quakes to report on the effects on buildings in the area. Of particular interest are the effects on recently built structures, because these buildings are, in effect, full-scale tests of the validity of our most recent design techniques. Each new edition of the building codes usually reflects some of the results of this cumulative growth of knowledge culled from the latest disasters.

2.2 GENERAL EFFECTS OF EARTHQUAKES

The ground movements caused by earthquakes can have several types of damaging effects. Some of the major effects are:

Direct Movement of Structures. Direct movement is the motion of the structure caused by its attachment to the ground. The two primary effects of this motion are a general destabilizing effect due to the shaking and to the impelling force caused by the inertia of the structure's mass.

Ground Surface Faults. Surface faults may consist of cracks, vertical shifts, general settlement of an area, landslides, and so on.

Tidal Waves. The ground movements can set up large waves on the surface of bodies of water that can cause major damage to shoreline areas.

Flooding, Fires, Gas Explosions, and So On. Ground faults or movements may cause damage to dams, reservoirs, river banks, buried pipelines, and so on, which may result in various forms of disaster.

Although all these possible effects are of concern, we deal in this book only with the first effect: the direct motion of structures.

Concern for this effect motivates us to provide for some degree of dynamic stability (general resistance to shaking) and some quantified resistance to energy loading of the structure.

The force effect caused by motion is generally directly proportional to the dead weight of the structure—or more precisely, to the dead weight borne by the structure. This weight also partly determines the character of dynamic response of the structure. The other major influences on the structure's response are its fundamental period of vibration and its efficiency in energy absorption. The vibration period is basically determined by the mass, the stiffness, and the size of the structure. Energy efficiency is determined by the elasticity of the structure and by various factors such as the stiffness of supports, the number of independently moving parts, the rigidity of connections, and so on.

A relationship of major concern is that which occurs between the period of the structure and that of the earthquake. Figure 2.2 shows a set of curves, called *spectrum curves*, that represent this relationship as derived from a large number of earthquake "playbacks" on structures with different periods. The upper curve represents the major effect on a structure with no damping. Damping results in a lowering of the magnitude of the effects, but a general adherence to the basic form of the response remains.

The general interpretation of the spectrum effect is that the earthquake has its major direct force effect on buildings with short periods. These tend to be buildings with stiff lateral resistive systems, such as shear walls and X-braced frames, and buildings that are small in size and/or squat in profile.

For very large, flexible structures, such as tall towers and high-rise buildings, the fundamental period may be so long that the structure develops a whiplash effect,

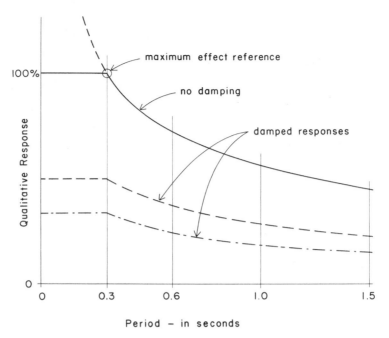

FIGURE 2.2 Spectrum response graphs.

with different parts of the structure moving in opposite directions at the same time, as shown in Fig. 2.3. Analysis for this behavior requires the use of dynamic methods that are beyond the scope of this book.

The three general cases of structural response are illustrated by the three cases shown in Fig. 2.4. Referring to the spectrum curves, for buildings with a period below that representing the upper cutoff of the curves (approximately 0.3 sec), the response is that of a rigid structure, with virtually no flexing. For buildings with a period slightly higher, there is some reduction in the force effect caused by the slight "giving" of the building and its using up some of the energy of the motion-induced force in its own motion. As the building period increases, the behavior approaches that of the slender tower, as shown in Fig. 2.3.

In addition to the movement of the structure as a whole, there are independent movements of individual parts. These each have their own periods of vibration, and the total motion occurring in the structure can thus be quite complex if it is composed of a number of relatively flexible parts.

2.3 EARTHQUAKE EFFECTS ON BUILDINGS

The principal concern in structural design for earthquake forces is for the laterally resistive system of the building. In most buildings this system consists of some combination of horizontally distributing elements (usually roof and floor diaphragms) and vertical bracing elements (shear walls, rigid frames, trussed bents, etc.). Failure of any part of this system, or of connections between the parts, can result in major damage to the building, including the possibility of total collapse.

It is well to remember, however, that an

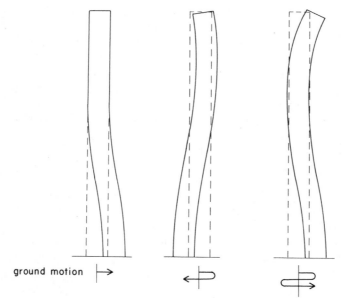

ground motion

FIGURE 2.3 Motion of a tall building during an earthquake.

earthquake shakes the whole building. If the building is to remain completely intact, the potential movement of all its parts must be considered. The survival of the structural system is a limited accomplishment if suspended ceilings fall, windows shatter, plumbing pipes burst, and elevators are derailed.

A major design consideration is that of tying the building together so that it is quite literally not shaken apart. With regard to the structure, this means that the various separate elements must be positively secured to one another. The detailing of construction connections is a major part of the structural design for earthquake resistance.

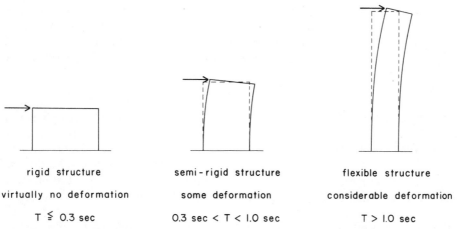

rigid structure

virtually no deformation

$T \lesssim 0.3$ sec

semi-rigid structure

some deformation

0.3 sec $< T < 1.0$ sec

flexible structure

considerable deformation

$T > 1.0$ sec

FIGURE 2.4 Fundamental seismic response of buildings.

In some cases it is desirable to allow for some degree of independent motion of parts of the building. This is especially critical in situations where a secure attachment between the structure and various nonstructural elements, such as window glazing, can result in undesired transfer of force to the nonstructural elements. In these cases use must be made of connecting materials and details that allow for the holding of the elements in place while still permitting relative independence of motion.

When the building form is complex, various parts of the building may tend to move differently, which can produce critical stresses at the points of connection between the parts of the building. The best solution to this sometimes is to provide connections (or actually in some cases nonconnections) that allow for some degree of independent movement of the parts. This type of connection is called a *seismic separation joint*, and its various problems are discussed in Sec. 6.5.

Except for the calculation and distribution of the loads, the design for lateral loads from earthquakes is generally similar to that for the horizontal forces that result from wind. In some cases the code requirements are the same for the two loading conditions. There are many special requirements for seismic design in the *UBC*, however, and the discussion in the next section, together with the examples in Part III, deal with the use of the code for analysis and design for earthquake effects.

2.4   GENERAL DESIGN CONSIDERATIONS FOR EARTHQUAKE FORCES

The influence of earthquake considerations on the design of building structures tends to be the greatest in the zones of highest probability of quakes. This fact is directly reflected in the *UBC* by the *Z* factor, which varies from 0.075 to 0.4, or by a ratio of more than 5 to 1. As a result, wind factors often dominate the design in the zones of lower seismic probability.

A number of general considerations in the design of lateral resistive systems were discussed at the end of Chapter 1. Most of these also apply to seismic design. Some additional considerations are discussed in the following paragraphs.

Influence of Dead Load.   Dead load is in general a disadvantage in earthquakes, because the lateral force is directly proportional to it. Care should be exercised in developing the construction details and in choosing materials for the building in order to avoid creating unnecessary dead load, especially at upper levels in the building. Dead load is useful for overturn resistance and is a necessity for the foundations that must anchor the building.

Advantage of Simple Form and Symmetry.   Buildings with relatively simple forms and with some degree of symmetry usually have the lowest requirements for elaborate or extensive bracing or for complex connections for lateral loads. Design of plan layouts and of the building form in general should be done with a clear understanding of the ramifications in terms of structural requirements when wind or seismic forces are high. When complex form is deemed necessary, the structural cost must be acknowledged.

Following Through with Load Transfers.   It is critical in design for lateral loads that the force paths be complete. Forces must travel from their points of origin through the whole system and into the ground. Design of the connections between elements and of the necessary drag struts, collectors, chords, blocking, hold downs, and so on is highly important to the integrity of the whole lateral resistive system. The ability to visualize the load paths and a reasonable understanding of building construction details and processes are prerequisites to this design work.

Use of Positive Connections. Earthquake forces often represent the most severe demands on connections because of their dynamic, shaking effects. Many means of connection that may be adequate for static force resistance fail under the jarring, loosening effects of earthquakes. Failures of a number of recently built buildings and other structures in earthquakes have been due to connection failures, even though the structures were designed in accordance with current code requirements and accepted practices. Increasing attention is being paid to this problem in the development of recommended details for building construction.

2.5 BUILDING CODE REQUIREMENTS FOR EARTHQUAKE EFFECTS

The model building code that generally presents the most up-to-date, complete guidelines for design for earthquake effects is the *Uniform Building Code.* This code is reissued every three years and its section on seismic design requirements is continually revised to reflect current developments in the field.

The 1988 edition of the *UBC* contained a considerable revision of the seismic design requirements. While it represents a major revision in comparison to the previous codes, it actually merely reflects developments that had been in preparation for some time, as reflected in various research reports and recommendations by engineering groups and government-sponsored agencies.

The following is a brief digest of the current *UBC* requirements. Applications of much of this material is illustrated in the building design examples in Part III.

Considerations for seismic effects are included in several chapters of the code—such as requirements for foundations, reinforced concrete, and so on. The basic requirements for investigation of seismic effects, however, are presented in Division III of Chapter 16. This section begins with a list of definitions of many of the special terms that are used in the section, and it is recommended that this list be studied for the purpose of a clear understanding of the terminology.

A critical determination for seismic design is what is called the *base shear*, which is essentially the total lateral force assumed to be delivered to the building at its base. For structural design, the force is actually assumed to be distributed vertically in the building, but the total force thus distributed is visualized as the base shear. As an equivalent static force effect, the base shear is expressed as some percent of the building weight. In the earliest codes this took the simple form of

$$V = 0.1W$$

or simply 10% of the building weight. This simple formula was embellished over the years to incorporate variables that reflected such issues as potential risk, dynamic response of the building, potential building/site interaction, and relative importance of the building's safety.

The changes in the 1988 *UBC* included considerations for an increased number of variations of the building form and the dynamic characteristics of the building. The formula for base shear now takes the form

$$V = \frac{ZIC}{R_W}W$$

The terms in this formula are as follows:

Z is a factor that adjusts for probability of risk; values are given in *UBC* Table 16-I for zones established by a map in *UBC* Fig. 16-2.

I is an *importance factor*; a token acknowledgment of the fact that in the event of a major disaster, the destruction of some buildings is more critical than that of others. In this view, three such buildings are those that house essential emergency facilities (police and fire stations, hospitals, etc.), potentially hazardous materials (toxic, explosive, etc.), or simply a lot of people in one place. I factors are given in *UBC* Table 16-K and the categories of the occupancies on which they are based are described in detail in the table.

C is a general factor that accounts for the specific, fundamental character of the building's dynamic response, as related to the general dynamic nature of major, recorded seismic events (translation: big earthquakes). Determination of C and the various factors that affect it are described in the following section.

R_W is the principal new factor in the 1988 *UBC*, although it relates to some issues dealt with previously by use of other factors. This factor generally accounts for considerations of the building's materials, type of construction, and type of lateral bracing system; a lot for one factor to deal with, but a considerable amount of code data goes into the establishing of this factor. Section 2.6 discusses the considerations for the determination of this factor.

W is the weight (read: *mass*) of the building. This is simply, basically, the dead load of the building construction, but may also include some considerations for the weight of some contents of the building: notably heavy equipment, furnishings, stored materials, and so on. The basic consideration is simply: What is the mass that is impelled by the seismic motion?

It should be noted that building code criteria in general are developed with a particular concern in mind. This concern is not for the preservation of the building's appearance, protection of the general security of the property as a financial investment, or the assurance that the building will remain functional after the big earthquake—or, at the least, be feasibly repairable for continued use. The building codes are concerned essentially—and pretty much only—with life safety: the protection of the public from injury or death. Designers or building owners with concerns beyond this basic one should consider the building code criteria to be really minimal and generally not sufficient to assure the other protections mentioned previously, relating to the security of the property.

Although it is of major importance, the determination of the base shear is only the first step in the process of structural design or investigation for seismic effects. Many other factors must be dealt with, some of which are discussed in later sections of this chapter. However, the process of design goes beyond merely a response to commands of the codes, and many issues relating to a general consideration of earthquake-resistive construction are discussed in Chapters 5 through 8. These include concerns for the building's lateral bracing system as well as general concern for the integrity of the whole of the building construction.

Although the basic considerations for seismic effects are dealt with in *UBC* Chapter 16, there are numerous requirements scattered throughout the code, many of which will be discussed in relation to the design of specific types of structures, use of particular materials, and development of various forms of construction.

2.6 CONSIDERATIONS FOR DYNAMIC RESPONSE

Seismic effects are essentially dynamic in character, as compared to gravity forces, which may generally be dealt with as static in nature. It is important to understand some of the critical differences between static and dynamic actions, and a short digest of dynamic concepts is given in Appendix A of this book. Readers with little background should use this reference—or others—to gain some understanding of basic principles of dynamics.

Much of the material presented for use in computations here and in the *UBC* uses the equivalent static force method, consisting of translating the true dynamic effects into ones for which basic static analyses can be used. Several issues to be considered in this process are discussed in this section.

Specific Dynamic Properties

The predictable response of an individual building to an earthquake can be generalized on the basis of many factors having to do with the type of building. These general considerations are largely dealt with in deriving the R_W factor for the base shear. Each building also has various specific properties, however, which must be determined individually for each case. The actual building weight, for example, is such a property, and the value for W in the equation for base shear is specific to a single building.

Additional specific properties include the following:

Fundamental period (T) of the building, as described in Sec. 2.2

Fundamental period of the building site

Interaction of the building and site, due to the matching of their individual periods and their linkage during seismic movements

The general incorporation of these considerations is through the use of the C factor, which is expressed as follows:

$$C = \frac{1.25S}{T^{2/3}}$$

where T is the fundamental period of the building and S is the site coefficient. Provisions are made in the code for limiting the value of C and the value of the C/R_W combination. Provisions are also made for the determination of both T and S on either an approximate basis, using generalized data, or an analytical basis, using more precise dynamic investigative methods.

Although the derivation of the C and R_W factors incorporates many considerations for dynamic effects, it must be noted that their use is still limited to producing shear force (V) for the equivalent static investigation. As much as this process may be refined, it is still not a real dynamic analysis using energy and work instead of static force.

One further adjustment required for the equivalent static method involves the distribution of the total lateral force effect to the various elements of the lateral bracing system. This issue is discussed further in Sec. 2.7.

Applications of much of the code criteria for the equivalent static method are illustrated in the examples in Part III.

General Categories for Dynamic Response

Many of the aspects of a building's response to an earthquake can be predicted on the basis of the general character of the building in terms of form, materials, and general construction. In previous editions of the *UBC*, this was accounted for by use of a *K* factor, with data providing for the differentiation of six different basic forms of construction. A major change in the 1988 *UBC* was the use of the R_W factor, generally replacing the *K* factor and providing for 14 categories of the construction of the lateral load-resisting system.

Another major change in the 1988 *UBC* was the identification of several cases of *structural irregularity*, which can limit the use of the equivalent static method and require a more rigorous, dynamic investigation. Two types of irregularity are defined: those which are related to vertical form and relationships, and those that are related to plan (horizontal) form and relationships. These issues are discussed more fully in Chapter 3.

Previous editions of the *UBC* were somewhat vague about what specifically constitutes a building that cannot be reliably investigated or designed by the equivalent static method. The 1988 edition had considerable definition of the categories of buildings that require a true dynamic investigation for design. These requirements include considerations for the risk zone, building size or number of stories, bracing system, and degree of either vertical or plan irregularity.

Dynamic Response of the Whole Building

A general effect of the 1988 edition of the *UBC* has been to cause a much greater concern for general architectural features of the building. While the real concern in many cases is for the response of the lateral load-resisting system, the form of that system is often substantially defined by the form of the building in general. The various forms of irregularity are usually derived from architectural features of the building plans and vertical profile. Beyond this are many special requirements based on concerns for damage to various nonstructural elements, such as ceilings, parapets and cornices, signs, heavy suspended light fixtures or equipment, and freestanding partitions, shelving, or other items subject to lateral movement, overturning, or detachment during an earthquake.

As mentioned previously, the main concern of codes is for life safety. Building collapse is the principal worry, but flying or falling elements of the building construction or furnishings can also represent major hazards. Various factors used to modify or regulate the design of the structure may be partly derived from concerns for nonstructural damage. Restrictions on lateral deflection (drift) are largely based on this concern.

2.7   SPECIAL PROBLEMS

In this section we present some of the special considerations for seismic response.

Acceptable Response

The meaning of safety and acceptable response becomes quite complex when disastrous force actions are considered. Few buildings can really be expected to endure violent windstorms or earthquakes with no damage whatsoever, or no discomfort on the part of occupants. For design it becomes necessary to define a particular level of qualified response (minor damage versus total collapse, for example) and then to quantify the magnitude of force that should be used to relate to that response.

For earthquakes, this translates into determining just how big a shock (dynamic effect) should be used in establishing specific levels of the building's response. Economic considerations make it unlikely that we will use the biggest earthquake imaginable as the quantified effect relating to minor damage.

Engineers are generally more aware of this aspect of design, as it is often quite basic to any design work relating to disastrous effects. Bomb shelters, for example, are always designed for a specific proximity to a particular-sized bomb. Bank vaults are rated on this basis and their structures designed accordingly.

Translating this concept into specific design data and practical criteria is not so easy, however. What frequently drops through the cracks and does not get real design effort is the construction detailing for various so-called "nonstructural" items: for example, glazing details for windows, supports for signs, and various elements of the HVAC, wiring, and piping systems.

Comparatively, it is relatively easy to design the major lateral resistive structure for resistance to full collapse under a specific size of seismic shock; and less easy to assure no glass breakage, broken piping, or dropped ceiling elements in a moderate earthquake. All of this gets more complicated when the value system for design goals puts life safety as the top concern. Rightly so, of course, but building owners may not often be aware that the building code minimum requirements are not doing much for protection of their property investment.

Distribution of Seismic Base Shear

The total horizontal force computed as the base shear (V) must be distributed both vertically and horizontally to the elements of the lateral load-resisting system. This begins with a consideration of the actual distribution of the building mass, which essentially develops the actual inertial forces. However, for various purposes in simulating dynamic response, the distribution of forces for the actual investigation of the structure may be modified.

Section 1628.4 of the *UBC* requires a redistribution of the lateral forces at the various levels of multistory buildings. As visualized vertically, these forces are assumed to be applied at the levels of the horizontal diaphragms, although the redistribution is intended to modify the loading to the vertical bracing system. The purpose of this modification is to move some of the lateral load to upper levels of the building, simulating more realistically the nature of the response of the vertical cantilever to the dynamic loads. Use of this criterion is illustrated in the multistory building examples in Part III.

In a horizontal direction, the total shear at any level of the building is generally considered to be distributed to the vertical elements of the system in proportion to their stiffness (resistance to lateral deflection). If the lateral bracing elements are placed symmetrically, and their centroid corresponds to the center of gravity of the building mass, this simple assumption may be adequate. However, two considerations may alter this simple distribution. The first has to do with the coincidence of location of the centroid of the bracing system (usually called the center of resistance) and the center of gravity of the building mass. If there is a major discrepancy in the location of these, there will be a horizontal torsional effect which will produce shears that must be added to those produced by the direct shear force. Even when no actual theoretical eccentricity of this type occurs, the code may require the inclusion of a so-called accidental eccentricity as a safety measure.

The second modification of horizontal distribution has to do with the relative horizontal stiffnes of the horizontal diaphragm (in effect, its deflection resistance in the diaphragm/beam action). The two major considerations that affect this are the aspect ratio of the diaphragm in plan (length-to-width ratio), and the basic construction of the diaphragm. Wood and formed steel deck diaphragms tend to be flexible, while concrete decks are stiff. The actions of diaphragms in this respect are discussed in Chapter 6.

In some cases it may be possible to manipulate the distribution of forces by alterations of the construction. Seismic separation joints represent one such alteration (see the discussion in Sec. 6.5). Another technique is to modify the stiffnesses of various vertical bracing elements to cause them to resist more or less of the total lateral load; thus a building may have several vertical bracing elements, but a few may take most of the load if they are made very stiff.

3

RESISTANCE OF BUILDINGS
TO LATERAL FORCES

In this chapter we discuss the general nature of the response of buildings to lateral force effects. Considerations in this regard include those for the force actions as discussed in the preceding two chapters. How the building responds to those actions and what can be done to enhance or improve its responses is the theme of this chapter.

3.1 APPLICATION OF WIND AND SEISMIC FORCES

To understand how a building resists the lateral load effects of wind and seismic force it is necessary to consider the manner of application of the forces and then to visualize how these forces are transferred through the lateral resistive structural system and into the ground.

Wind Forces

The application of wind forces to a closed building is in the form of pressures applied normal to the exterior surfaces of the building. In one design method the total effect on the building is determined by considering the vertical profile, or silhouette, of the building as a single vertical plane surface at right angles to the wind direction. A direct horizontal pressure is assumed to act on this plane.

Figure 3.1 shows a simple rectangular building under the effect of wind normal to one of its flat sides. The lateral resistive structure that responds to this loading consists of the following:

Wall surface elements on the windward side are assumed to take the total wind pressure and are typically designed to span vertically between the roof and floor structures.

Roof and floor decks, considered as rigid planes (called diaphragms), receive the edge loading from the windward wall and distribute the load to the vertical bracing elements.

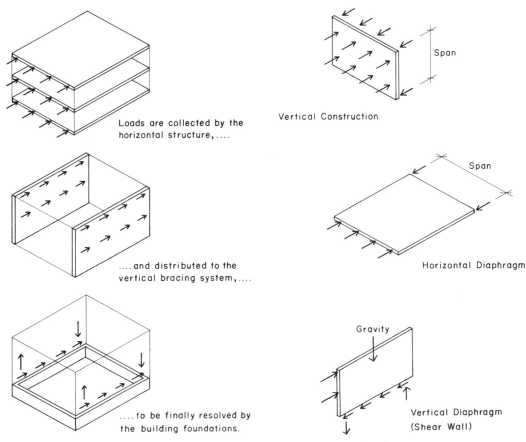

FIGURE 3.1   Propagation of wind forces and basic functions of elements in a box system.

Vertical frames or shear walls, acting as vertical cantilevers, receive the loads from the horizontal diaphragms and transfer them to the building foundations.

The foundations must anchor the vertical bracing elements and transfer the loads to the ground.

The propagation of the loads through the structure is shown on the left in Fig. 3.1 and the functions of the major elements of the lateral resistive system are shown on the right in the figure. The exterior wall functions as a simple spanning element loaded by a uniformly distributed pressure normal to its surface and delivering a reaction force to its supports. In most cases, even though the wall may be continuous through several stories, it is considered as a simple span at each story level, thus delivering half of its load to each support. Referring to Fig. 3.1, this means that the upper wall delivers half of its load to the roof edge and half to the edge of the second floor. The lower wall delivers half of its load to the second floor and half to the first floor.

This may be a somewhat simplistic view of the function of the walls themselves, depending on their construction. If they are framed walls with windows or doors, there may be many internal load transfers within the wall. Usually, however, the external load delivery to the horizontal structure will be as described.

The roof and second-floor diaphragms function as spanning elements loaded by the edge forces from the exterior wall and spanning between the end shear walls, thus producing a bending that develops tension on the leeward edge and compression on the windward edge. It also produces shear in the plane of the diaphragm that becomes a maximum at the end shear walls. In most cases the shear is assumed to be taken by the diaphragm, but the tension and compression forces due to bending are transferred to framing at the diaphragm edges. The means of achieving this transfer depends on the materials and details of the construction.

The end shear walls act as vertical cantilevers that also develop shear and bending. The total shear in the upper story is equal to the edge load from the roof. The total shear in the lower story is the combination of the edge loads from the roof and second floor. The total shear force in the wall is delivered at its base in the form of a sliding friction between the wall and its support. The bending caused by the lateral load produces an overturning effect at the base of the wall as well as the tension and compression forces at the edges of the wall. The overturning effect is resisted by the stabilizing effect of the dead load on the wall. If this stabilizing moment is not sufficient, a tension tie must be made between the wall and its support.

If the first floor is attached directly to the foundations, it may not actually function as a spanning diaphragm but rather will push its edge load directly to the leeward foundation wall. In any event, it may be seen in this example that only three-quarters of the total wind load on the building is delivered through the upper diaphragms to the end shear walls.

This simple example illustrates the basic nature of the propagation of wind forces through the building structure, but there are many other possible variations with more complex building forms or with other types of lateral resistive structural systems.

Seismic Forces

Seismic loads are actually generated by the dead weight of the building construction. In visualizing the application of seismic forces, we look at each part of the building and consider its weight as a horizontal force. The weight of the horizontal structure, although actually distributed throughout its plane, may usually be dealt with in a manner similar to the edge loading caused by wind. In the direction normal to their planes, vertical walls will be loaded and will function structurally in a manner similar to that for direct wind pressure. The load propagation for the box-shaped building in Fig. 3.1 will be quite similar for both wind and seismic forces.

If a wall is reasonably rigid in its own plane, it tends to act as a vertical cantilever for the seismic load in the direction parallel to its surface. Thus, in the example building, the seismic load for the roof diaphragm would usually be considered to be caused by the weight of the roof and ceiling construction plus only those walls whose planes are normal to the direction being considered. These different functions of the walls are illustrated in Fig. 3.2. If this assumption is made, it will be necessary to calculate a separate seismic load in each direction for the building.

For determination of the seismic load, it is necessary to consider all elements that are permanently attached to the structure. Ductwork, lighting and plumbing fixtures,

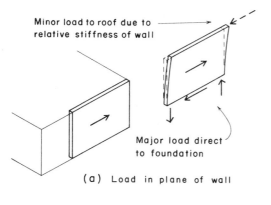

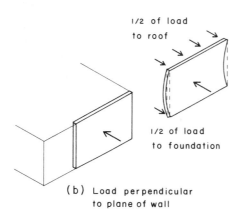

FIGURE 3.2   Seismic loads caused by wall weight.

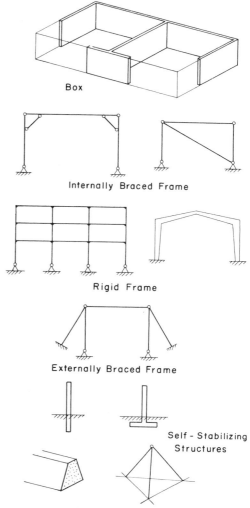

FIGURE 3.3   Types of lateral load resistive systems.

supported equipment, signs, and so on, will add to the total dead weight for the seismic load. In buildings such as storage warehouses and parking garages it is also advisable to add some load for the building contents.

3.2   TYPES OF LATERAL RESISTIVE SYSTEMS

The building in the preceding example illustrates one type of lateral resistive system: the box or panelized system. As shown in Fig. 3.3, the general types of systems are

those discussed in the following paragraphs.

Box or Panelized System

The box or panelized system is usually of the type shown in the preceding example, consisting of some combination of horizontal and vertical planar elements. Actually, most buildings use horizontal diaphragms simply because the existence of roof and floor construction provides them as a matter of course. The other types of systems usually consist of variations of the vertical bracing elements. An occasional exception is a roof structure that must be braced by trussing or other means when there are a large number of roof openings or a roof deck with little or no diaphragm strength.

Internally Braced Frames

The typical assemblage of post and beam elements is not inherently stable under lateral loading unless the frame is braced in some manner. Shear wall panels may be used to achieve this bracing, in which case the system functions as a box even though there is a frame structure. It is also possible, however, to use diagonal members, X-bracing, knee-braces, struts, and so on, to achieve the necessary stability of the rectangular frame. The term *braced frame* usually refers to these techniques.

Rigid Frames

Although the term *rigid frame* is a misnomer since this technique usually produces the most *flexible* lateral resistive system, the term refers to the use of moment-resistive joints between the elements of the frame.

Externally Braced Frames

The use of guys, struts, buttresses, and so on, that are applied externally to the structure or the building results in externally braced frames.

Self-Stabilizing Elements and Systems

Retaining walls, gravity dams, flagpoles, pyramids, tripods, and so on, in which stability is achieved by the basic form of the structure are examples of self-stabilizing elements and systems.

Each of these systems has variations in terms of materials, form of the parts, details of construction, and so on. These variations may result in different behavior characteristics, although each of the basic types has some particular properties. An important property is the relative stiffness or resistance to deformation, which is of particular concern in evaluating energy effects, especially for response to seismic loads. A box system with diaphragms of poured-in-place concrete is usually very rigid, having little deformation and a short fundamental period. A multistory rigid frame of steel, on the other hand, is usually quite flexible and will experience considerable deformation and have a relatively long fundamental period. In seismic analysis these properties are used to modify the percentage of the dead weight that is used as the equivalent static load to simulate the seismic effect.

Elements of the building construction developed for the gravity load design, or for the general architectural design, may become natural elements of the lateral resistive system. Walls of the proper size and in appropriate locations may be theoretically functional as shear walls. Whether they can actually serve as such will depend on their construction details, on the materials used, on their height-to-width ratio, and on the manner in which they are attached to the other elements of the system for load transfer. It is also possible, of course, that the building construction developed only for gravity load resistance and architectural planning considerations may *not* have the necessary attributes for lateral load resistance, thus requiring some replanning or the addition of structural elements.

Many buildings consist of mixtures of the basic types of lateral resistive systems. Walls existing with a frame structure, al-

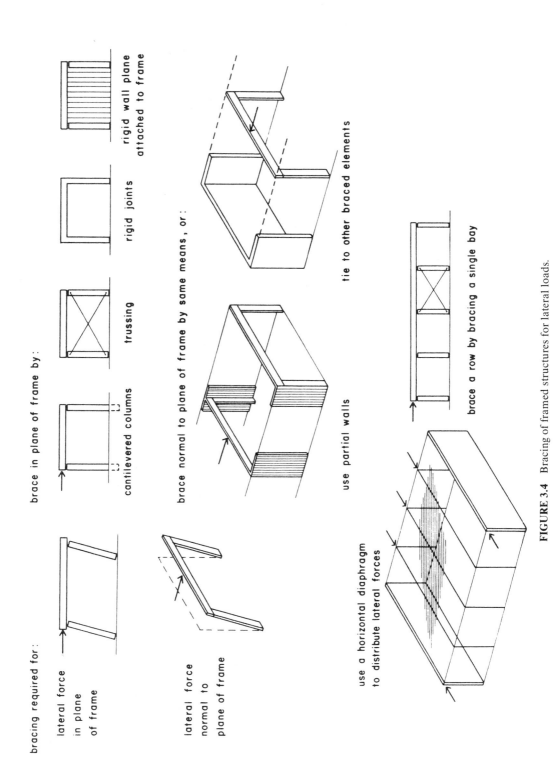

FIGURE 3.4 Bracing of framed structures for lateral loads.

bracing required for:

lateral force in plane of frame

lateral force normal to plane of frame

brace in plane of frame by:

cantilevered columns

trussing

rigid joints

rigid wall plane attached to frame

brace normal to plane of frame by same means, or:

use partial walls

tie to other braced elements

use a horizontal diaphragm to distribute lateral forces

brace a row by bracing a single bay

though possibly not used for gravity loads, can still be used to brace the frame for lateral loads. Shear walls may be used to brace a building in one direction whereas a braced frame or rigid frame is used in the perpendicular direction.

Multistory buildings occasionally have one type of system, such as a rigid frame, for the upper stories and a different system, such as a box system or braced frame, for the lower stories to reduce deformation and take the greater loads in the lower portion of the structure.

In many cases it is neither necessary nor desirable to use every wall as a shear wall or to brace every bay of the building frame. Figure 3.4 shows various situations in which the lateral bracing of the building is achieved by partial bracing of the system. This procedure does require that there be some load-distributing elements, such as the roof and floor diaphragms, horizontal struts, and so on, that serve to tie the unstabilized portions of the building to the lateral resistive elements.

There is a possibility that some of the elements of the building construction that are not intended to function as bracing elements may actually end up taking some of the lateral load. In frame construction, surfacing materials of plaster, drywall, wood paneling, masonry veneer, and so on, may take some lateral load even though the frame is braced by other means. This is essentially a matter of relative stiffness, although connection for load transfer is also a consideration. What can happen in these cases is that the stiffer finish materials take the load first, and if they are not strong enough, they fail and the intended bracing system then goes to work. Although collapse may not occur, there can be considerable damage to the building construction as a result of the failure of the supposed nonstructural elements.

The choice of the type of lateral resistive system must be related to the loading con-ditions and to the behavior characteristics required. It must also, however, be coordinated with the design for gravity loads and with the architectural planning considerations. Many design situations allow for alternatives, although the choice may be limited by the size of the building, by code restrictions, by the magnitude of lateral loads, by the desire for limited deformation, and so on.

3.3 LATERAL RESISTANCE OF ORDINARY CONSTRUCTION

Even when buildings are built with no consideration given to design for lateral forces, they will have some natural capacity for lateral force resistance. It is useful to understand the limits and capabilities of ordinary construction as a starting point for the consideration of designing for enhanced levels of lateral force resistance.

Wood Frame Construction

Wood structures can be categorized broadly as light frame or heavy timber. Light frames—using mostly 2 × dimension lumber for wall studs, floor joists, and roof rafters—account for the vast majority of small, low-rise buildings in the United States. In most cases the frames are covered on both surfaces by some type of surfacing material. Many of these surfacing systems have a usable, quantifiable, capacity for diaphragm resistance. Thus, without any major alteration of the basic structure, most light wood frames can be made resistive to lateral forces through the use of a combination of horizontal and vertical diaphragms.

Many of the ordinary elements of light wood frames can be utilized as parts of the lateral resistive system, serving as diaphragm chords, collectors, and edge transfer members. Wall studs, posts, sills and plates, and roof and floor framing members, occurring routinely in the structure, are often able to be utilized for these functions. Alterations necessary to make them more functional are often limited to moderate increases in sizes or to the use of some additional fastening or anchoring. When members are long and not able to be installed as a single piece (as with the top plate on a long wall), it may be necessary to use stronger splicing than is ordinarily required for gravity resistance alone.

Common practices of carpentry result in a considerable amount of fastening between members of a light wood frame. Building codes often specify minimum requirements for such fastening. Before undertaking to design such a structure for seismic loads, it well behooves the designer to become thoroughly familiar with these requirements as well as current local practices of contractors and workers. References with suggested details for construction become rapidly out of date as codes and construction practices and the availability of materials and equipment change.

In recent times there has been a trend toward the extensive use of sheet metal fastening devices for the assemblage of light wood structures. In general, these tend to increase resistances to lateral loads because the continuity of the frame is greater and the anchorage of members is more positive.

There is a considerable range in the diaphragm shear capability of various surfacing materials. The following are some widely used products:

Plywood. This may be used as the structural backup for a variety of finishes or may be used with a special facing as the complete surfacing. Most plywoods offer considerable potential for shear resistance. With an increase in structural quality, greater thickness than the minimum required for

other functions, and a greater number of nails, it is possible to develop considerable strength for either horizontal or vertical diaphragm use.

Board Sheathing. Boards of 1-in. nominal thickness, with shiplap or tongue-and-groove edges, were once quite commonly used for sheathing. When applied in a position diagonal to the frame, they can produce some diaphragm action. Capacities are nowhere near that of plywood, however, so that for this and other reasons this type of structural covering is not much used at present.

Plaster. Portland cement plaster, applied over wire-reinforced backing and adequately secured to the framing, produces a very stiff diaphragm with load capacities equal to that of the thinner plywoods. On the exterior it is called stucco, which due to its popularity is the definitive vertical bracing material for light structures in southern and western United States.

Miscellaneous Surfacing. Gypsum drywall, gypsum plaster, nonstructural plywood, and particleboard can develop some diaphragm capacity that may be sufficient for low stress situations. Stiffness is minimal, thus narrow diaphragms should be avoided.

When the same material is applied to both sides of a wall, the code permits the use of the sum of the resistances of the two surfaces. For interior walls this is quite common and permits the utilization of low-strength surfacing. However, for exterior walls the two surfaces are seldom the same, thus the stronger (usually the exterior) must be used alone.

In the past a widely used method for bracing light wood frames consisted of diagonal bracing, typically in the form of 1-in.-nominal-thickness boards with their outside surfaces made flush with the face of the framing by cutting notches in the framing members (called *let-in bracing*). The acceptance of rated load capacities for a wider range of popular surfacing materials has made this practice largely redundant. When diagonal bracing is used today, it often consists of thin steel straps applied to the faces of the framing members.

Some of the problems encountered in developing seismic resistance with light wood frame construction follow.

1. Lack of adequate solid walls to serve as shear walls. This may be due to the building planning, with walls insufficient at certain locations or in a particular direction. Walls may also simply be too broken up in short lengths by doors and windows. For multistory buildings there may be a problem where upper level walls do not occur above walls in lower levels.

2. Lack of adequate diaphragm surfacing. Many types of surfacing have rated capacities for shear. Each, however, has its limits and some materials are not rated for code-acceptable loadings.

3. Lack of continuity of framing. Because it is often not required for gravity load conditions, members that could function as chords or collectors may consist of separate pieces that are not spliced for tension or compression continuity.

4. Lack of adequate connections. Load transfers—most notably those from horizontal to vertical diaphragms—may not be possible without modification of the construction details, involving additional framing, increased nailing, or use of special anchorage devices. Ordinary code-required minimum wall sill bolting is generally not acceptable for overturn resistance and is often not adequate for sliding resistance.

Wood post-and-beam structures can sometimes be made to function as braced frames or moment-resistive frames, the latter usually being somewhat more difficult to achieve. Often, however, these structures occur in combination with wall construction of reinforced masonry or wood frame plus surfacing so that the post and beam frame need not function for lateral bracing. Floor and roof decks are often the same as for light frames and thus are equally functional as horizontal diaphragms.

A problem with post-and-beam construction is often the lack of ability for load transfers between members required for resistance to lateral loads. At present this is less the case due to the increasing use of metal framing devices for beam seats, post caps and bases, and so on.

When heavy timber frames are exposed to view, a popular choice for roof or floor decks is a timber deck of 1.5 in. or greater thickness. Although such a deck has a minimal shear capacity, it is usually not adequate for diaphragm development except for small buildings. The most common and economical means for providing the necessary diaphragm resistance is to simply nail a continuous plywood surface to the top of the timber deck.

Structural Masonry

For seismic zones 3 and 4 the only masonry structural construction permitted is *reinforced masonry*. These comments are confined to structural masonry walls constructed with hollow concrete units (concrete blocks), with the voids partly or wholly reinforced and filled with grout (see Fig. 10.15).

Structural masonry walls have considerable potential for utilization as shear walls. There are, however, a number of problems that must be considered.

1. *Increased Load.* Due to their weight, stiffness, and brittleness, masonry walls must be designed for higher lateral seismic forces.
2. *Limited Stress Capacity.* The unit strength and the mortar strength must be adequate for the required stress resistances. In addition, both vertical and horizontal reinforcing is required for major shear wall functions.
3. *Cracks and Bonding Failures.* Walls not built to the specifications usually used for seismic-resistive construction often have weakened mortar joints and cracking. These reduce seismic resistance, especially in walls with minimal reinforcing.

Code specifications for reinforced concrete block walls result in a typical minimum construction that has a particular limit of shear wall capacity. This limit is beyond the limit for the strongest of the wood-framed walls with plywood on a single side, and so the change to a masonry wall is a significant step. Beyond the minimum value the load capacity is increased by adding additional reinforcing and filling more of the block voids. At its upper limits the reinforced masonry wall approaches the capacity of a reinforced concrete wall.

Anchorage of masonry walls to their supports is usually simply achieved. Resistance to vertical uplift and horizontal sliding can typically be developed by the usual doweling of the vertical wall reinforcing. The anchorage of horizontal diaphragms to masonry walls is another matter and typically requires the use of more "positive" anchoring methods than are ordinarily used when seismic risk is low.

Reinforced Concrete Construction

Poured concrete elements for most structures are ordinarily quite extensively reinforced, thus providing significant compensation for the vulnerability of the tension-weak material. Even where structural demands are not severe, minimum two-way reinforcing is required for walls and slabs to absorb effects of shrinkage and fluctuation of temperature. This form of construction has considerable natural potential for lateral force resistance.

Subgrade building construction most often consists of thick concrete walls, in many cases joined to horizontal concrete structures with solid cast-in-place slabs. The typical result is a highly rigid, strong boxlike structure. Shears in the planes of the walls and slabs can be developed to considerable stress levels with minimum required reinforcing. Special attention must be given to the maintaining of continuity through pour joints and control joints and to the anchorage of reinforcing at wall corners and intersections and at the joints between slabs and walls. This does not always result in an increase in the amount of reinforcing but may alter some details of its installation.

Structures consisting of concrete columns used in combination with various concrete spanning systems require careful study for the development of seismic resistance as rigid frame structures. The following are some potential problems:

1. *Weight of the Structure.* This is ordinarily considerably greater than that of wood or steel construction, with the resulting increase in the total seismic force.
2. *Adequate Reinforcing for Seismic Effects.* Of particular concern are the shears and torsions developed in framing elements and the need for

continuity of the reinforcing or anchorage at the intersections of elements. A special problem is that of vertical shears developed by vertical accelerations, most notably punching shear in slab structures.

3. *Ductile Yielding of Reinforcing.* This is the desirable first mode of failure, even for gravity load resistance. With proper design it is a means for developing a yield character in the otherwise brittle, tension-weak structure.

4. *Detailing of Reinforcing.* Continuity at splices and adequate anchorage at member intersections must be assured by careful layout of reinforcing installation.

5. *Tying of Compression Bars.* Column and beam bars should be adequately tied in the region of the column-beam joint.

When concrete walls are used in conjunction with concrete frames, the result is often similar to that of the plywood-braced wood frame, the walls functioning to absorb the major portion of the lateral loads due to their relative stiffness. At the least, however, the frame members function as chords, collectors, drag struts, and so on. The forces at the intersections of the walls and the frame members must be carefully studied to assure proper development of necessary force transfers.

As with masonry structures, considerable cracking is normal in concrete structures; much of it is due to shrinkage, temperature expansion and contraction, settlement or deflection of supports, and the normal development of internal tension forces. In addition, built-in cracks of a sort are created at the cold joints that are unavoidable between successive, separate pours. Under the back-and-fourth swaying actions

of an earthquake, these cracks will be magnified, and a grinding action may occur as stresses reverse. The grinding action can be a major source of energy absorption but can also result in progressive failures or simply a lot of pulverizing and flaking off of the concrete. If reinforcing is adequate, the structure may remain safe, but the appearance is sure to be affected.

It is virtually impossible to completely eliminate cracking from masonry and cast-in-place concrete buildings. Good design, careful construction detailing, and quality construction can reduce the amount of cracking and possibly eliminate some types of cracking. However, the combination of shrinkage, temperature expansion, settlement of supports, creep, and flexural stress is a formidable foe.

Steel Frame Construction

Structures with frames of steel can often quite readily be made resistive to lateral loads, usually by producing either a braced (trussed) frame or a moment-resistive frame. Steel has the advantage of having a high level of resistance to all types of stress and is thus not often sensitive to multi-directional stresses or to rapid stress reversals. In addition, the ductility of ordinary structural steel provides a toughness and a high level of energy absorption in the plastic behavior mode of failure.

The high levels of stress obtained in steel structures are accompanied by high levels of strain, resulting often in considerable deformation. The actual magnitudes of the deformations may affect the building occupants or contents or may have undesirable results in terms of damage to non-structural elements of the building construction. Deformation analysis is often a critical part of the design of steel structures, especially for moment-resistive frames.

The ordinary post and beam steel frame is essentially unstable under lateral loading. Typical framing connections have some minor stiffness and moment resistance but are not effective for development of the rigid joints required for a moment-resistive frame. Frames must therefore either be made self-stable with diagonal bracing or with specially designed moment-resistive connections or be braced by shear walls.

Steel frames in low-rise buildings are often braced by walls, with the steel structure serving only as the horizontal spanning structure and vertical gravity load resisting structure. Walls may consist of masonry or of wood or metal frames with various shear-resisting surfacing. For the wall-braced structure, building planning must incorporate the necessary solid wall construction for the usual shear wall braced building. In addition, the frame will usually be used for chord and collector actions; therefore, the connections between the decks, the walls, and the frame must be designed for the lateral load transfers.

The trussed steel structure is typically quite stiff, in a class with the wall-braced structure. This is an advantage in terms of reduction of building movements under load, but it does mean that the structure must be designed for as much as twice the total lateral force as a moment-resistive frame. Incorporating the diagonal members in vertical planes of the frame is often a problem for architectural planning, essentially similar to that of incorporating the necessary solid walls for a shear wall braced structure.

In the past steel frames were mostly used in combination with decks of concrete or formed sheet steel. A popular construction for low-rise buildings at present—where fire-resistance requirements permit its use—is one that utilizes a wood infill structure of joists or trusses with a plywood deck. A critical concern for all decks is the ade-quate attachment of the deck to the steel beams for load transfers to the vertical bracing system. Where seismic design has not been a factor, typical attachments are often not sufficient for these load transfers.

Buildings of complex unsymmetrical form sometimes present problems for the development of braced or moment-resistive frames. Of particular concern is the alignment of the framing to produce the necessary vertical planar bents. Randomly arranged columns and discontinuities due to openings or voids can make bent alignment or continuity a difficult problem.

Precast Concrete Construction

Precast concrete structures present unique problems in terms of lateral bracing. Although they share many characteristics with cast-in-place concrete structures, they lack the natural member-to-member continuity that provides considerable lateral stability. Precast structures must therefore be dealt with in a manner similar to that for post and beam structures of wood or steel. This problem is further magnified by the increased dead weight of the structure, which results in additional lateral force.

Separate precast concrete members are usually attached to each other by means of steel devices that are cast into the members. The assemblage of the structure thus becomes a steel-to-steel connection problem. Where load transfer for gravity resistance is limited to simple bearing, connections may have no real stress functions, serving primarily to hold the members in position during construction. Under lateral load, however, all connections will likely be required to transfer shear, tension, bending, and torsion. Thus for seismic resistance many of the typical connections used for gravity resistance alone will be inadequate.

Because of their weight, precast concrete

spanning members may experience special problems due to vertical accelerations. When not sufficiently held down against upward movement, members may be bounced off their supports (a failure described as dancing.)

Precast concrete spanning members are often also prestressed, rather than simply utilizing ordinary steel reinforcing. This presents a possible concern for the effects of the combined loading of gravity and lateral forces or for upward movements due to vertical acceleration. Multiple loading conditions and stress reversals tend to greatly complicate the design of prestressing.

As with frames of wood or steel, those of precast concrete must be made stable with trussing, moment connections, or in-fill walls. If walls are used, they must be limited to masonry or concrete. Connections between the frame and any bracing walls must be carefully developed to assure the proper load transfers.

Miscellaneous Construction

Foundations. Where considerable below-grade construction occurs—with heavy basement walls, large bearing footings, basement or sublevel floor construction of reinforced concrete, and so on—the below-grade structure as a whole usually furnishes a solid base for the above-grade building. Not many extra details or elements are required to provide for seismic actions. Of principal concern is the tying together of the base of the building, which is where the seismic movements are transmitted to the building. If the base does not hold together as a monolithic unit, the result will be disastrous for the supported building. Buildings without basements or those supported on piles or piers may not ordinarily be sufficiently tied together for this purpose, thus requiring some additional construction.

Freestanding Structures. Freestanding structures include exterior walls used as fences as well as large signs, water towers, and detached stair towers. The principal problem is usually the large overturning effect. Rocking and permanent soil deformations that result in vertical tilting must be considered. It is generally advisable to be quite conservative in the design for soil pressure due to the overturning effects. When weight is concentrated at the top—as in the case of signs or water towers—the dynamic rotational effect is further increased. These concerns apply also to the elements that may be placed on the roof of a building.

3.4 LIMITS OF MATERIALS AND ELEMENTS

Although the range of possibilities for development of lateral resistive structures is considerable, it is necessary to be aware of the limitations that exist. Limitations may be real, established by stress-level capacities of materials or the available sizes of ordinary construction elements. Limits may also be somewhat artificial or ambiguous, established by code requirements or rules of thumb in design practice. The latter are not developed arbitrarily but rather are evolved from the collective experience and judgments of generations of professional researchers and designers. However, in time, changes in construction methods, design practices, and use of materials and products make old rules lose touch with reality.

The limits of basic types of construction are discussed in Sec. 3.3. The limits of typical elements of lateral resistive structural systems are discussed in Part II. The following are some basic types of limits that should be recognized, although the specific quantification of data and the relative status of particular materials or systems tend to exist in a constant state of flux.

1. *Limits of Materials.* Common materials have specific stress limits; variation is typical over some range of quality or type (stress grade of wood, f'_c of concrete, etc.).
2. *Available Sizes.* Concrete is unique in its ability to form elements of virtually any size; most other materials and products have practical limits in terms of commonly available shapes and sizes.
3. *Functional Limits.* For various reasons particular materials, products, construction techniques, or details of construction are often limited to specific uses or excluded from specific uses. Unfinished bolts, which are mostly used only for temporary or minor connections, are excluded from use in braced frames. Wood-framed shear walls should not be used to brace a concrete frame. For the bolts, the limits are due to their inability to maintain a tight (nonslipping) joint. For the wood-framed wall it is simply not reasonable to brace the heavy, rigid, system with a light, deformable one.
4. *Size Limits.* Upper and lower limits of size exist—sometimes for actual reasons of the availability of products, sometimes merely for practicality, as in the case of the limits of minimum size for concrete members (wall thickness, beam width, column diameter, etc.).
5. *Aspect Ratios.* Examples of aspect ratios are the practical, or in some cases the code-specified, limits for the h/t of columns or L/d of beams. A major type of limitation is the length-to-width ratio of wood diaphragms (expressed as span-to-width for hori-zontal diaphragms and as height-to-width for vertical diaphragms by *UBC* Table 23-I-I).
6. *Structural Behavior.* Performance of the structure is mostly limited by the assigning of values for allowable stresses (in the working stress method) or load factors (for the ultimate strength method). Except for vertical deflection of beams and lateral drift (horizontal deflection) of individual stories, codes do not provide much criteria for limitation of deformations.

3.5 DESIGN CONSIDERATIONS FOR LATERAL RESISTIVE SYSTEMS

The design of the lateral load resistive system for a building involves a great number of factors. The principal considerations are the following.

Determination of the Loading

Determining the loading is usually established by the satisfaction of the requirements of the building code with jurisdiction. Critical load values, as well as various requirements for the form of the structural analysis and design, are determined by the degree of local concern for extremes of wind storms or earthquakes. This concern is primarily based on the history of disasters in the area.

Selection and Planning of the Lateral Resistive System for the Building

As previously discussed, this selection and planning must be coordinated with the gravity load design and the architectural design in general. In some cases the design for lateral loads may be a major factor in establishing the building form and detail, in selection of materials, and so on. In other cases it may consist essentially of assuring the proper construction of ordinary elements of the construction.

Detailed Analysis and Design of the Elements of the Lateral Resistive System

With the loading established and the system defined the performance of the individual parts and of the system as a whole must be investigated. An important aspect of the investigation is the complete following through of the loads from their origin to their final resolution in the ground. With the internal forces and stresses determined the design of the parts of the system is usually a matter of routine, using code specifications and data from the code or from other reference sources.

Development of Structural Construction Details and Specifications

Such development constitutes essentially the documentation of the design and is a task of major importance. A thorough analytical investigation and complete set of structural calculations will be useless unless the results are translated into directives usable by the builders of the building.

Convincing the Authorities Who Grant Building Permits That the Structure Is Adequate

In most cases someone employed or retained as a consultant by the code enforcing body will review the structural calculations and the construction drawings and specifications for compliance with the local code requirements and with acceptable practices of design and construction. Although it should be expected that a competent and thorough design effort will receive a good review, there is usually some room for individual judgment and personal preference so that the potential exists for some conflict between the designer and the reviewer.

In some ways the easiest part of this process is that of the structural analysis and design. The analysis and design may be laborious if the building is large or complex, but it is usually routine in nature, with considerable information and guidance available from codes, texts, industry brochures, and so on. Some degree of training in engineering mechanics and basic structural analysis and design is necessary, but most of the work is "cookbook" in nature.

Determination of the lateral loads is also reasonably simple, at least with the use of the equivalent static load methods. One possible complication is due to the fact that the structure must be defined in some detail before the loads can be determined and their propagation through the system investigated, which is somewhat like needing to know the answer before a person can formulate the question.

Some things must be known about the structure before one can analyze its behavior. As a result, the early stages of design often consist of some guessing and trying—approximating a structure and then analyzing it to see if it works. This process is easier, of course, when the designer has worked on similar problems before, or if he or she has the results of previous similar designs as a basis for a more educated first guess.

The more difficult aspects of design for lateral loads are the development of the basic systems for the lateral resistive structure and the development of the necessary construction drawings and specifications to assure proper construction. This work requires considerable understanding of the problems of the building design and construction in general because decisions about the basic structural scheme and some of the details of the structure may have considerable influence on the general form and detail of the building and on the economics and general feasibility of the construction.

As the potential for the influence or ramifications of design decisions broadens, the concerns and value systems of other areas of the building design must be considered. If the person doing the basic structural design is not capable of dealing with all of those other areas, he or she should—at the least—have some level of awareness of them. Seismic response is a serious concern, but is only one of many concerns for building design. The structural designer must become as aware as possible of the total building design process; otherwise, the most skillfully executed, optimal structural design work may be at odds with the general design development and cause some embarrassment to the structural designer.

4

BUILDING PLANNING FOR LATERAL RESISTANCE

The planning of buildings must respond to many concerns besides that for resistance to lateral forces. However, where violent windstorms or earthquakes are a real likelihood, it is possible to give them some major consideration—or at least to avoid going the other direction, making general building forms that are as stupid as possible for good lateral resistance. In this chapter we treat some issues that relate building form to lateral resistance, providing some basis for judgment in basic planning.

4.1 ARCHITECTURAL DESIGN ISSUES

When the need to develop resistance to seismic forces is kept in mind throughout the entire process of the building design, it will have bearing on many areas of the design development. This chapter deals with various considerations that may influence the general planning as well as choices for materials systems, and construction details.

When lateral design is dealt with as an afterthought rather than being borne in mind in the earliest decisions on form and planning of the building, it is quite likely that optimal conditions will not be developed. Some of the major issues that should be kept in mind in the early planning stages follow.

1. The need for *some kind* of lateral bracing system. In some cases, because of the building form or size or the decision to use a particular structural material or system, the choice may be highly limited. In other situations there may be several options, with each having different required features (alignment of columns, incorporation of solid walls, etc.). The particular system to be used should be established early,

although it may require considerable exploration and development of the options in order to make an informed decision.

2. Implications of architectural design decisions. When certain features are desired, it should be clearly understood that there are consequences in the form of problems with regard to lateral design. Some typical situations that commonly cause problems are the following:

> General complexity and lack of symmetry in the building form
>
> Random arrangement of vertical elements (walls and columns), resulting in a haphazard framing system in general
>
> Lack of continuity in the horizontal structure due to openings, multiplane roofs, split-level floors, or open spaces within the building
>
> Building consisting of aggregates of multiple, semidetached units, requiring considerations for linking or separation for seismic interaction
>
> Special forms (curved walls, sloping floors, etc.) that limit the performance of the structure
>
> Large spans, heights, or wall openings that limit placement of structural elements and result in high concentrations of load
>
> Use of nonstructural materials and construction details that result in high vulnerability to damage caused by lateral movements

3. Allowance for lateral design work. Consideration should be given to the time, cost, and scheduling for the lateral investigation and design development. This is most critical when the building is complex or when an extensive dynamic analysis is required. Sufficient time should be allowed for a preliminary investigation of possible alternatives for the lateral bracing system, since a shift to another system at late stages

of the architectural design work will undoubtedly cause problems.

4. Design styles not developed with seismic effects in mind. In many situations popular architectural design styles or features are initially developed in areas where seismic effects are not of concern. When these are imported to regions with high risk of seismic activity, a mismatch often occurs. Early European colonizers of Central and South America and the west coast of North America learned this the hard way. The learning goes on.

4.2 RELATIONS OF BUILDING FORM TO LATERAL RESISTANCE

The form of a building has a great deal to do with the determination of the effects of seismic activity on the building. This chapter discusses various aspects of building form and the types of problem commonly experienced.

Most buildings are complex in form. They have plans defined by walls that are arranged in complex patterns. They have wings, porches, balconies, towers, and roof overhangs. They are divided vertically by multilevel floors. They have sloping roofs, arched roofs, and multiplane roofs. Walls are pierced by openings for doors and windows. Floors are pierced by stairways, elevators, ducts, and piping. Roofs are pierced by skylights, vent shafts, and chimneys. The dispersion of the building mass and the overall response of the building to seismic effects can thus be complicated; difficult to visualize, let alone to quantitatively evaluate.

Despite this typical complexity, investigation for seismic response may often be simplified by the fact that we deal mostly with those elements of the building that are directly involved in the resistance of lateral forces; what we refer to as the *lateral resistive system.* (The *UBC* uses the term *lateral load-resisting system.*) Thus most of the building construction, including parts of the structure that function strictly for resistance of gravity loads, may have only minimal involvement in seismic response. These nonstructural elements contribute to the load (generated by the building mass) and may offer damping effects to the structure's motion, but may not significantly contribute to the development of resistance to lateral force.

A discussion of the issues relating to building form must include consideration of two separate situations: the form of the building as a whole and the form of the lateral resistive system. Figure 4.1 shows a simple one-story building, with the general exterior form illustrated in the upper figure. The lower figure shows the same building with the parapet, canopy, window wall, and other elements removed, leaving the essential parts of the lateral resistive system. This system consists primarily of the horizontal roof surface and the portions of the vertical walls that function as shear walls. The whole building must be considered in determining the building mass for the lateral load, but the stripped down structure must be visualized in order to investigate the effects of lateral forces.

In developing building plans and the building form in general, architectural designers must give consideration to many issues. Seismic response has to take its place in line with the needs for functional interior spaces, control of traffic, creation of acoustic privacy, separation for security, energy efficiency, and general economic and technical feasibility. In this book we dwell primarily on the problems of lateral response, but it must be kept in mind that the architect must deal with all of these other concerns.

Development of a reasonable lateral resistive structural system within a building may be easy or difficult, and for some proposed plan arrangements may be next to impossible. Figure 4.2 shows a building plan in the upper figure for which the potentiality for development of shear walls in the north–south direction is quite reasonable but in the east–west direction is not so good as there is no possibility for shear walls on the south side. If the modification shown in the middle figure is acceptable, the building can be adequately braced by shear walls in both directions. If the open south wall is really essential, it may be possible to brace this wall by using a column and beam structure that is braced by trussing or by rigid connections, as shown in the lower figure.

In the plan shown in Fig. 4.3*a* the column layout results in a limited number of possible bents that may be developed as moment-resistive frames. In the north–south direction the interior columns are either offset from the exterior columns or the bent is interrupted by the floor opening; thus the two end bents are the only ones

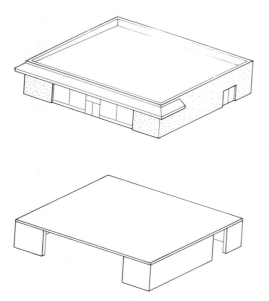

FIGURE 4.1 Building and lateral load resistive system.

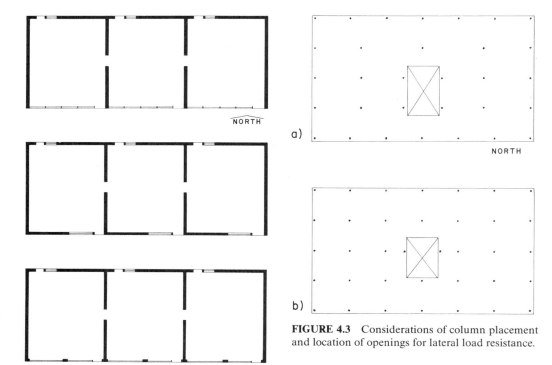

a)

NORTH

b)

FIGURE 4.3   Considerations of column placement and location of openings for lateral load resistance.

FIGURE 4.2   Plan development of lateral load resistive systems.

usable. In the east-west direction the large opening interrupts two of the three interior bents, leaving only three usable bents that are not disposed symmetrically in the plan. The modification shown in Fig. 4.3b represents an improvement in lateral response, with six usable bents in the north–south direction and four symmetrically placed bents in the east–west direction. This plan, however, has more interior columns, smaller open spaces, and a reduced size for the opening—all of which may present some drawbacks for architectural reasons.

In addition to planning concerns the vertical massing of the building has various implications on its seismic response. The three building profiles shown in Fig. 4.4a, b, and c represent a range of potential response with regard to the fundamental

period of the building and the concerns for lateral deflection. The short, stiff building shown in (a) tends to absorb a larger jolt from an earthquake because of its quick response (short period of natural vibration). The tall, slender building, on the other hand, responds slowly, dissipating some of the energy of the seismic action in its motion. However, the tall building may develop some multimodal response, a whiplash effect, or simply so much actual deflection that it may have problems of its own.

The overall inherent stability of a building may be implicit in its vertical massing or profile. The structure shown in Fig. 4.4d has considerable potential for stability with regard to lateral forces, whereas that shown in Fig. 4.4e is highly questionable. Of special concern is the situation in which abrupt change in stiffness occurs in the vertical massing. The structure shown in Fig.

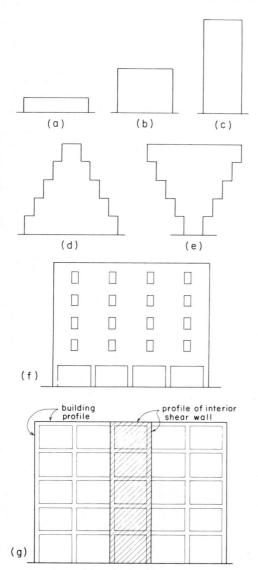

FIGURE 4.4 Considerations of building profile and lateral load resistance.

As with the building plan, consideration of the vertical massing must include concerns for the form of the lateral resistive system as well as the form of the whole building. Figure 4.4g shows a building whose overall profile is quite stout. However, if the building is braced by a set of interior shear walls, as shown in the section, it is the profile of the shear walls that must be considered. In this case the shear wall is quite slender in profile.

Investigation of the seismic response of a complex building is, in the best of circumstances, a difficult problem. Anything done to simplify the investigation will not only make the analysis easier to perform but will tend to make the reliability of the results more certain. Thus, from a seismic design point of view, there is an advantage in obtaining some degree of symmetry in the building massing and in the disposition of the elements of the lateral resistive structure.

When symmetry does not exist, a building tends to experience severe twisting as well as the usual rocking back and forth. The twisting action often has its greatest effects on the joints between elements of the bracing system. Thorough investigation and careful detailing of these joints for construction are necessary for a successful design. The more complex the seismic response and the more complicated and unusual the details of the construction, the more difficult it becomes to assure a thorough and careful design.

Most buildings are not symmetrical, being sometimes on one axis, often not on any axis. However, real architectural symmetry is not necessarily the true issue in seismic response. Of critical concern is the alignment of the net effect of the building mass (or the centroid of the lateral force) with the center of stiffness of the lateral resistive system—most notably the center of stiffness of the vertical elements of the system. The more the eccentricity of the

4.4f has an open form at its base, resulting in the so-called *soft story*. While this type of system may be designed adequately by the general requirements of the equivalent static force method, a true dynamic analysis will indicate serious problems, as borne out by some recent serious failures.

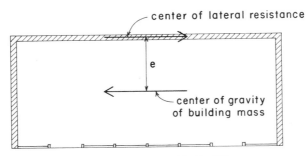

FIGURE 4.5 Three-sided building.

centroid of the lateral force from the center of stiffness of the lateral bracing system, the greater the twisting effect on the building.

Figure 4.5 shows an extreme example—the so-called *three-sided building*. In this situation the lack of resistive vertical elements on one side of the building requires that the opposite wall take all of the direct effect of the lateral force that is parallel to it.

Assuming the centroid of the building mass to be approximately at the center of the plan, this results in a large eccentricity between the load and the resisting wall. The twisting action that results will be partly resisted by the two end walls that are at right angles to the load, but the general effect on the building is highly undesirable. This type of structure is presently highly

FIGURE 4.6 Common example of a three-sided structure: an open store front with closed, heavy walls on the sides and back. End shear walls on the front were not sufficient here to prevent major lateral movement in the plane of the front wall, resulting in loss of all the large front windows. In this case, the excessive torsion also resulted in failure of the end panels of the roof diaphragm in shear at the wall header and full collapse of the end framing.

restricted for use in regions of high seismic risk (Fig. 4.6).

When a building is not architecturally symmetrical, the lateral bracing system must either be adjusted so that its center of stiffness is close to the centroid of the mass or it must be designed for major twisting effects on the building. As the complexity of the building form increases, it may be necessary to consider the building to be multimassed.

Many buildings are multimassed rather than consisting of a single geometric form. The building shown in Fig. 4.7 is multimassed, consisting of an L-shaped tower that is joined to an extended lower portion. Under lateral seismic movement the various parts of this building will have different responses. If the building structure is developed as a single system, the building movements will be very complex, with extreme twisting effects and considerable strain at the points of connection of the discrete parts of the mass.

If the elements of the tower of the building in Fig. 4.7 are actually separated, as shown in Fig. 4.8a or c, the independent movements of the separated elements will be different due to their difference in stiffness. It may be possible to permit these in-

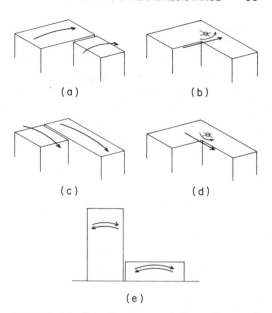

FIGURE 4.8 Seismic movements in multimassed buildings.

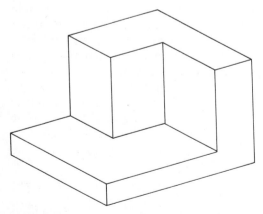

FIGURE 4.7 Multimassed building form.

dependent movements by providing structural connections that are detailed to tolerate the type and magnitude of the actual deformations. Thus the twisting effects on the building and the strain at the joints between the elements of the mass may be avoided.

There is also the potential for difference in response movements of the tower and the lower portion of the building, as shown in Fig. 4.8e. Actual separation may be created at this connection of the masses to eliminate the need for investigation of the dynamic interaction of the separate parts. However, it may not be feasible or architecturally desirable to make the provisions necessary to achieve either of the types of separation described. The only other option is therefore to design for the twisting effects and the dynamic interactions that result from having a continuous, single structural system for the entire building. The advisability or feasibility of one option over the other is often difficult to establish and may require considerable study of alternate designs.

Figure 4.9a shows an L-shaped building in which the architectural separation of the masses is accentuated. The linking element, although contiguous with the two parts, is unlikely to be capable of holding them together under seismic movements. If

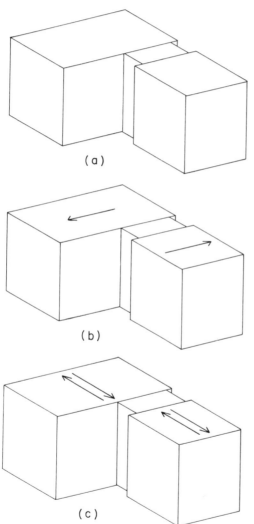

(a)

(b)

(c)

FIGURE 4.9   Relative movements in linked buildings.

it cannot, there are two forms of differential movement that must be provided for, as shown in Fig. 4.9b and c. In addition to providing for these movements, it is also necessary to consider the bracing of the linkage element. If it is not capable of being independently braced, it must be attached to one or the other of the larger elements for support, making for a quite complex study of actions at the connection of the masses.

Stress concentrations will occur at all points of discontinuity, such as window and door openings, changes in shape of a wall, or connections between any separate parts of a building mass. The construction at all these locations must be studied for potential damage. Damage may be prevented by reinforcement or other strengthening, or by providing relief joints (control joints) in the construction at the points of potential stress (Fig. 4.10).

When individual parts of multimassed buildings are joined, there are many potential problems, some of which were just described. Figure 4.11 shows three types of action that must often be considered for such structures. When moving at the same time, as shown in Fig. 4.11a, a problem for the separate masses becomes the actual dimension of the separation that must be provided to prevent them from bumping each other (called battering or hammering). If they are not actually separated, their independent deflections may be a basis for consideration of the forces that must be considered in preventing them from being torn apart (Fig. 4.12).

Another potential action of separately moving parts is that shown in Fig. 4.11b. This involves a shearing action on the joint similar to that which occurs in laminated elements subjected to bending. For the vertically cantilevered elements shown, both the shear and lateral deflection effects vary from zero at the base to a maximum at the top. The taller the structure, the greater the

FIGURE 4.10 A tall shear wall at the joint between two separate but joined building masses is pushed and pulled, causing damage at its top, bottom, and the connection to the lower mass. Damage here is essentially "cosmetic" and could probably have been prevented by use of relief joints in the stucco at the locations of the cracks.

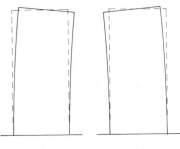

(a) differential deflection

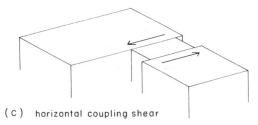

(C) horizontal coupling shear

FIGURE 4.11 (*Continued*)

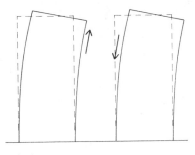

(b) vertical coupling shear

FIGURE 4.11 Coupling shears in linked elements.

actual dimension of the critical movements near the top (Fig. 4.13).

A third type of action is the horizontal shearing effect illustrated in Fig. 4.11c. This is probably the most common type of problem that must be dealt with, as it occurs frequently in one-story structures, whereas the vertical shear and lateral deflection problems are usually severe only in taller structures.

Individual joined masses are sometimes so different in size or stiffness that the in-

FIGURE 4.12   Earthquake damage to a church with a connected tower. The tower was strong enough but flexible, so that its movement battered the lower, stiffer church structure, causing extensive damage, visible here on the exterior (*upper photo*) and interior (*lower photo*).

dicated solution is to simply attach the smaller part to the larger and let it tag along. Such is the case for the buildings shown in Fig. 4.14a and b in which the smaller lower portion and the narrow stair tower would be treated as attachments.

In some instances the tag-along relationship may be a conditional one, as shown

FIGURE 4.13 Movement of the separate large masses of this apartment building caused major fracture of the light stair structure between the masses.

in Fig. 4.14c, where the smaller element extends a considerable distance from the larger mass. In this situation the movement of the smaller part to and away from the larger may be adequately resisted by the attachment. However, some bracing would probably be required at the far end of the smaller part to assist resistance to movements parallel to the connection of the two parts.

The tag-along technique is often used for stairs, chimneys, entries, and other elements that are part of a building, but are generally outside the main mass. It is also possible, of course, to consider the total structural separation of such elements in some cases.

Another classic problem of joined elements is that of coupled shear walls. These are shear walls that occur in sets in a single wall plane and are connected by the continuous construction of the wall. Figure 4.15 illustrates such a situation in a multistory building. The elements that serve to link such walls—in this example the spandrel panels beneath the windows—are

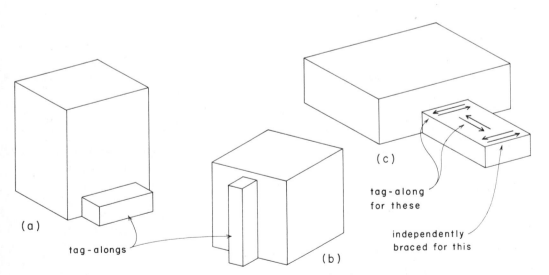

FIGURE 4.14 External elements, linked for lateral bracing.

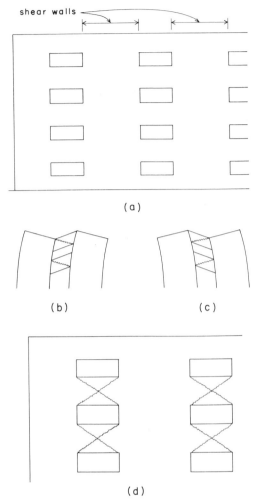

FIGURE 4.15   Effects of movements in continuous wall construction.

Forces applied to buildings must flow with some direct continuity through the elements of the structure, be transferred effectively from element to element, and eventually be resolved into the ground. Where there are interruptions in the normal flow of the forces, problems will occur. For example, in a multistory building the resolution of gravity forces requires a smooth, vertical path; thus columns and bearing walls must be stacked on top of each other. If a column is removed in a lower story, a major problem is created, requiring the use of a heavy transfer girder or other device to deal with the discontinuity.

A common type of discontinuity is that of openings in horizontal and vertical diaphragms. These can be a problem as a result of their location, size, or even shape. Figure 4.17 shows a horizontal diaphragm with an opening. The diaphragm is braced by four shear walls, and if it is considered to be uninterrupted, it will distribute its load to the walls in the manner of a continuous beam. (See the discussion of flexibility of horizontal diaphragms in Sec. 6.1.) If the relative size of the opening is as shown in Fig. 4.17a, this assumption is a reasonable one. What must be done to assure the integrity of the continuous diaphragm is to reinforce the edges and corners of the opening and to be sure that the net diaphragm width at the opening is adequate for the shear force.

If the opening in a horizontal diaphragm is as large as that shown in Fig. 4.17b, it is generally not possible to maintain the continuity of the whole diaphragm. In the example the best solution would be to consider the diaphragm as consisting of four individual parts, each resisting some portion of the total lateral force. For openings of sizes between the ones shown in Fig. 4.17 judgment must be exercised as to the best course.

wracked by the vertical shearing effect illustrated in Fig. 4.11b. As the building rocks back and forth, this effect is rapidly reversed, developing the diagonal cracking shown in Fig. 4.15b and c. This results in the X-shaped crack patterns shown in Fig. 4.15d, which may be observed on the walls of many masonry, concrete, and stucco-surfaced buildings in regions of frequent seismic activity (Fig. 4.16).

FIGURE 4.16 Diagonal cracking in stiff, brittle construction takes an X-form as the building moves back and forth in an earthquake. The example shown here is a pierced concrete shear wall with the usual vulnerable piers and spandrel beams between window openings.

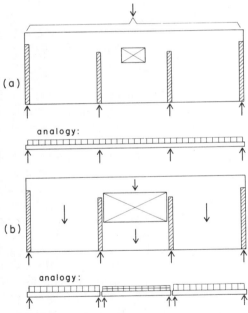

FIGURE 4.17 Effects of openings on the character of distribution by a horizontal diaphragm. (*a*) Minor effect; diaphragm functions as a continuous beam. (*b*) Major effect; diaphragm functions as a set of simple beams with subdiaphragms at openings.

Another discontinuity that must sometimes be dealt with is that of the interrupted multistory shear wall. Figure 4.18*a* shows such a situation, with a wall that is not continuous down to its foundation. In this example it may be possible to utilize the horizontal structure at the second level to redistribute the horizontal shear force to other shear walls in the same plane. The overturn effect on the upper shear wall, however, cannot be so redirected, thus requiring that the columns at the ends of the shear wall continue down to the foundation.

It is sometimes possible to redistribute the shear force from an interrupted wall, as shown in Fig. 4.18*b*, with walls that are sidestepped rather than in the same vertical plane of the upper wall. Again, however, the overturn on the upper wall must be accommodated by continuing the structure at the ends of the wall down to the foundation.

Figure 4.19 shows an X-braced frame structure with a situation similar to that of

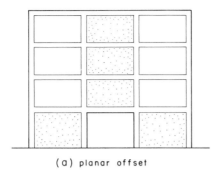

(a) planar offset

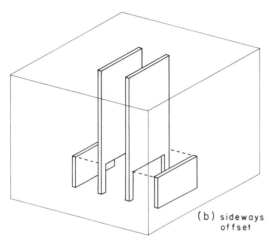

(b) sideways offset

FIGURE 4.18   Offsets in the lateral bracing system: (*a*) vertical offset; (*b*) offset in plan.

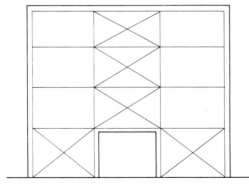

FIGURE 4.19   Vertical offset in a braced frame system.

the shear wall in Fig. 4.18*a*. The individual panels of X-bracing are sufficiently similar in function to the individual panels of the shear wall to make the situation have the same general options and requirements for a solution.

Discontinuities are usually inevitable in multistory and multimassed buildings. They add to the usual problems of dissymmetry to create many difficult situations for analysis and design and require careful study for the proper assumptions of behavior and the special needs of the construction.

4.3   BUILDING FORM AND THE *UBC*

A major effect of the changes in seismic design criteria in the 1988 edition of the *UBC* was a significant advancement of seismic design of buildings as an architectural design concern. This is not entirely a new concern, but has been given considerably more attention in this edition of the code.

There is a steadily growing body of evidence—obtained mostly from post-quake inspections of buildings following major seismic events (translation: big earthquakes) which indicates that building form and choice of materials has considerable influence on the response of buildings to earthquake effects. A major response to this is represented by the addition of the definition of the *regular* structure (and by corollary, the *irregular* structure). The qualifications of irregularity described in *UBC* Tables 16-L and 16-M—while describing structural properties—are generally created by architectural planning of the building. By general inference, the building form conditions described in these tables are all negative factors with regard to the seismic resistance of buildings.

There is really very little in the new seismic design requirements in the 1988 *UBC* that is unfamiliar to experienced structural engineers. The *UBC* essentially simply reflects what is well established in engineering and research circles, and quite frankly follows—rather than leads—what is common practice by the leading structural design practitioners. Publication of these materials in the 1988 *UBC* merely brings the architectural designer more specifically into the circle of participants in the development of intelligent designs for seismic-resistive buildings.

We do not advocate that all buildings should be "regular" and that any irregularity is intolerable. However, it behooves all architectural designers who work on buildings in zones of high seismic risk to understand the significance and the consequences of various types of irregularity and how structural irregularity relates to architectural design factors.

The following is a brief discussion of some of the building form considerations that relate to the requirements of the *UBC* Tables 16-L and 16-M.

UBC Table 16-L, Vertical Irregularities

There are five cases—labeled 1 through 5—described in the table, as follows:

1: *Soft Story.* This situation is described in Sec. 4.4. The common ways in which this occurs are when an individual story (often the ground-level story) is made taller and/or more open in construction. This form is quite commonly used with commercial multistory buildings, so the condition must frequently be considered.

2: *Weight Irregularity.* This often occurs in conjunction with other irregularities—such as setbacks or interior open spaces. However, another common occurrence is a very heavy roof with rooftop HVAC equipment above a lightweight framed building.

3: *Vertical Geometric Irregularity.* This refers to setbacks of more than a specified amount in the lateral resistive structure. If core bracing or some other system that does not follow the building's exterior profile is used, this irregularity may be avoided, even when the building itself has major setbacks. If a general perimeter bracing system is used, however, this irregularity may be unavoidable.

4: *In-Plane Discontinuity in Vertical Bracing.* This refers to offsets of the form illustrated in Fig. 4.18*a*, limiting them to a specified distance. This is generally less of a problem on the building exterior than on the interior, where changes in building usage (or occupancy) on different levels requires the reorganization of spaces.

5: *Weak Story.* This is also described briefly in Sec. 4.4. One way that this occurs is when code requirements for minimum construction result in major redundancy of structural capacity: for example, a continuous, long wall with plywood sheathing with code-required minimum nailing, possibly not intended as a shear wall, but obviously having a great resistance to lateral force. If this construction occurs above a more open-walled story, both a soft story and a weak story may result. Minimum masonry or concrete wall construction are similar situations.

UBC Table 16-M, Plan Irregularities

Again, there are five cases, labeled 1 through 5, as follows:

1: *Torsional Irregularity.* This occurs when there is a major discrepancy between the centroid of the building mass and the centroid of the whole lateral resistive system. The limits in the table refer to disproportionate deformations of opposite ends of the building. This occurs most commonly when the plan of the building is unsymmetrical—probably the general, rather than a special, case for all buildings. The problem is not essentially one of maintaining symmetry in the building, although that may be a simple way to deal with the problem. What is essential is to distribute bracing elements with stiffnesses that match the distribution of the building mass.

The key word here is *stiffness*, not strength. A case of this irregularity may occur when mixed bracing systems are used (shear walls plus frames, etc.). This generally requires a very careful analysis of the relative deformations of the vertical bracing elements when plan dissymmetry or other complexity of the building form occurs.

2. *Reentrant Corners.* This has to do with situations such as that for the building with an L-shaped plan, whose actions are illustrated in Fig. 4.9. The

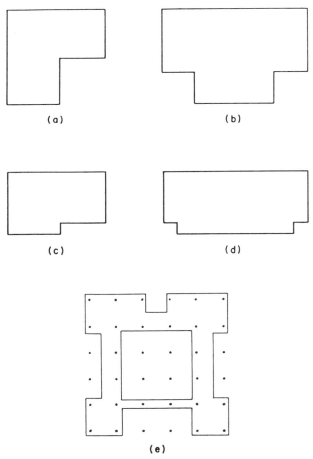

FIGURE 4.20 Reentrant corner; *UBC* qualification.

table defines a limit for extension of the building on both sides of the corner, so that the irregular classification is not given for minor corners, such as those shown in Fig. 4.20c and d. It should also be noted that the table considers this case only when both the building and its bracing system have the reentrant corner condition. If perimeter bracing is used, this may be the case. However, it may be possible to develop a bracing system that does not follow the building plan in this manner; such a case is that for the system shown in Fig. 4.20e.

3: *Diaphragm Discontinuity.* In most cases we consider a roof or floor diaphragm to be a contiguous, rigid planar element, capable of distributing lateral forces to the vertical elements without significant concern for its own deformation. However, either plan form discrepancies or changes in construction may modify the behavior of the diaphragm sufficiently to make this not the case. Large openings for skylights or enclosed atrium spaces may create this situation.

4: *Out-of-Plane Offsets.* This refers to the situation illustrated in Fig. 4.18b, which is considered to be an irregularity without qualification.

5: *Nonparallel Systems.* This generally refers to shear walls or braced bents that are curved or angled in plan, with respect to a simple x- and y-axis system.

Again, we state that having an irregularity is not necessarily an irredeemable disgrace, but—at the least—calls some attention to special concerns for seismic response. And quite possibly, a collection of several irregularities in the same building may be cause for serious reflection on the quality of the building design.

4.4 SPECIAL PROBLEMS

Vulnerable Elements

There are many commonly used elements of buildings that are especially vulnerable to damage due to earthquakes. Most of these are nonstructural, that is, not parts of the structural system for resistance of gravity or lateral loads. Because of their nonstructural character, they do not routinely receive thorough design study by the structural designer; thus in earthquake country they constitute major areas of vulnerability. Some typical situations are the following:

1. *Suspended Ceilings.* These are subject to horizontal movement. If not restrained at their edges, or hung with elements that resist horizontal movement, they will swing and bump other parts of the construction. Another common failure consists of the dropping of the ceiling due to downward acceleration if the supports are not resistive to a jolting action.

2. *Cantilevered Elements.* Balconies, canopies, parapets, and cornices should be designed for significant seismic force in a direction perpendicular to the cantilever. In most cases, codes provide criteria for consideration of these forces.

3. *Miscellaneous Suspended Objects.* Lighting fixtures, signs, HVAC equipment, loudspeakers, catwalks, and other items that are supported by hanging should be studied for the effects of pendulumlike movements. Supports should tolerate the movement or should be designed to restrain it.

4. *Piping.* Building movements during seismic activity can cause the rupture of piping that is installed in a conventional manner. In addition to the usual allowances for thermal expansion, provisions should be made for

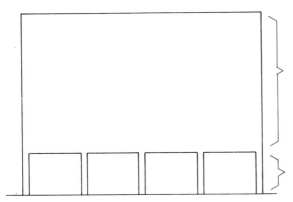

May be
· Solid wall
· Wall with few small windows
· Rigid frame with shorter story height or closer column spacing

Relatively flexible story

FIGURE 4.21 Soft story.

the flexing of the piping or for isolation from the structure that is sufficient to prevent any damage. This is obviously most critical for piping that is pressurized.

5. *Stiff, Weak Elements.* Any parts of the building construction that are stiff but not strong are usually vulnerable to damage. This includes window glazing, plastered surfaces, wall and floor tile (especially of ceramic or cast materials), and any masonry (especially veneers of brick, tile, precast concrete, or stone). Reduction of damage and of hazard to occupants requires careful study of installation details for attachment. In most cases extensive use of control joints is advised to permit movements without fractures.

Soft Story

Any discontinuity that constitutes an abrupt change in the structure is usually a source of some exceptional distress. This is true for static load conditions as well, but is especially critical for dynamic loading conditions. Any abrupt increase or decrease in stiffness will result in some magnification of deformation and stress in a structure subjected to energy loading. Open-

FIGURE 4.22 Classic form of the soft story, with a highly rigid structure above a more flexible one. Failure here is not significant in the wood frame, but could not be tolerated by the stiff stucco covering. This type of failure is one of poor coordination between the structural designer and the architect who details the stucco. Relief joints at the location of these cracks would probably have resulted in no visible damage.

ings, notches, necking-down points, and other form variations produce these abrupt changes in either the horizontal or vertical structure. An especially critical situation is the so-called *soft story*, shown in Fig. 4.21.

The soft story could—and indeed sometimes does—occur at an upper level. How-

ever, it is more common at the ground-floor level between a rigid foundation system and some relatively much stiffer upper level system. The tall, open ground floor has both historical precedent and current popularity as an architectural feature. This is not always strictly a matter of design

FIGURE 4.23 Soft story failure in a large apartment complex. Individual units extended on the rear of the apartment block had open parking at the ground level with only steel pipe column supports. These units were wagged like the tail of a dog until the columns failed and the rear end of the units dropped.

style, as it is often required for functional reasons.

Reference to *story* here is only figurative, as the condition is essentially one of change of mass or relative stiffness. A situation of this type can occur with a one-story building where an open portion occurs below a relatively tall solid portion above (Fig. 4.22). However, the classic case is a very open, lightly structured space beneath a solid form of construction, a typical example being apartments with parking in a ground-level space created by merely lifting the apartments up and providing light columns for support (Fig. 4.23).

If the tall, relatively open ground floor is necessary, Fig. 4.24 presents some possibilities for having this feature with a reduction of the soft-story effect. The methods shown consist of the following:

1. Bracing some of the open bays (Fig. 4.24*a*). If designed adequately for the forces, the braced frame (truss) should have a class of stiffness closer to a rigid shear wall, which is the usual upper structure in these situations. However, the soft story effect can also occur in rigid frames where the "soft" story is simply significantly less stiff.

2. Keeping the building plan periphery open while providing a rigidly braced interior (Fig. 4.24*b*).

3. Increasing the number and/or stiffness of the ground-floor columns for an all-rigid frame structure (Fig. 4.24*c*).

4. Using tapered or arched forms for the ground-floor columns to increase their stiffness (Fig. 4.24*d*).

5. Developing a rigid first story as an upward extension of a heavy foundation structure (Fig. 4.24*e*).

The soft story is actually a method for providing critical damping or major energy

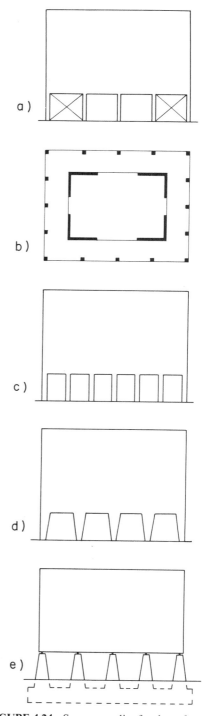

FIGURE 4.24 Some remedies for the soft story.

absorption, which could be a *positive* factor in some situations. However, the major stress concentrations and deformations must be carefully provided for, and a true dynamic analysis is certainly indicated.

Although the concept of and concerns for the soft story have been known for some time, the 1988 *UBC* for the first time presented a code definition of what constitutes a soft story. This is given in *UBC* Table 16-L as one of the conditions that constitutes a vertical structural irregularity.

Weak Story

Another category of vertical discontinuity described in *UBC* Table 16-L is that of the *weak story*, qualified as a discontinuity in capacity. The table defines this condition as that of any story of a structure in which the total lateral story strength is less than 80% of that of the story above. The table

entry also refers to the code, which further stipulates that structures be limited to two stories or a total height of 30 ft where a weak story has a strength of less than 65% of the story above. An exemption to the preceding limitation is given if the weak story has a strength of $3(R_W/8)$ times that required by the usual code requirements for the story shear.

It is essential to understand the distinction between a soft story and a weak story, although it is possible for a single story to be both. The soft-story qualification is based on stiffness, or simply the relative resistance to lateral deformation or story drift. The weak-story qualification is based on strength in terms of force resistance (statics) or energy capacity (dynamics).

A form of weak story may be created by the overstiffening or overstrengthening of a lower floor. Correction for the potential soft story situation of an open ground floor

FIGURE 4.25 The second-floor columns in this five-story concrete structure collapsed, dropping the upper portion and triggering massive failures at the building ends, which had large masonry shear walls. The open first floor was a potential soft story, but its highly stiffened structure remained intact, thus shifting the real base for seismic shear to the second-floor level, where both the lateral and vertical seismic forces worked on the second-story columns. The abrupt change in both strength and stiffness between the first and second stories might have caused a soft story or weak story condition for the second story—or maybe both.

FIGURE 4.26   Vertical collapse of a portion of a large apartment complex. The unit in front was the same height as that still standing behind it before its first story collapsed. A possible cause here was a soft or weak story because of a more open first floor, although the first story may simply not have been strong enough for the base shear.

may result in a concentrated effect on the second floor, which may need more stiffness or strength than a simple static investigation would reveal (Fig. 4.25).

In some cases both a weak story and a soft story may occur. This sometimes results in changes in construction at different levels or in changes in plans that cause different distribution of bracing in different levels (Fig. 4.26).

The general purpose of the provisions for irregularity is to call attention to potential problems. Providing good engineering design for these cases often involves using something beyond the simple static investigation procedures.

PART II

LATERAL RESISTIVE
ELEMENTS AND SYSTEMS

5

LATERAL-LOAD-RESISTING SYSTEMS

In this chapter we treat the basic forms of lateral bracing systems, as presently used for most buildings. Development of these basic systems also involves the use of various special elements, which are discussed in Chapter 6. The problems of planning and utilizing these systems for buildings are treated in general in Chapter 7 and illustrated more fully in the many design case examples in Part III.

5.1 BOX SYSTEMS

The term *box system* was coined to describe a building consisting of a connected set of horizontal and vertical elements consisting of stiffened planar construction (ordinarily walls, roofs, and floors). If the elements themselves, plus their connections, are capable of developing resistance to lateral forces, the result is typically a relatively stiff bracing system.

The box system most often uses sets consisting of horizontal planar construction (called horizontal diaphragms) and vertical planar construction (called vertical diaphragms, or shear walls). However, any planar construction (trussed bent, rigid frame bent) could be used in either a horizontal or a vertical plane to substitute for the usual wall, roof, or floor plane in a box system.

Trussed bents are often quite stiff in resistance to deflections, and thus commonly qualify for use in a box system. The general nature of the stiff box system is thus frequently achieved with some mixture of solid wall, roof, or floor planes and planar bents with triangulated, trussed framing.

Rigid frames, despite their common name, are typically the *least* resistive to lateral deflection; that is, they are the least rigid of the basic forms of bracing. This is due to the bending of the frame members and rotations of the frame joints. However, if the rigid frame members are individually very stiff, the frame may become quite stiff overall, and the nature of a pierced wall, versus a flexible frame, takes over. Since many walls in the box-shaped building are pierced by door or window openings, there is a range in relative stiffness that extends from the solid wall to the truly open, flexible, rigid frame, as shown in Fig. 5.1.

While examples can be found of all the possible variations of the basic box system just described, the most common usage

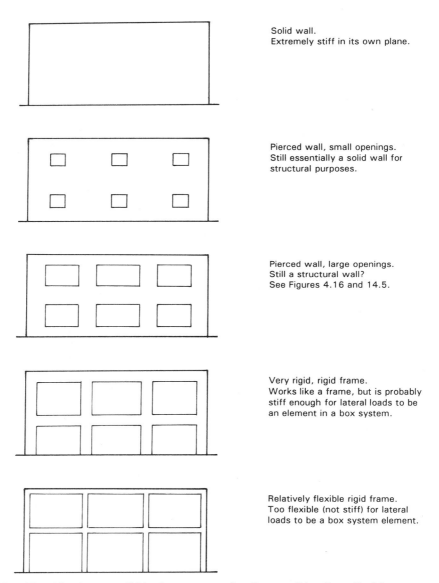

Solid wall.
Extremely stiff in its own plane.

Pierced wall, small openings.
Still essentially a solid wall for structural purposes.

Pierced wall, large openings.
Still a structural wall?
See Figures 4.16 and 14.5.

Very rigid, rigid frame.
Works like a frame, but is probably stiff enough for lateral loads to be an element in a box system.

Relatively flexible rigid frame.
Too flexible (not stiff) for lateral loads to be a box system element.

FIGURE 5.1 Transition in a monolithic planar construction, from a solid wall to a flexible, moment-resisting frame (rigid frame).

remains that with solid planar elements, typically formed by ordinary wall, roof, and floor constructions. Design of various examples of this construction, with the many possible combinations of materials for the individual elements are presented in the design cases in Part III.

The horizontal diaphragm, consisting essentially of the continuous structural deck of a roof or floor, is utilized in most buildings; simply because most buildings have continuous roof and floor surfaces. Variations in lateral-resistive systems are thus mostly achieved with different elements for the vertical components.

Where walls of adequately solid, strong construction are used for the general building construction, they can often be utilized as shear walls, producing a box system with all solid planar elements. Where a major framework of steel or concrete is used, or major open interior space is required, or extensive openings are required in exterior walls or roofs, the general lateral resistive system will usually be developed with trussing or a moment-resisting frame.

The general problems of developing horizontal and vertical diaphragms are discussed in Chapter 6. The remaining sections in this chapter treat the other forms of general bracing systems.

5.2 BRACED FRAMES

Although there are actually several ways to brace a frame against lateral loads, the term *braced frame* is used to refer to frames that utilize trussing as the primary bracing technique. In buildings, trussing is mostly used for the vertical bracing system in combination with the usual horizontal diaphragms. It is also possible, however, to use a trussed frame for a horizontal system, or to combine vertical and horizontal trussing in a truly three-dimensional trussed framework. The latter is more common for

open tower structures, such as those used for large electrical transmission lines and radio and television transmitters.

Use of Trussing for Bracing

Post and beam systems, consisting of separate vertical and horizontal members, may be inherently stable for gravity loading, but they must be braced in some manner for lateral loads. The three basic ways of achieving this are through shear panels moment-resistive joints between the members, or by trussing. The trussing, or triangulation, is usually formed by the insertion of diagonal members in the rectangular bays of the frame.

If single diagonals are used, they must serve a dual function: acting in tension for the lateral loads in one direction and in compression when the load direction is reversed (see Fig. 5.2a). Because long tension members are more efficient than long compression members, frames are often braced with a crisscrossed set of diagonals (called *X-bracing*) to eliminate the need for the compression members. In any event the trussing causes the lateral loads to induce only axial forces in the members of the frame, as compared to the behavior of the rigid frame. It also generally results in a frame that is stiffer for both static and dynamic loading, having less deformation than the rigid frame.

While the stiffness of a truss represents an advantage in some regards (notably in less movement that causes damage to the nonstructural elements of the building) it means that the truss lacks the potential for resiliency and energy absorption that exists with more flexible structures. Thus significant deflection of the truss can occur only with buckling of compression members, tensile yielding of tension members, or major deformation of joints, none of which is really desirable. Joints in particular should be made to resist any loosening,

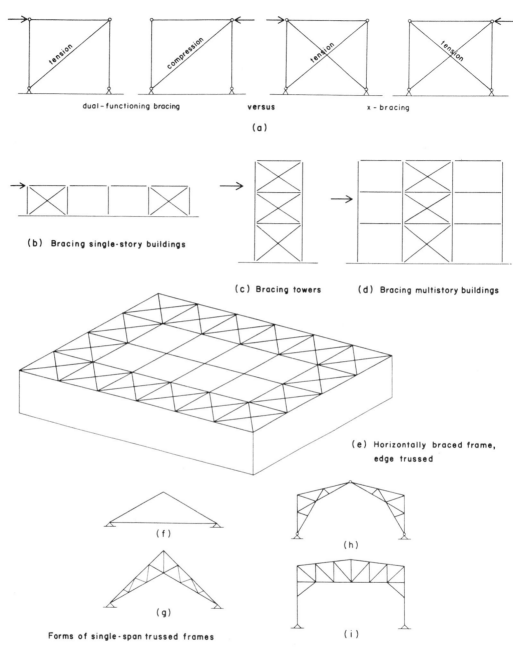

dual-functioning bracing   **versus**   x-bracing

(a)

(b) Bracing single-story buildings

(c) Bracing towers   (d) Bracing multistory buildings

(e) Horizontally braced frame, edge trussed

(f)

(g)

(h)

(i)

Forms of single-span trussed frames

FIGURE 5.2 Considerations of braced frames.

brittle fracture, tearing, or other undesired forms of failure, and should preferably be stronger than the members they connect. This is all to say that the stiffness of a braced frame can be an advantage, but the structure pays a penalty to develop it. You have to accept the penalty to get the bonus.

Single-story, single-bay buildings may

be braced as shown in Fig. 5.2a. Single-story, multibay buildings may be braced by bracing less than all of the bays in a single plane of framing, as shown in Fig. 5.2b. The continuity of the horizontal framing is used in the latter situation to permit the rest of the bays to tag along. Similarly, a single-bay, multistoried, tower-like structure, as shown in Fig. 5.2c, must have its frame fully braced, whereas the more common type of frame for the multistored building, as shown in Fig. 5.2d, is usually only partly braced. Since either the single diagonal or the criss-crossed X-bracing causes obvious problems for interior circulation and for openings for doors and windows, building planning often makes the limited bracing a necessity.

Just about any type of floor construction used for multistored buildings usually has sufficient capacity for diaphragm action in the lateral bracing system. Roofs, however, often utilize light construction or are extensively perforated by openings, so that the basic construction is not capable of the usual horizontal, planar diaphragm action. For such roofs or for floors with many openings, it may be necessary to use a trussed frame for the horizontal part of the lateral bracing system. Figure 5.2e shows a roof for a single-story building in which trussing has been placed in all the edge bays of the roof framing in order to achieve the horizontal structure necessary. As with vertical trussed frames, the horizontal trussed frame may be partly trussed, as shown in Fig. 5.2e, rather than fully trussed.

For single-span structures, trussing may be utilized in a variety of ways for the combined gravity- and lateral-load-resistive system. Figure 5.2f shows a typical gable roof with the rafters tied at their bottom ends by a horizontal member. The tie, in this case, serves the dual functions of resisting the outward thrust due to gravity loads and of one of the members of the single triangle, trussed structure that is rigidly resistive to lateral loads. Thus the wind force on the sloping roof surface, or the horizontal seismic force caused by the weight of the roof structure, is resisted by the triangular form of the rafter–tie combination.

The horizontal tie shown in Fig. 5.2f may not be architecturally desirable in all cases. Some other possibilities for the single-span structure—all producing more openness beneath the structure—are shown in Fig. 5.2g, h, and i. Figure 5.2g shows the so-called *scissors truss*, which can be used to permit more openness on the inside or to permit a ceiling that has a form reflecting that of the gable roof. Figure 5.2h shows a trussed bent that is a variation on the three-hinged arch. The structure shown in Figure 5.2i consists primarily of a single-span truss that rests on end columns. If the columns are pin-jointed at the bottom chord of the truss, the structure lacks basic resistance to lateral loads and must be separately braced. If the column in Fig. 5.2i is continuous to the top of the truss, it can be used in rigid frame action for resistance to lateral loads. Finally, if the knee-braces, shown in the figure are added, the column is further stiffened, and the structure has more load resistance and less deflection under lateral loading.

Planning of Bracing. Some of the problems to be considered in using braced frames are the following:

1. Diagonal members must be placed so as not to interfere with the action of the gravity-resistive structure or with other building functions. If the bracing members are designed essentially as axial stress members, they must be located and attached so as to avoid loadings other than those required for their bracing functions. they must also be located so as not to interfere with door, window, or roof

openings or with ducts, wiring, piping, light fixtures, and so on.

2. As mentioned previously, the reversibility of the lateral loads must be considered. As shown in Fig. 5.2a, such consideration requires that diagonal members be dual functioning (as single diagonals) or redundant (as X-bracing) with one set of diagonals working for load from one direction and the other set working for the reversal loading.

3. Although the diagonal bracing elements usually function only for lateral loading, the vertical and horizontal elements must be considered for the various possible combinations of gravity and lateral load. Thus the total frame must be analyzed for all the possible loading conditions, and each member must be designed for the particular critical combinations that represent its peak response conditions.

4. Long, slender bracing members, especially in X-braced systems, may have considerable sag due to their own dead weight, which requires that they be supported by sag rods or other parts of the structure.

5. The trussed structure should be "tight." Connections should be made in a manner to assure that they will be initially free of slack and will not loosen under the load reversals or repeated loadings. This means generally avoiding connections that tend to loosen or progressively deform such as those that use nails, loose pins, and unfinished bolts.

6. To avoid loading on the diagonals, the connections of the diagonals are sometimes made only after the gravity-resistive structure is fully assembled and at least partly loaded by the building dead loads.

7. The deformation of the trussed structure must be considered, and it may relate to its function as a distributing element, as in the case of a horizontal structure, or to the establishing of its relative stiffness, as in the case of a series of vertical elements that share loads. It may also relate to some effects on nonstructural parts of the building, as was discussed for shear walls.

8. In most cases it is not necessary to brace every individual bay of the rectangular frame system. In fact, this is often not possible for architectural reasons. As shown in Fig. 5.2b, walls consisting of several bays can be braced by trussing only a few bays, or even a single bay, with the rest of the structure tagging along like cars in a train.

The braced frame can be mixed with other bracing systems in some cases. Figure 5.3a shows the use of a braced frame for the vertical resistive structure in one direction and a set of shear walls in the other direction. In this example the two systems act independently, except for the possibility of torsion, and there is no need for a deflection analysis to determine the load sharing.

Figure 5.3b shows a structure in which the end bays of the roof framing are X-braced. For loading in the direction shown, these braced bays take the highest shear in the horizontal structure, allowing the deck to be designed for a lower shear stress.

Figure 5.4 shows a low-rise office building in which X-braced steel bents are used in combination with wood-framed shear walls for the lateral bracing system. The detail shown in Fig. 5.4 illustrates the typical use of steel gusset plates welded to the vertical and horizontal framing members for attachment of the diagonal braces. In this case the diagonal members consist of

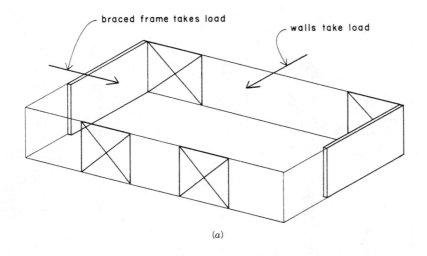

braced frame takes load

walls take load

(a)

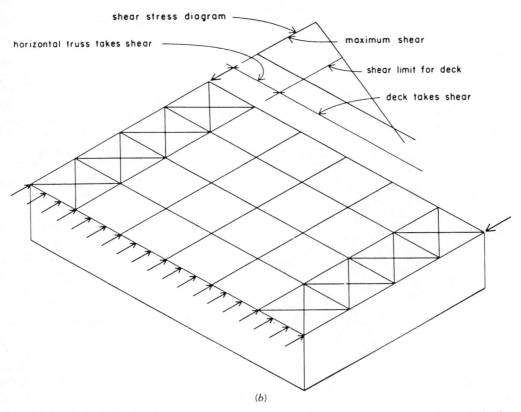

shear stress diagram

maximum shear

horizontal truss takes shear

shear limit for deck

deck takes shear

(b)

FIGURE 5.3 Use of braced systems: (*a*) mixed vertical elements for lateral resistance; (*b*) mixed horizontal diaphragm and braced frame.

77

(c)

(d)

FIGURE 5.4 *Upper:* X-braced bents in a mixed steel and wood system. *Lower:* joint detail for the steel bent, showing use of back-to-back steel channels for the diagonals.

single steel channel sections turned back-to-back to form the X-braces.

Although buildings and their structures are often planned and constructed in two-dimensional components (horizontal floor and roof planes and vertical wall or framing bent planes), it must be noted that the building is truly three dimensional. Bracing against lateral forces is thus a three-dimensional problem, and although a single horizontal or vertical plane of the structure may be adequately stable and strong, the whole system must interact appropriately. While the single triangle is the basic unit for a planar truss, the three-dimensional truss may not be truly stable just because its component planes are braced.

In a purely geometric sense the basic unit for a three-dimensional truss is the four-sided figure called a tetrahedron. However, since most buildings consist of spaces that are rectangular boxes, the three-dimensional trussed building structure usually consists of rectangular units rather than multiples of the pyramidal tetrahedral form (Fig. 5.5). When so used, the single planar truss unit is much the same as a solid planar wall or deck unit, and general reference to the box-type system typically includes both forms of construction.

Typical Construction

Development of the details of construction for trussed bracing is in many ways similar to the design of spanning trusses. The materials used (generally wood or steel), the form of individual truss members, the type of jointing (nails, bolts, welds, etc.), and the magnitudes of the forces are all major considerations. Since many of the members of the complete truss serve dual roles for gravity and lateral loads, member selection is seldom based on truss action alone. Quite often trussed bracing is produced by simply adding diagonals (or X-bracing) to a system already conceived for the gravity loads and for the general development of the desired architectural forms and spaces.

Figure 5.6 shows some details for wood framing with added diagonal members. Wood-framing members are most often rectangular in cross section and metal connecting devices of various form are used in the assembly of frameworks. Figure 5.6a shows a typical beam and column assembly with diagonals consisting of pairs of wood

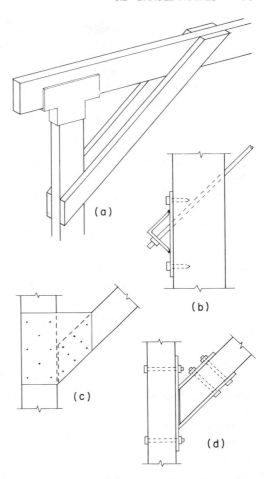

FIGURE 5.6 Framing of trussed bents in wood.

members bolted to the frame. When X-bracing is used, and members need take only tension forces, slender steel rods may be used; a possible detail for this is shown in Fig. 5.6b. For the wood diagonal an alternative to the bolted connection is the type of joint in Fig. 5.6c employing a gusset plate to attach single members all in a single plane. If architectural detailing makes the protruding members shown in Fig. 5.6a or even the protruding gussets in Fig. 5.6c undesirable, a bolted connection like that shown in Fig. 5.6d may be used.

As discussed in next section, a con-

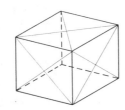

single tetrahedron   the trussed box

FIGURE 5.5 Three-dimensional trussing.

tributing factor in the deformation of the bracing under loading may be movements within the connections. Bolted connections are especially vulnerable when used in shear resistance, since both oversizing of the holes and shrinkage of the wood contribute to a lack of tightness in the joints. In some cases it may be possible to increase the tightness of the joints by using some form of shear developer such as steel split rings.

Gusset plates ordinarily consist of plywood, sheet steel, or steel plate, depending mostly on the magnitude of the loads. Plywood joints should be glued or the nails should be ring or spiral shafted to increase the joint tightness. Steel plate gussets are usually attached by either lag screws or through bolts. Thin sheet metal gussets are either nailed or screwed in place, the latter being preferred for maximum tightness.

Figure 5.7 shows some details for the incorporation of diagonal bracing in steel frames. As with wood structures, bolt loosening is a potential problem. For bolts used in tension—or for the threaded ends of round steel rods—loosening of nuts can be prevented by welding them in place or by scarring the threads. For shear-type connections, highly tensioned, high-strength bolts are preferred over ordinary, unfinished bolts. A completely welded connection will produce the stiffest joint, but on-site bolting in the field is usually preferred over welding.

Various steel elements can be used for diagonal members, depending on the magnitude of loads, the problems of incorporating or exposing the members in the construction, and the requirements for attachment to the structural frame. Figure 5.8 shows an interior view of a building in which a system of exposed truss bents is used for the roof structure as well as the lateral bracing system. Columns are round steel pipes and truss members are mostly

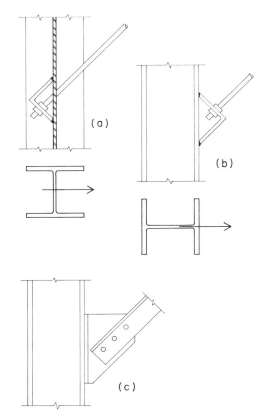

FIGURE 5.7   Use of trussed steel framing details for wide-flange columns.

double angles with welded gusset plate joints, as shown in the detail in Fig. 5.8.

Use of round steel pipe diagonals is shown in Fig. 5.19b. Steel channels are used for X-bracing in the building shown in Fig. 5.4.

Stiffness and Deflection.   As has been stated previously, the braced frame is typically a relatively stiff structure. This is based on the assumption that the major contribution to the overall deformation of the structure is the shortening and lengthening of the members of the frame as they experience the tension and compression forces due to the truss action. However, the

(d)

(e)

FIGURE 5.8 Building with an exposed steel structure consisting of a two-way trussed bent system with round pipe columns. (Public Library for City of Thousand Oaks, California; Albert C. Martin Office, Architects.)

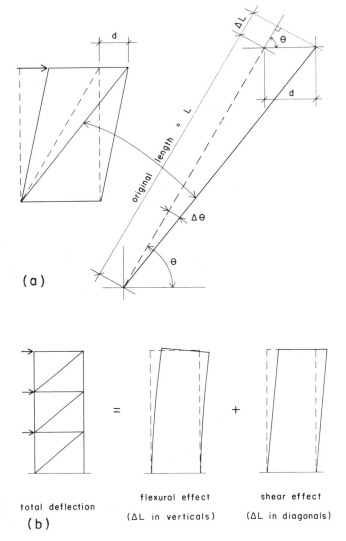

FIGURE 5.9 Lateral deflection of trussed bents.

two other potentially significant contributions to the movement of the braced frame, with either or both of major concern, are:

1. *Movement of the Supports.* This includes the possibilities of deformation of the foundations and yielding of the anchorage connections. If the

foundations rest on compressible soil, there will be some movement due to soil stress. Deformation of the anchorage may be due to a combination of lengthening of anchor bolts and bending of column base plates.

2. *Deformation in the Frame Connections.* This is a complex problem having to do with the general nature of the con-

nection (e.g., welds or glue versus bolts or nails) as well as its form and layout and the deformation of the parts being connected.

It is good design practice in general to study the connection details for braced frames with an eye toward reduction of deformation within the connections. As has been mentioned previously, this generally favors the choice of welding, gluing, high-strength bolts, wood screws, and other fastening techniques that tend to produce stiff, tight joints. It may also favor the choice of materials or form of the frame members as these choices may affect the deformations within the joints or the choice of connecting methods.

The deflection, or drift, of single-story X-braced frames is usually caused primarily by the tension elongation of the diagonal X members. As shown in Fig. 5.9*a*, the elongation of one diagonal moves the rectangular-framed bay into a parallelogram form. The approximate value of the deflection, *d* in Fig. 5.9*a*, can be derived as follows.

Assuming the change in the angle of the diagonal, $\Delta\theta$ in the figure, to be quite small, the change in length of the diagonal may be used to approximate one side of the triangle of which *d* is the hypotenuse. Thus

$$d = \frac{\Delta L}{\cos\theta} = \frac{TL/AE}{\cos\theta} = \frac{TL}{AE\cos\theta}$$

where

T = tension in the X caused by the lateral load

A = cross-sectional area of the X

E = elastic modulus of the X

θ = angle of the X from the horizontal

The deflection of multistory X-braced frames has two components, both of which may be significant. As shown in Fig. 5.9*b*, the first effect is caused by the change in length of the vertical members of the frame as a result of the overturning moment. The second effect is caused by the elongation of the diagonal X, as discussed for the single-story frame. These deflections occur in each level of the frame and can be calculated individually and summed up for the whole frame. Although this effect is also present in the single-story frame, it becomes more pronounced as the frame gets taller with respect to its width. These deflections of the cantilever beam can be calculated using standard formulas such as those given in the beam diagrams and formulas in Part 2 of the *Manual of Steel Construction* (Ref. 11).

5.3 MOMENT-RESISTIVE FRAMES

There is some confusion over the name to be used in referring to frames in which interactions between members of the frame include the transfer of moments through the connections. In years past the term most frequently used was *rigid frame*. This term came primarily from the classification of the connections or joints of the frame as *fixed* (or rigid) versus *pinned*, the latter term implying a lack of capability to transfer moment through the joint. As a general descriptive term, however, the name was badly conceived, since the frames of this type were generally the most deformable under lateral loading when compared to trussed frames or those braced by vertical diaphragms. The *UBC* (Ref. 1) uses the specific term *moment-resisting space frame* and gives various qualifications for such a frame when it is used for seismic resistance. With apologies to the *UBC*, although we will assume the type of frames they thus define, we prefer not to use this rather cumbersome mouthful of a term, so will use the simpler term of rigid frame in our discussions.

General Behavior

In rigid frames with moment-resistive connections, both gravity and lateral loads produce interactive moments between the members. Design must consider both load conditions.

In most cases rigid frames are actually the most flexible of the basic types of lateral resistive systems. This deformation character, together with the required ductility, makes the rigid frame a structure that absorbs energy loading through deformation as well as through its sheer brute strength. The net effect is that the structure actually works less hard in force resistance because its deformation tends to soften the loading. This is somewhat like rolling with a punch instead of bracing oneself to take it head on.

Most moment-resistive frames consist of either steel or concrete. Steel frames have either welded or bolted connections between the linear members to develop the necessary moment transfers. Frames of concrete achieve moment connections through the monolithic concrete and the continuity and anchorage of the steel reinforcing. Because concrete is basically brittle and not ductile, a ductile character is essentially produced by the ductility of the reinforcing. The type and amount of reinforcing and the details of its placing become critical to the proper behavior of rigid frames of reinforced concrete.

A complete presentation of the design of moment-resistive ductile frames for seismic loads is beyond the scope of this book. Such design can be done only by using plastic design for steel and ultimate strength design for reinforced concrete. We present only a brief discussion of this type of structure. For wind loading the analysis and design may be somewhat more simplified. However, if the structure is considerably indeterminate, an accurate analysis requires a complex and laborious calculation. We show some examples in Chapter 12, but limit the analysis to approximate methods.

For lateral loads in general, the rigid frame offers the advantage of a high degree of freedom in architectural terms. Walls and interior spaces are freed of the necessity for solid diaphragms or diagonal members. For building planning as a whole, this is a principal asset. Walls, even where otherwise required to be solid, need not be of a construction qualifying them as shear walls.

When seismic force governs as the critical lateral load, the moment-resistive frame has the advantage of having the highest value for the R_W factor, and thus obtains the lowest design value for base shear (see *UBC* Table 16-N). This is only true, of course, for frames qualified as "special" by the *UBC* criteria whether used alone or in a dual system. This favored status has to do with the dynamic behavior of the relatively flexible frames, but also reflects preferences with regard to the nature of behavior of the frames. Some considerations are the toughness resulting from formation of plastic hinges, the response to continued, cyclic loading, and the redundancy of capacity of the typically highly indeterminate systems.

Deformation analysis is a critical part of the design of rigid frames because such frames tend to be relatively deformable when compared to other lateral resistive systems. The deformations have the potential of causing problems in terms of movements of a disturbing nature that can be sensed by the building occupants or of damage to nonstructural parts of the building, as previously discussed. The need to limit deformations often results in the size of vertical elements of the frame being determined by stiffness requirements rather than by stress limits.

Loading Conditions

Unlike shear walls or X-bracing, rigid frames are not generally able to be used for lateral bracing alone. Thus their structural actions induced by the lateral loads must always be combined with the effects of gravity loads. These combined loading conditions may be studied separately in order to simplify the work of visualizing and quantifying the structural behavior, but it should be borne in mind that they do not occur independently.

Figure 5.10*a* shows the form of deformation and the distribution of internal bending moments in a single-span rigid frame, as induced by vertical gravity loading. If the frame is not required to resist lateral loads, the singular forms of these responses may be assumed, and the various details of the structure may be developed in this context. Thus the direction of rotation at the column base, the sign of moment at the beam-to-column joint, the sign of the bending moment and nature of the corresponding stresses at midspan of the beam, and

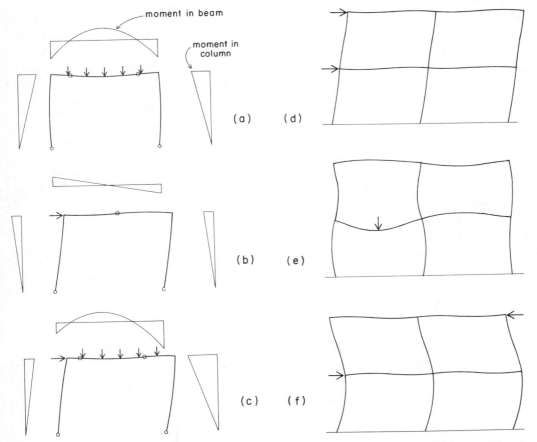

FIGURE 5.10 Behavior of rigid frames: (*a*) under gravity load; (*b*) under lateral load; (*c*) under combined gravity and lateral loads; (*d*) lateral load on a multistory bent; (*e*) effect of single gravity load in a multiunit bent; (*f*) effects of rapid reversals of lateral loads.

the location of inflection points in the beam may all relate to choices for the form and details of the members and development of any connection details. If the frame is reasonably symmetrical, the only concerns for deflection are the outward bulging of the columns and the vertical sag of the beam.

Under action of lateral loading due to wind or seismic force, the form of deformation and distribution of internal bending moment will be as shown in Fig. 5.10b. If the gravity and lateral loadings are combined, the net effect will be as shown in Fig. 5.10c. Observing the effects of the combined loading, we note the following:

1. Horizontal deflection at the top of the frame (called drift) must now be considered, in addition to the deflections mentioned previously for gravity load alone.

2. The maximum value for the moment at the beam-to-column connection is increased on one side and reduced on the other side of the bent. The increased moment requires that the beam, the column, and the connection at the joint must all be stronger for the combined loading.

3. If the lateral load is sufficient, the minimum value for the moment at the beam-to-column joint may be one of opposite sign from that produced by gravity loading alone. The form of the connection, and possibly the design of the members may need to reflect this reversal of the sense of the moment.

4. The direction of the lateral load shown in Fig. 5.10b is reversible so that two combinations of load must be considered: gravity plus lateral load to the right and gravity plus lateral load to the left.

While single-span rigid frames are often used for buildings, the multispan or multistory frame is the more usual case. Figure 5.10d and e show the response of a two-bay, two-story frame to lateral loads and to a gravity-type load applied to a single beam. The response to lateral loads is essentially similar to that for the single bent in Fig. 5.9. For gravity loads the multiunit frame must be analyzed for a more complex set of potential combinations, because the live load portion of the gravity loads must be considered to be random, and thus may or may not occur in any given beam span.

Lateral loads produced by winds will generally result in the loading condition shown in Fig 5.10d. Because of its relative flexibility and size, however, a multistory building frame may quite likely respond so slowly to seismic motions that upper levels of the frame experience a whiplashlike effect; thus separate levels may be moving in opposite directions at a single moment. Figure 5.10f illustrates a type of response that may occur if the two-story frame experiences this action. Only a true dynamic analysis can ascertain whether this action occurs and is of critical concern for a particular structure.

Approximate Analysis for Gravity Loads. Most rigid frames are statically indeterminate and require the use of some method beyond simple statics for their analysis. Simple frames of few members may be analyzed by some hand method using handbook coefficients, moment distribution, and so on. If the frame is complex, consisting of several bays and stories, or having a lack of symmetry, the analysis will be quite laborious unless performed with some computer-aided method.

For preliminary design it is often useful to have some approximate analysis, which can be fairly quickly performed. Internal forces, member sizes, and deflections thus

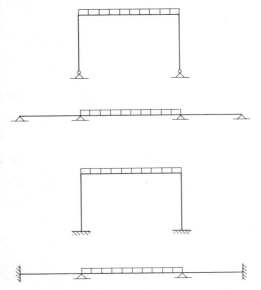

FIGURE 5.11 Single-span rigid frame; continuous beam analogy.

bined as required for the various critical combinations of dead load, live load, wind load, and seismic load (see *UBC* Sec. 1603.6 in Appendix B).

Approximate Analysis for Lateral Loads. Various approximate methods may be used for the analysis of rigid frames under statically applied lateral loading. For ordinary frames, whether single-bay, multibay, or even multistory, approximate methods are commonly used for loading due to wind or as obtained from an equivalent static load analysis for seismic effects. More exact analyses are possible, of course, especially when performed by computer-aided methods.

For frames that are complex—due to irregularities, lack of symmetry, tapered members, and so on—analysis is hardly feasible without the computer. This is also true for analyses that attempt to deal with the true dynamic behavior of the structure under seismic load. With the increasing availability of the software, and the accumulation of experience with its use, this type of analysis is becoming more widespread in use. For quick approximations for preliminary design, however, approximation methods are likely to continue in use for some time.

For the simple bent shown in Fig. 5.12, the effects of the single lateral force may be quite simply visualized in terms of the deflected shape, the reaction forces, and the variation of moment in the members.

If the columns are assumed to be pin based and of equal stiffness, it is reasonable to assume that the horizontal reactions at the base of the columns are equal, thus permitting an analysis by statics alone. If the column bases are assumed to be fixed, the frame is truly able to be analyzed only by indeterminate methods, although an approximate analysis can be made with an assumed location of the inflection point in

determined may be used for a quick determination of the structural actions and the feasibility of some choices of systems and components. For the simple, single-bay bent shown in Fig. 5.11, the analysis for gravity loads is quite simple, since a single-loading condition exists (as shown) and the only necessary assumption is that of the relative stiffnesses of the beam and columns. For the frame with pin-based columns the analogy is made to a three-span beam on rotation-free supports. If the column bases are fixed, the end supports of the analogous beam are assumed fixed.

For multibayed, multistoried frames, an approximate analysis may be performed using techniques such as that described in Chapter 8 of the *ACI Code* (Ref. 14). This is more applicable to concrete frames, of course, but can be used for quick approximation of welded steel frames as well. Even when using approximate methods, it is advisable to analyze separately for dead and live loads. The results can thus be com-

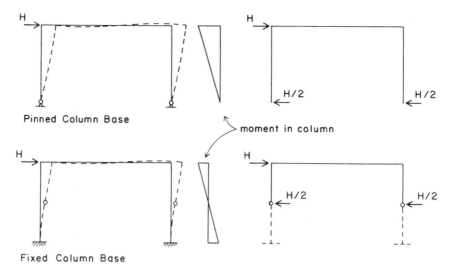

FIGURE 5.12 Effects of column base conditions on a single-span frame.

the column. In truth, the column bases will most likely be somewhere between these two idealized conditions. Approximate designs are sometimes done by combining the results from both idealized conditions (pinned and fixed bases) and designing for both. Adjustments are made when the more precise-nature of the condition is established by the detailed development of the base construction.

For multibayed frames, such as those shown in Fig. 5.13, an approximate analysis may be done in a manner similar to that for the single-bay frame. If the columns are all of equal stiffness, the total load is simply divided by the number of columns. Assumptions about the column base condition would be the same as for the single-bay frame. If the columns are not all of the same stiffness, an approximate distribution can be made on the basis of relative stiffness.

Figure 5.13c illustrates the basis for an approximation of the horizontal shear forces in the columns of a multistory building. As for the single-story frame, the in-

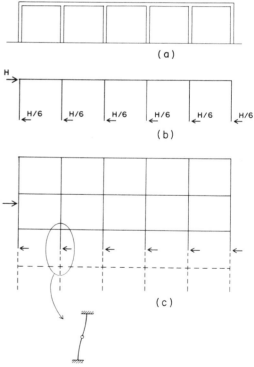

FIGURE 5.13 Distribution of lateral load in a multiunit frame.

dividual column shears are distributed on the basis of assumed column stiffnesses. For the upper columns the inflection point is assumed to occur at mid-height, unless column splice points are used to control its location.

General Design Considerations

From a purely structural point of view, there are many advantages of the rigid frame as a lateral bracing system for resisting seismic effects. The current codes tend to favor it over shear walls or braced (trussed) frames by requiring less design load due to the higher assigned R_W value. For large frames the combination of slow reaction time (because of long fundamental period of vibration) and various damping effects means that forces tend to dissipate rapidly in the remote portions of the frame.

Architecturally, the rigid frame offers the least potential for interference with planning of open spaces within the building and in exterior walls. It is thus highly favored by architects whose design style includes the use of ordered rectangular grids and open spaces of rectangular or cubical form. Geometries other than rectangular ones are possible also; what remains as the primary advantage is the absence of the need for diagonal bracing or solid walls at determined locations.

Some of the disadvantages of rigid frames are the following:

1. Lateral deflection, or drift, of the structure is likely to be a problem. As a result, it is often necessary to stiffen the frame—mostly by increasing the number and/or size of the columns.
2. Connections must be stronger, especially in steel frames. Addition of heavily welded steel joints, of additional reinforcing for bar anchorage, shear, and torsion in concrete frames, and for cumbersome moment-resis-

tive connections in wood frames can add appreciably to the construction cost and time.
3. For large frames dynamic behavior is often quite complex, with potential for whiplash, resonance, and so on. Dynamic analysis is more often required for acceptable design when a rigid frame is used. In time, dynamic analysis may be routine and easily performed with available aids on all structures; at the present it is quite expensive and time consuming.

When rigid frames are to be used for the lateral bracing system for a building, there must be a high degree of coordination between the planner of the building form and the structural designer. As illustrated in the examples in Chapter 12, the building plan must accommodate some regularity and alignment of the columns that constitute the vertical members of the rigid frame bents. This may not include *all* of the building columns, but those that function in the rigid frames must be aligned to form the planes of the bents. Openings in floors and roofs must be planned so as not to interrupt beams that occur as horizontal members of the bents.

There must, of course, be a coordination between the designs for lateral and gravity loads. If a rigid frame is used for bracing, it will in most cases be used in conjunction with some type of framed floor and roof systems. Horizontal members of the floor and roof systems designed for gravity loads will also be used as horizontal elements of the rigid frames. However, in some cases, much of the horizontal structure may be used only for gravity loads. In the system that utilizes only the bents that occur in exterior walls, none of the interior portion of the framing is involved in the rigid frame actions; thus its planning is free of concerns for the bent development.

Selection of members and construction details for rigid frames depends a great deal on the materials used. The following discussion deals separately with the problems of bents of steel and reinforced concrete.

Steel Frames. Steel frames with moment-resistive connections were used for early skyscrapers. Fasteners consisted of rivets, which were widely used until the development of high-strength bolts. Today,

rigid frame construction in steel mostly utilizes welded joints, although bolting is sometimes used for temporary connection during erection. Figure 5.14 shows the erected frame for a low-rise office building, using a steel frame for both gravity and lateral load resistance. The principal members of the frame are wide flange (I-shaped) rolled steel sections, and moment connections are welded. This is the most common form of steel rigid frame for building construction.

FIGURE 5.14 Steel frame structure with a perimeter moment-resisting frame.

Another form of steel rigid frame is the trussed bent. A single-span bent is illustrated in Fig. 5.2i, and the use of a two-span bent is described in the building design study in Fig. 11.13.

For tall buildings a currently popular system is one that uses a perimeter bracing arrangement with closely spaced, stiff steel columns and heavy spandrel beams. This type of structure offers a major advantage in its overall stiffness in resistance to lateral drift (horizontal deflection). Figure 5.15 shows the erected frame for such a structure. Note that the exterior rigid bents are discontinuous at the corners, thus avoiding

FIGURE 5.15   Use of closely spaced columns and stiff spandrels in a perimeter bent system.

the high concentration of forces on the corner columns, especially those due to torsional action of the building.

As a result of experience in the January, 1994, Northridge earthquake, this popular form of bracing for seismic forces is now seriously in question. Several recently-built buildings suffered damage in the form of fractured connections of the steel rigid frames. At the time of writing of this book research is still inconclusive with regard to what modifications may be necessary to improve the construction of the steel rigid frame with welded joints.

Reinforced Concrete Frames.   Cast-in-place frames with monolithic columns and beams have a natural rigid frame action. For seismic resistance both columns and beams must be specially reinforced for the shears and torsions at the member ends. Beams in the column-line bents ordinarily use continuous top and bottom reinforcing with continuous loop ties that serve the triple functions of resisting shear, torsion, and compression bar buckling.

Figure 5.16 shows two buildings with exposed concrete frames that are of this type of construction. It is possible, of course, to brace such a building with concrete shear walls and to use the frame strictly for gravity resistance, except for collector and chord functions.

Precast concrete structures are often difficult to develop as rigid frames, unless the precast elements are developed as individual bent units instead of the usual, single, linear members. Moment-resistive joints for these structures are usually quite difficult to develop, the more common solution being to use a shear wall system for lateral bracing.

Failures of concrete structures in recent earthquakes have pointed out various vulnerabilities of this construction. This has resulted in increased requirements which have benefitted new construction but leave older structures with questionable resistance. Retrofitting (bringing up to present standards) of older buildings has become a major industry in areas of high risk for earthquakes. Some of the issues involved in this are discussed in Part IV.

FIGURE 5.16   Exposed concrete frames.

5.4 ECCENTRICALLY BRACED FRAMES

As shown in Fig. 5.17, there are three basic types of systems that use diagonals. The first is the traditional trussing, with all members connected at common intersec-tions. This is the good old, pin-connected system with two–force members (either tension or compression, but no direct shear or bending) (see Fig. 5.17a). Because the bracing members all connect to frame joints, this is called *concentric bracing*. This system works, although a more common usage for lateral loading is the X-braced system because the loading direction is reversible.

Various systems employ members with one end not connected to a frame joint (see Fig. 5.17b). The connection made within the length of a member is called an *eccentruic connection* and the system is described as *semi–concentric bracing*. Knee–bracing was used in various forms in ancient times, although it involves development of some shear and bending in the columns. It was probably the means for development of the first rigid bents. Variations on this were the K-bracing and V-bracing (also called chevron bracing if the V is upside down) developed with early highrise steel con-struction, and still widely used for wind

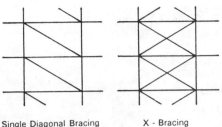

Single Diagonal Bracing   X - Bracing

(*a*)

Concentric, Traditional Bracing

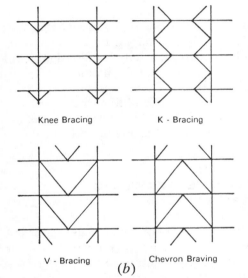

Knee Bracing   K - Bracing

V - Bracing   Chevron Braving

(*b*)

Semi - Concentric Bracing, Beam - to - Column

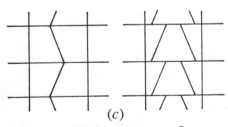

(*c*)

Eccentric Bracing, Beam - to - Beam

FIGURE 5.17 Forms of eccentric bracing.

bracing. All of these arrangements allow for some more opening of the braced bay than systems using full diagonals or Xs.

A more recent development is *fully eccentric bracing* in which neither of the ends of the diagonal are connected to a beam/column joint. While possible only marginally constituting trussing, this system is still capable of considerably stiffening a rigid frame with members otherwise arranged in rectilinear patterns.

Use of eccentric bracing results in some combined form of truss and rigid frame actions, as shown in Fig. 5.18. The trussing action often produces a relative stiffness for the bents that is close to that for a regular truss, and its use is sometimes significant for this stiffening effect on the otherwise more flexible, purely rigid frame structure. On the other hand, the addition of bending in some members adds a redundant be-

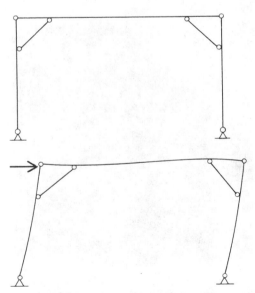

FIGURE 5.18 Knee-braced bent.

FIGURE 5.19 (*Continued*)

havior and considerably more energy capacity to an ordinary trussed system. This added energy capacity is quite significant for seismic resistance, making the eccentrically braced frame presently quite popular for tall steel building structures.

Figure 5.19 shows some examples of the use of eccentric bracing in steel structures.

5.5 MIXED FRAME AND WALL SYSTEMS

Most buildings consist of combinations of walls and some framing of wood, steel, or concrete. The planning and design of the lateral resistive structure require some judgments and decisions regarding the roles of the frame and the walls. In this section we discuss some of the issues relating to this aspect of design.

FIGURE 5.19 Use of eccentric bracing: (*a*) knee braces in a light steel bent; (*b*) inverted V-bracing in a low-rise steel frame; (*c*) complex bent with combinations of eccentric and ordinary trussing.

Coexisting, Independent Elements

Most buildings have some solid walls, that is, walls with continuous surfaces free of openings. When the gravity load-resistive structure of the building consists of a frame, the relationship between the walls and the frame has several possibilities with regard to action caused by lateral loads.

The frame may be a braced frame or a moment-resistive frame designed for the total resistance of the lateral loads, in which case the attachment of walls to the frame must be done in a manner that prevents the walls from absorbing lateral loads. Because solid walls tend to be quite stiff in their own planes, such attachment often requires the use of separation joints or flexible connections that will allow the frame to deform as necessary under the lateral loads.

The frame may be essentially designed for gravity resistance only, with lateral load resistance provided by the walls acting as shear walls (see Fig. 5.20). This method requires that some of the elements of the frame function as collectors, stiffeners, shear wall end members, or diaphragm chords. If the walls are intended to be used strictly for lateral bracing, care must be exercised in the design of construction details to assure that beams that occur above the walls are allowed to deflect without transferring loads to the walls.

Load Sharing

When walls are firmly attached to vertical elements of the frame, they usually provide continuous lateral bracing in the plane of the wall, thus permitting the vertical frame elements to be designed for column action using their stiffness in the direction perpendicular to the wall. Thus 2×4 studs may be designed as columns using h/d ratios based on their larger dimension.

In some cases both walls and frames may be used for lateral load resistance at different locations or in different directions. Figure 5.21 shows four such situations. In Fig. 5.21a a shear wall is used at one end of the building and a frame at the other end for the wind from one direction. In Fig. 5.21b walls are used for the lateral loads from one direction and frames for the load from the other direction. In both cases the walls and frames do not actually interact; that is, they act independently with regard to load sharing.

Figure 5.21c and d show situations in which walls and frames interact to share a direct load. The walls and frames share the total load from a single direction. If the horizontal structure is a rigid diaphragm, the load sharing will be on the basis of the relative stiffness of the vertical elements. This relative stiffness must be established by the calculated deflection resistance of the elements, as discussed previously.

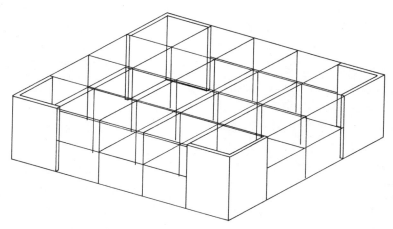

FIGURE 5.20 Wall-braced frame structure.

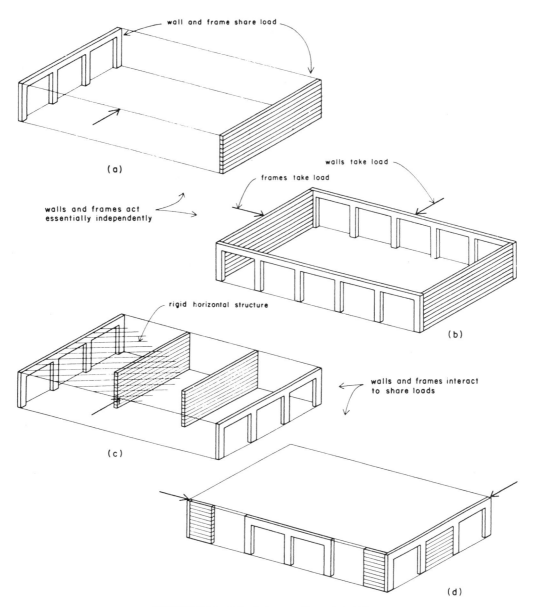

FIGURE 5.21 Mixed shear walls and rigid frames.

Dual Systems. The *UBC* Sec. 1627.6.5 defines a dual system as one in which a special moment-resistive space frame interacts with shear walls or eccentric bracing to form a combined resistance to lateral loads. The code requires that the two systems be designed to resist loads in proportion to their relative stiffnesses, but that the space frame be designed for not less than 25% of the total load. Use of a dual system permits the obtaining of the various benefits of behavior of the rigid frame, while having a much stiffer structure, thus reducing the problems of major movements experienced with some lighter steel frames or the extensive cracking in some concrete frames.

5.6 SPECIAL SYSTEMS

The preceding sections in this chapter treat the ordinary means for bracing buildings for lateral force. There are, of course, many possibilities for special building forms or situations in which some other forms of bracing might be used. A few examples are the following.

1. *Arches and Domes.* In modern construction, arches are typically constituted as rigid elements having some bending capacity as well as the ability to resist the natural arch compression forces. Thus an arch in its own plane usually has some rigid frame capacity and is quite stable. In a direction perpendicular to the arch, however, some bracing is required, just as for a series of beams, planar trusses, trussed bents, and so on: This cross-bracing may be achieved by any of the basic means: shear walls, trussing, or rigid frame action. Domes, on the other hand, have a natural three-dimensional stability, if it can be developed in the overall structure.

2. *Tension Structures.* Lateral stability of tension structures must usually be developed as part of the overall design. There are many variations of tension structures and few common techniques. A special problem of draped tension elements is sometimes encountered because of their lack of bending or shear capabilities; thus loads other than pure gravity may induce difficult conditions. Secondary structures must sometimes be used to stabilize highly flexible tension cable or surface structures.

3. *Large Abutments or Piers.* Long-span structures sometimes use very large support elements and the support elements may be constituted as massive structures with their own inherent stability. These elements may thus become individual, freestanding structures, subjected to the combined gravity and lateral loads.

4. *Pole Structures.* A pole-type structure is one that is analogous to a nail that sustains lateral force. The pole is inserted in the ground and the combination of the surrounding earth and the bending capacity of the pole produce a lateral resistance in the form of a vertical cantilever. Fenceposts, driven piles, and other structures function in this manner. The general problems of pole structures are discussed in Sec. 8.7 and 13.3.

This book is intended primarily as a reference for common and simple systems, so we do not intend to try to illustrate all of the possibilities for lateral bracing. No matter how exotic the structure, however, the basic principles demonstrated here must still apply, and both investigations and design must essentially be pursued in a similar manner.

6

ELEMENTS OF LATERAL RESISTIVE SYSTEMS

The basic systems used for lateral force resistance in buildings were presented in Chapter 5. Development of these systems, and the general development of the whole building as a force resistive entity, requires the use of various special elements and devices. In this chapter we present some of the component elements commonly used in lateral resistive systems.

6.1 HORIZONTAL DIAPHRAGMS

Most lateral resistive structural systems for buildings consist of some combination of vertical elements and horizontal elements. The horizontal elements are most often the roof and floor framing and decks. When the deck is of sufficient strength and stiffness to be developed as a rigid plane, it is called a *horizontal diaphragm.*

General Behavior

A horizontal diaphragm typically functions by collecting the lateral forces at a particular level of the building and then distributing them to the vertical elements of the lateral resistive system. For wind forces the lateral loading of the horizontal diaphragm is usually through the attachment of the exterior walls to its edges. For seismic forces the loading is partly a result of the weight of the deck itself and partly a result of the weights of other parts of the building that are attached to it.

The particular structural behavior of the horizontal diaphragm and the manner in which loads are distributed to vertical elements depend on a number of considerations that are best illustrated by various example cases in Part III. Some of the general issues of concern are treated in the following discussions.

Relative Stiffness of the Horizontal Diaphragm. If the horizontal diaphragm is relatively flexible, it may deflect so much that its continuity is negligible and the distribution of load to the relatively stiff vertical elements is essentially on a load periphery basis. If the deck is quite rigid, on the other hand, the distribution to vertical elements will be essentially in proportion to their relative stiffness with respect to each other. The possibility of these two situations is illustrated for a simple box system in Fig. 6.1.

Torsional Effects. If the centroid of the lateral forces in the horizontal diaphragm does not coincide with the centroid of the stiffness of the vertical elements, there will be a twisting action (called *rotation effect* or *torsional effect*) on the structure as well as the direct force effect. Figure 6.2 shows a structure in which this effect occurs because of a lack of symmetry of the structure. This effect is usually of significance only if the horizontal diaphragm is relatively stiff. This stiffness is a matter of the materials of the construction as well as

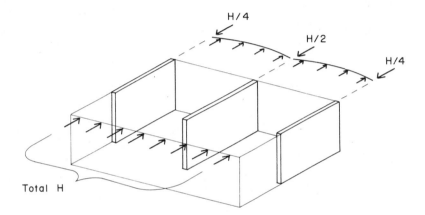

Peripheral distribution – flexible horizontal diaphragm

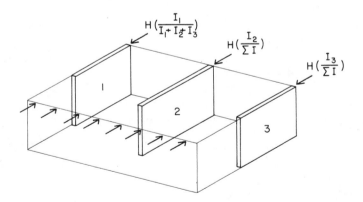

Proportionate stiffness distribution – rigid horizontal diaphragm

FIGURE 6.1 Peripheral distribution versus proportionate stiffness distribution.

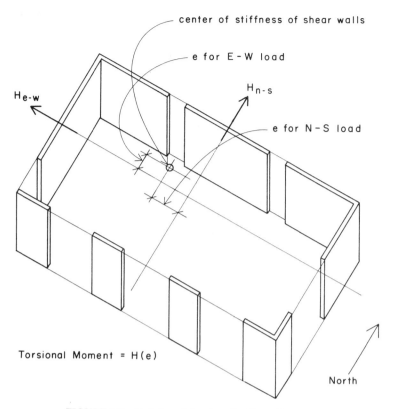

center of stiffness of shear walls

e for E-W load

H_{e-w}

H_{n-s}

e for N-S load

Torsional Moment = H(e)

North

FIGURE 6.2   Rotational (torsional) effect of lateral load.

the depth-to-span ratio of the horizontal diaphragm. In general, wood and metal decks are quite flexible, whereas poured concrete decks are very stiff.

Relative Stiffness of the Vertical Elements.

When vertical elements share load from a rigid horizontal diaphragm, as shown in the lower figure in Fig. 6.1, their relative stiffness must usually be determined in order to establish the manner of the sharing. The determination is comparatively simple when the elements are similar in type and materials such as all plywood shear walls. When the vertical elements are different, such as a mix of plywood and masonry shear walls or of some shear walls and some braced frames, their actual deflec-

tions must be calculated in order to establish the distribution, and this may require laborious calculations.

Use of Control Joints.

The general approach in design for lateral loads is to tie the whole structure together to assure its overall continuity of movement. Sometimes, however, because of the irregular form or large size of a building, it may be desirable to control its behavior under lateral loads by the use of structural separation joints. In some cases these joints function to create total separation, allowing for completely independent motion of the separate parts of the building. In other cases the joints may control movements in a single direction while achieving connection for load transfer in other directions. A general discussion of separation joints is given in Sec. 6.5.

Design and Usage Considerations

In performing their basic tasks, horizontal diaphragms have a number of potential stress problems. A major consideration is that of the shear stress in the plane of the diaphragm caused by the spanning action of the diaphragm as shown in Fig. 6.3. This

spanning action results in shear stress in the material as well as a force that must be transferred across joints in the deck when the deck is composed of separate elements such as sheets of plywood or units of formed sheet metal. The sketch in Fig. 6.4 shows a typical plywood framing detail at the joint between two sheets. The stress in the deck at this location must be passed from one sheet through the edge nails to the framing member and then back out through the other nails to the adjacent sheet.

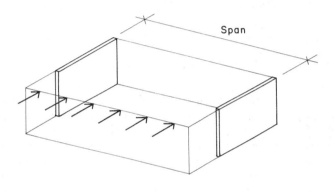

Beam Analogy

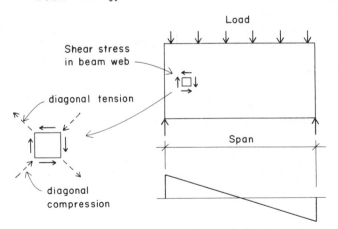

FIGURE 6.3 Functions of a horizontal diaphragm.

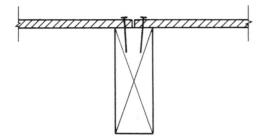

FIGURE 6.4 Load continuity in the plywood diaphragm.

As is the usual case with shear stress, both diagonal tension and diagonal compression are induced simultaneously with the shear stress. The diagonal tension becomes critical in materials such as concrete. The diagonal compression is a potential source of buckling in decks composed of thin sheets of plywood or metal. In plywood decks the thickness of the plywood relative to the spacing of framing members must be considered, and it is also why the plywood must be nailed to intermediate framing members (not at edges of the sheets) as well as at edges. In metal decks the gauge of the sheet metal and the spacing of stiffening ribs must be considered. Tables of allowable loads for various deck elements usually incorporate some limits for these considerations.

Diaphragms with continuous deck surfaces are usually designed in a manner similar to that for webbed steel beams. The web (deck) is designed for the shear, and the flanges (edge-framing elements) are designed to take the moment, as shown in Fig. 6.5. The edge members are called *chords*, and they must be designed for the tension and compression forces at the edges. With diaphragm edges of some length, the latter function usually requires that the edge members be spliced for some continuity of the forces. In many cases there are ordinary elements of the framing system, such as spandrel beams or top plates of stud walls, that have the potential to function as chords for the diaphragm.

In some cases the collection of forces into the diaphragm or the distribution of loads to vertical elements may induce a stress beyond the capacity of the deck alone. Figure 6.6 shows a building in which a continuous roof diaphragm is connected to a series of shear walls. Load collection and force transfers require that some force be dragged along the dotted lines shown in the figure. For the outside walls it is possible that the edge framing used for chords can do double service for this purpose. For the interior shear wall, and possibly for the edges if the roof is cantilevered past the walls, some other framing elements may be necessary to reinforce the deck.

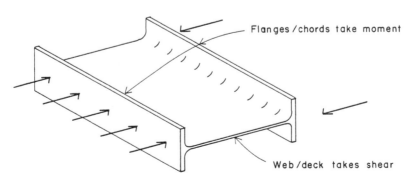

Flanges/chords take moment

Web/deck takes shear

FIGURE 6.5 Functioning of a horizontal diaphragm; beam analogy.

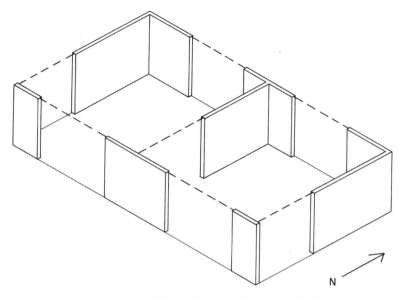

FIGURE 6.6 Collector functions in a box system.

The diaphragm shear capacities for commonly used decks of various materials are available from the codes or from load tables prepared by deck manufacturers. Loads for plywood decks are given in the *UBC* (see Appendix B). Other tabulations are available from product manufacturers, although care should be exercised in their use to be certain that they are acceptable to the building code of jurisdiction.

A special situation is a horizontal system that consists partly or wholly of a braced frame. Care may be required when there are a large number of openings in the roof deck, or when the diaphragm shear stress is simply beyond the capacity of the deck. In the event of a deck with no code-accepted rating for shear, the braced frame may have to be used for the entire horizontal system.

The horizontal deflection of flexible decks, especially those with high span-to-depth ratios, may be a critical factor in their design. Calculation of actual deflection dimensions may be required to determine the effect on vertical elements of the build-

ing construction or to establish positively whether the deck must be considered as essentially flexible or rigid, as discussed previously.

The use of subdiaphragms may also be required in some cases, necessitating the design of part of the whole system as a separate diaphragm, even though the deck may be continuous.

Typical Construction

The most common horizontal diaphragm is the plywood deck for the simple reason that wood frame construction is so popular and plywood is mostly used for roof and floor decks. For roofs the deck may be as thin as $\frac{3}{8}$ in., but for flat roofs with water-proof membranes decks are usually $\frac{1}{2}$ in. or more. Attachment is typically by nailing, although glued floor decks are used for their added stiffness and to avoid nail popping and squeaking. Mechanical devices for nailing may eventually become so common that shear capacities will be based on some other fastener; at present the common wire nail is still the basis for load rating.

Attachment of plywood to chords and collectors and load transfers to vertical shear walls are also mostly achieved by nailing. Code-acceptable shear ratings are based on the plywood type and thickness, the nail size and spacing, and features such as size and spacing of framing and use of blocking. Load capacities for plywood decks are given in *UBC* Table 23-I-J-1 which is reprinted in Appendix B.

In general, plywood decks are quite flexible and should be investigated for deflection when spans are large or span-to-depth ratios are high.

Decks of boards or timber, usually with tongue-and-groove joints, were once popular but are given low rating for shear capacity at present. Where the exposed plank-type deck is desired, it is not uncommon to use a thin plywood deck on top of it for lateral force development.

Steel decks offer possibilities for use as diaphragms for either roof or floors. Acceptable shear capacities should be obtained from the supplier for any particular product, as capacities vary considerably and code approval is not consistent. Stiffnesses are generally comparable with those of plywood decks. Floor decks receive concrete fill, which significantly increases the stiffness of the deck. In some situations it is practical to use the concrete as the basic shear-resisting element.

Poured-in-place concrete decks provide the strongest and stiffest diaphragms. Precast concrete deck units, as well as the slab portions of precast systems, can be used for diaphragms. Precast units must be adequately attached to each other and to supporting members for diaphragm actions; if designed only for gravity, ordinary attachments will usually not be adequate for lateral forces. As with steel deck, concrete fill is sometimes placed on top of precast units,

which both stiffens and strengthens the system.

Many other types of roof deck construction may function adequately for diaphragm action, especially when the required unit shear resistance is low. Acceptability by local building code administration agencies should be determined if any construction other than those described is to be used.

Stiffness and Deflection. As spanning elements, the relative stiffness and actual dimensions of deformation of horizontal diaphragms depend on a number of factors such as:

The materials of the construction

The continuity of the spanning diaphragm over a number of supports

The span-to-depth ratio of the diaphragm

The effect of various special conditions, such as chord length changes, yielding of connections and influence of large openings

In general, wood and light-gauge metal decks tend to produce quite flexible diaphragms, whereas poured concrete decks tend to produce the most rigid diaphragms. Ranging between these extremes are decks of lightweight concrete, gypsum concrete, and composite constructions of lightweight concrete fill on metal deck. For true dynamic analysis the variations are more complex because the weight and degree of elasticity of the materials must also be considered.

With respect to their span-to-depth ratios, most horizontal diaphragms approach the classification of deep beams. As shown in Fig. 6.7, even the shallowest of diaphragms, such as the maximum 4 to 1 case allowed for a plywood deck by the *UBC*, tends to present a fairly stiff flexural mem-

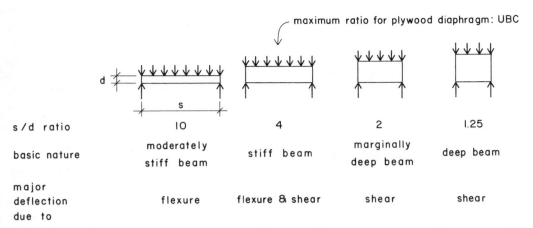

FIGURE 6.7 Behavior of horizontal diaphragms related to depth-to-span ratios.

ber. As the span-to-depth ratio falls below about 2, the deformation characteristic of the diaphragm approaches that of a deep beam in which the deflection is primarily caused by shear strain rather than by flexural strain. Thus the usual formulas for deflection caused by flexural strain become of limited use.

The following formula is used for the calculation of deflection of simple-span plywood diaphragms:

$$\Delta = \frac{5vL^3}{8EAb} + \frac{vL}{4Gt}$$

$$+ \ 0.094 Le_n + \Sigma \ \frac{\Delta_c X}{2b}$$

in which the four terms account for four different contributions to the deflection, as follows:

Term 1 accounts for the length change of the chords.

Term 2 accounts for the shear strain in the plywood panels.

Term 3 accounts for the lateral bending of the nails.

Term 4 accounts for additional change in the chord lengths caused by slip in the chord splices.

The deflection of steel deck diaphragms is discussed and illustrated in the *Inryco Lateral Diaphragm Data Manual 20-2* published by Inryco, Inc. The formula used for calculating the deflection of a simple span deck is

$$\Delta_t = \frac{5WL_s^4 \times 1728}{384E \times I}$$

$$+ \left(q \times F \times \frac{L_s}{2} \times 10^{-6} \right)$$

in which the first term accounts for flexural deflection caused by the length change of the chords and the second term for shear strain and panel distortion in the diaphragm web.

The quantities q and F vary as a function of the type and gauge of the deck, the fastening patterns and methods used, and the possible inclusion of concrete fill.

As with deck load capacities, deflections and the relative stiffness of deck systems are generally based on materials presented in the *Diaphragm Design Manual*, published by the Steel Deck Institute. For a specific product, however, designers should obtain information from the product supplier and verify that any data or procedures used are acceptable to the building permit-approving agency for the work.

6.2 VERTICAL DIAPHRAGMS (SHEAR WALLS)

Vertical diaphragms are usually the walls of buildings. As such, in addition to their shear wall function, they must fulfill various architectural functions and may also be required to serve as bearing walls for the gravity loads. The location of walls, the materials used, and some of the details of their construction must be developed with all these functions in mind.

The most common shear wall constructions are those of poured concrete, masonry, and wood frames of studs with surfacing elements. Wood frames may be made rigid in the wall plane by the use of diagonal bracing or by the use of surfacing materials that have sufficient strength and stiffness. Choice of the type of construction may be limited by the magnitude of shear caused by the lateral loads, but will also be influenced by fire code requirements and the satisfaction of the various other wall functions, as described previously.

General Behavior

Some of the structural functions usually required of vertical diaphragms are the following (see Fig. 6.8):

1. *Direct Shear Resistance.* This usually consists of the transfer of a lateral force in the plane of the wall from some upper level of the wall to a lower level or to the bottom of the wall. This results in the typical situation of shear stress and the accompanying diagonal

tension and compression stresses, as discussed for horizontal diaphragms.

2. *Cantilever Moment Resistance.* Shear walls generally work like vertical cantilevers, developing compression on one edge and tension on the opposite edge, and transferring an overturning moment (M) to the base of the wall.

3. *Horizontal Sliding Resistance.* The direct transfer of the lateral load at the base of the wall produces the tendency for the wall to slip horizontally off its supports.

The shear stress function is usually considered independently of other structural

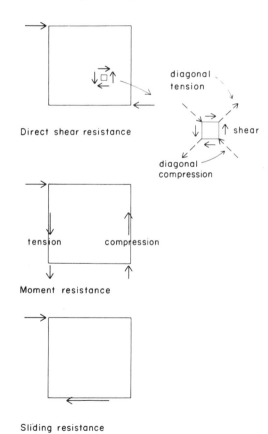

FIGURE 6.8   Functions of a shear wall.

functions of the wall. The maximum shear stress that derives from lateral loads is compared to some rated capacity of the wall construction, with the usual increase of one third in allowable stresses because the lateral load is most often a result of wind or earthquake forces. For concrete and masonry walls the actual stress in the material is calculated and compared with the allowable stress for the material. For structurally surfaced wood frames the construction as a whole is generally rated for its total resistance in pounds per foot of the wall length in plan. For a plywood-surfaced wall this capacity depends on the type and thickness of the plywood; the size, wood species, and spacing of the studs; the size and spacing of the plywood nails; and the inclusion or omission of blocking at horizontal plywood joints.

The analysis and design of shear walls of concrete and reinforced masonry are discussed in the examples in Chapter 10. For walls of concrete the procedures are based on the requirements of the American Concrete Institute (ACI) Code (Ref. 14). For walls of reinforced hollow concrete block construction the procedures are based on the requirements in the *UBC*.

For wood stud walls the *UBC* provides tables of rated load capacities for several types of surfacing, including plywood, diagonal wood boards, plaster, gypsum drywall, and particleboard. This material is included in the *UBC* reprints in Appendix B.

Although the possibility exists for the buckling of walls as a result of the diagonal compression effect, this is usually not critical because other limitations exist to constrain wall slenderness. The thickness of masonry walls is limited by maximum values for the ratio of unsupported wall height or length-to-wall thickness. Concrete thickness is usually limited by forming and pouring considerations, so that thin walls are not common except with precast construction. Slenderness of wood studs is limited by gravity design and by the code limits as a function of the stud size. Because stud walls are usually surfaced on both sides, the resulting sandwich-panel effect is usually sufficient to provide a reasonable stiffness.

As in the case of horizontal diaphragms, the moment effect on the wall is usually considered to be resisted by the two vertical edges of the wall acting as flanges or chords. In the concrete or masonry wall this results in a consideration of the ends of the wall as columns, sometimes actually produced as such by thickening of the wall at the ends. In wood-framed walls the end framing members are considered to fulfill this function. These edge members must be investigated for possible critical combinations of loading because of gravity and the lateral effects.

The overturn effect of lateral loads must be resisted for the building as a whole, as well as for individual elements of the vertical lateral bracing system. For wind, the overturning caused by the horizontal forces must be combined with the uplift caused by upward suction pressure on the roof. For buildings with a height-to-width ratio of 0.5 or less and a maximum height of 60 ft, the combination of the effects of overturning and uplift may be reduced by one-third. Weight of earth over footings may be used to calculate the dead-load-resisting moment. For both the entire building and its individual lateral bracing elements, the overturning moment must not exceed two-thirds of the dead-load-resisting moment (see *UBC* Sec. 1619.1).

For seismic effects, *UBC* Sec. 1631.1 specifies that only 85% of the dead load be used to resist uplift effects when using the working stress method for materials. This means that any anchorage elements that are required can be designed for their working stress resistance.

For an individual shear wall, the overturn investigation is summarized in Fig. 6.9. Specific applications are illustrated in the design examples in Chapter 10.

Resistance to horizontal sliding at the base of a shear wall is usually at least partly resisted by friction caused by the dead loads. For masonry and concrete walls with dead loads that are usually quite high, the frictional resistance may be more than sufficient. If it is not, shear keys must be provided. For wood-framed walls the friction is usually ignored and the sill bolts are designed for the entire load.

Design and Usage Considerations

An important judgment that must often be made in designing for lateral loads is that of the manner of distribution of lateral force between a number of shear walls that share the load from a single horizontal diaphragm. In some cases the existence of symmetry or of a flexible horizontal diaphragm may simplify this consideration. In many cases, however, the relative stiffnesses of the walls must be determined for this calculation.

If considered in terms of static force and elastic stress-strain conditions, the relative stiffness of a wall is inversely proportionate to its deflection under a unit load. Figure 6.10 shows the manner of deflection of a shear wall for two assumed conditions. In (a) the wall is considered to be fixed at its top and bottom, flexing in a double curve with an inflection point at midheight. This is the case usually assumed for a continuous wall of concrete or masonry in which a series of individual wall portions (called *piers*) are connected by a continuous upper wall or other structure of considerable stiffness. In (b) the wall is considered to be fixed at its bottom only, functioning as a vertical cantilever. This is the case for independent, freestanding

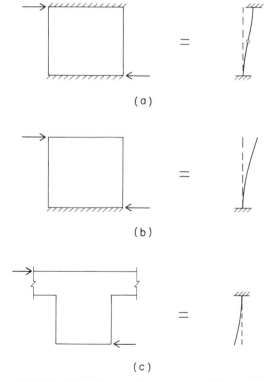

(a)

(b)

(c)

FIGURE 6.10 Shear wall support conditions: (a) fixed top and bottom; (b) and (c) cantilevered.

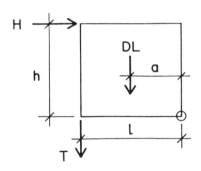

To determine T:

for wind –   DL(a) + T(l) = 1.5 [H(h)]
for seismic – 0.85 [DL(a)]+ T(l) = H(h)

FIGURE 6.9 Determination of stability and tiedown requirements for a shear wall; working stress method.

walls or for walls in which the continuous upper structure is relatively flexible. A third possibility is shown in (c) in which relatively short piers are assumed to be fixed at their tops only, which produces the same deflection condition as in (b).

In some instances the deflection of the wall may result largely from shear distortion, rather than from flexural distortion, perhaps because of the wall materials and construction or the proportion of wall height to plan length. Furthermore, stiffness in resistance to dynamic loads is not quite the same as stiffness in resistance to static loads. The following recommendations are made for single-story shear walls:

1. For wood-framed walls with height-to-length ratios of 2 or less, assume the stiffness to be proportional to the plan length of the wall.

2. For wood-framed walls with height-to-length ratios over 2 and for concrete and masonry walls, assume the stiffness to be a function of the height-to-length ratio and the method of support (cantilevered or fixed top and bottom). Use the values for pier rigidity given in Appendix C.

3. Avoid situations in which walls of significantly great differences in stiffness share loads along a single row. The short walls will tend to receive a small share of the loads, especially if the stiffness is assumed to be a function of the height-to-length ratio.

4. Avoid mixing of shear walls of different construction when they share loads on a deflection basis.

Item 4 in the preceding list can be illustrated by two situations as shown in Fig. 6.11. The first situation is that of a series of panels in a single row. If some of these pan-

els are of concrete or masonry and others of wood frame construction, the stiffer concrete or masonry panels will tend to absorb the major portion of the load. The load sharing must be determined on the basis of actual calculated deflections. Better yet is a true dynamic analysis, because if the load is truly dynamic in character, the periods of the two types of walls are of more significance than their stiffness.

In the second situation shown in Fig. 6.11, the walls share load from a rigid horizontal diaphragm. this situation also requires a deflection calculation for determining the distribution of force to the panels.

In addition to the various considerations mentioned for the shear walls themselves, care must be taken to assure that they are properly anchored to the horizontal diaphragms. Problems of this sort are illustrated in the examples in Chapter 10.

A final consideration for shear walls is that they must be made an integral part of the whole building construction. In long building walls with large door or window openings or other gaps in the wall, shear walls are often considered as entities (isolated, independent piers) for their design. However, the behavior of the entire wall under lateral load should be studied to be sure that elements not considered to be parts of the lateral resistive system do not suffer damage because of the wall distortions.

An example of this situation is shown in Fig 6.12. The relatively long solid portion is assumed to perform the bracing function for the entire wall and would be designed as an isolated pier. However, when the wall deflects, the effect of the movement on the shorter piers, on the headers over openings, and on the door and window framing must be considered. The headers must not be cracked loose from the solid wall portions or pulled off their supports.

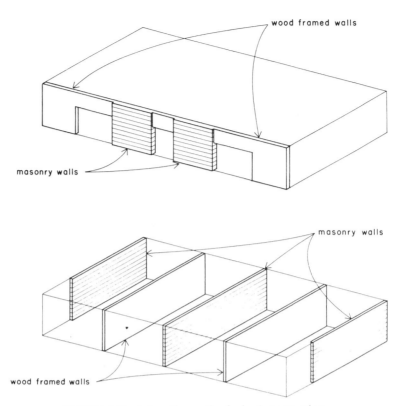

FIGURE 6.11 Interacting walls of mixed construction.

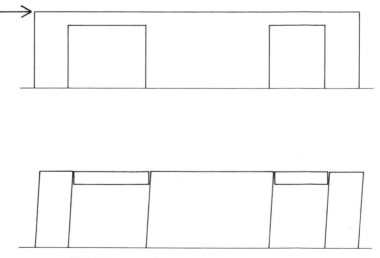

FIGURE 6.12 Effect of shear wall of mixed construction.

Typical Construction

The various types of common construction for shear walls mentioned in the preceding discussion are wood and steel frames with various surfacing, reinforced masonry, and concrete. The only frame wall used extensively in the past was the plywood covered wood stud wall. Experience and testing have established acceptable ratings for other surfacing, so that plywood is used somewhat less when shear loads are low (Fig 6.13).

For all types of walls there are various considerations (good carpentry, fire resistance, available products, etc.) that establish a certain minimum construction. In many situations this "minimum" is really adequate for low levels of shear loading, and the only extra requirements are in the area of attachments and joint load transfers. Increasing wall strength beyond the minimum usually requires increasing the size or quality of units, adding or strengthening attachments, developing supporting elements to function as chords or collectors, and so on. It well behooves the designer to find out the standards for basic construction to know what the minimum code-defined construction consists of so

FIGURE 6.13 Wood frame diaphragm. Garage with side wall of stucco fastened directly to studs, short end wall with plywood, and let-in bracing for stability of the frame during construction.

that added strength can be developed when necessary—but using methods consistent with the ordinary types of construction.

Load Capacity. Load capacities for ordinary wood-framed shear walls are given in the load tables in the *UBC*. For seismic actions the only masonry construction ordinarily acceptable is that of reinforced masonry, of which the most common form is one using hollow precast concrete units (concrete blocks). The design of masonry construction should be done with the references that present material acceptable to the building code of jurisdiction.

Most concrete design is based on the current edition of the ACI Code (*Building Code Requirements for Reinforced Concrete*, Ref. 14), although local codes are sometimes slow in accepting new changes in the ACI Code editions. The current edition of the ACI Code provides some criteria for seismic design, but recent developments—mostly in the form of suggested details for construction—are more stringent for areas of high seismic risk. Concrete design in general and seismic design in particular have become quite sophisticated and complex, and there are few simple guides. The shear wall is a relatively simple concrete element, but its design should be done in conformance with the latest codes and practices.

Stiffness and Deflection. As with the horizontal diaphragm, there are several potential factors to consider in the deflection of a shear wall. As shown in Fig. 6.14*a*, shear walls also tend to be relatively stiff in most cases, approaching deep beams instead of ordinary flexural members.

The two general cases for the vertical shear wall are the cantilever and the doubly fixed pier. The cantilever, fixed at its base, is the most commonly used. Fixity at both the top and bottom of the wall usually affects deflection only when the wall is rela-

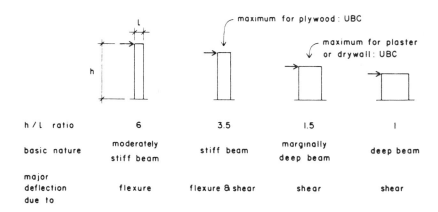

(a) Behavior of cantilvered elements related to height-to-length ratios

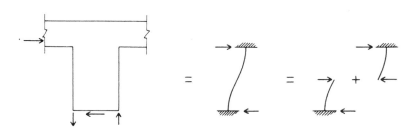

(b) Deflection assumption for a fully fixed masonry pier

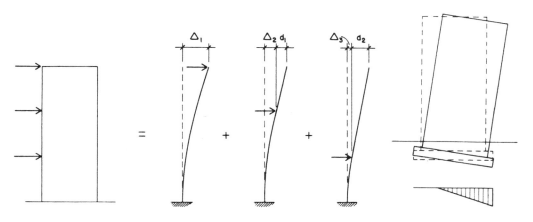

(c) Deflection of a multistory shear wall

(d) Shear wall tilt caused by uneven soil pressure

FIGURE 6.14 Aspects of lateral deflection of shear walls.

tively short in length with respect to its height. Walls with long lengths in proportion to their height fall into the deep beam category in which the predominant shear strain is not affected by the fixity of the support.

As shown in Fig. 6.14*b*, if the doubly fixed pier is assumed to have an inflection point at its midheight, its deflection can be approximated by considering it to be the sum of the deflections of two half-height cantilevered piers. Yielding of the supports and flexure in the horizontal structure will produce some rotation of the assumedly fixed ends, which will result in some additional deflection.

The following formula is used for calculating deflection of cantilevered plywood shear walls similar to that for the plywood horizontal diaphragm:

$$\Delta = \frac{8vh^3}{EAb} + \frac{vh}{Gt} + 0.376he_n + d_a$$

in which the four terms account for the following:

Term 1 accounts for the change in length of the chords (wall end framing).
Term 2 accounts for the shear strain in the plywood panels.
Term 3 accounts for the nail deformation.
Term 4 is a general term for including the effects of yield of the anchorage.

The formula can also be used for calculating the deflection of a multistory wall, as shown in Fig. 6.14*c*. For the loading as shown in the illustration, a separate calculation would be made for each of the three loads (Δ_1, Δ_2, and Δ_3). To these would be added the deflection at the top of the wall caused by the rotation effects of the lower loads (d_1 and d_2). Thus the total deflection

at the top of the wall would be the sum of the five increments of deflection.

Rotation caused by soil deformation at the base of the wall can also contribute to the deflection of shear walls (see Fig. 6.14*d*). This is especially critical for tall walls on isolated foundations placed on relatively compressible soils, such as loose sand and soft clay—a situation to be avoided if at all possible.

6.3 COLLECTORS AND TIES

Transfer of loads from horizontal to vertical elements in laterally resistive structural systems frequently involves the use of some structural members that serve the functions of struts, drags, ties, collectors, and so on. These members often serve two functions—as parts of the gravity-resistive system or for other functions in lateral load resistance.

Figure 6.15 shows a structure consisting of a horizontal diaphragm and a number of exterior shear walls. For loading in the north–south direction the framing members labeled A serve as chords for the roof diaphragm. In most cases they are also parts of the roof edge or top of the wall framing. For the lateral load in the east–west direction they serve as collectors. This latter function permits us to consider the shear stress in the roof diaphragm to be a constant along the entire length of the edge. The collector "collects" this constant stress from the roof and distributes it to the isolated shear walls, thus functioning as a tension/compression member in the gaps between the walls.

In the example in Fig. 6.15 the collector A must be attached to the roof edge to develop the transfer of the constant shear stress. The collector A must be attached to the individual shear walls for the transfer of the total load in each wall. In the gaps between walls the collector gathers the roof

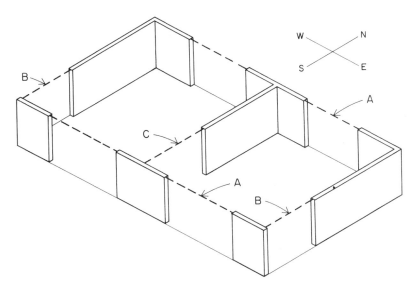

FIGURE 6.15 Collector functions in a box system.

edge load and functions partly as a compression member, pushing some of the load to the forward wall, and partly as a tension member, dragging the remainder of the collected load into the rearward wall.

Collectors B and C in Fig. 6.15 gather the edge load from the roof deck under the north–south lateral loading. Their function over the gap reverses as the load switches direction. They work in compression for load in the northerly direction, pushing the load into the walls. When the load changes to the southerly direction, they work in tension, dragging the load into the walls.

The complete functioning of a lateral resistive structural system must be carefully studied to determine the need for such members. As mentioned previously, ordinary members of the building construction will often serve these functions: top plates of the stud walls, edge framing of roofs and floors, headers over openings, and so on. If so used, such members should be investigated for the combined stress effects involved in their multiple roles.

6.4 ANCHORAGE ELEMENTS

The attachment of elements of the lateral resistive structure to one another, to collectors, or to supports usually involves some type of anchorage element. There is a great variety of these, encompassing the range of situations with regard to load transfer conditions, magnitude of the forces, and various materials and details of the structural members and systems.

Tiedowns

Resistance to vertical uplift is sometimes required for elements of a braced- or moment-resistive frame, for the ends of shear walls, or for light roof systems subject to the force of upward wind suction. For concrete and reinforced masonry structures, such resistance is most often achieved by doweling and/or hooking of reinforcing bars. Steel columns are usually anchored by the anchor bolts at their bases. Figure 6.16 shows some of the devices that are used for

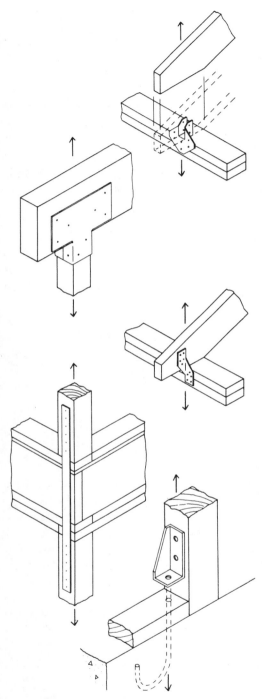

FIGURE 6.16 Anchoring of wood frames.

anchoring wood structural elements. In many cases these devices have been load tested and their capacities rated by their manufacturers. When using them, it is essential to determine whether the load ratings have been accepted by the building code agency with jurisdiction for a specific building design.

The term *tiedown* or *hold-down* is used mostly to describe the type of anchor shown in the lower right corner of Fig. 6.16.

Horizontal Anchors. In addition to the transfer of vertical gravity load and lateral shear load at the edges of horizontal diaphragms, there is usually a need for resistance to the horizontal pulling away of walls from the diaphragm edge. In many cases the connections that are provided for other functions also serve to resist this action. Codes usually require that this type of anchorage be a "positive" one, not relying on such things as the withdrawal of nails or lateral force on toe nails. Figure 6.17 shows some of the means used for achieving this type of anchorage.

Shear Anchors. The shear force at the edge of a horizontal diaphragm must be transferred from the diaphragm into a collector or some other intermediate member, or directly into a vertical diaphragm. Except for cast-in-place concrete structures, this process usually involves some means of attachment. For wood structures the transfer is usually achieved through the lateral loading of nails, bolts, or lag screws for which the codes or industry specifications provide tabulated load capacities. For steel deck diaphragms the transfer is usually achieved by welding the deck to supporting steel framing.

If the vertical system is a steel frame, these members are usually parts of the frame system. If the vertical structure is

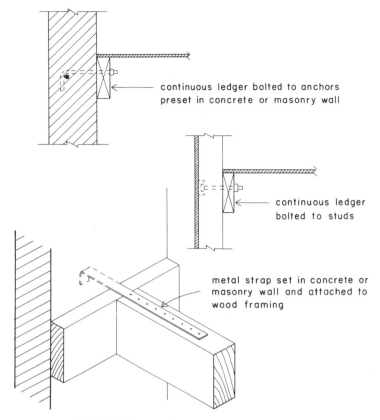

continuous ledger bolted to anchors
preset in concrete or masonry wall

continuous ledger
bolted to studs

metal strap set in concrete or
masonry wall and attached to
wood framing

FIGURE 6.17   Horizontal anchorage for walls.

concrete or masonry, the edge transfer members are usually attached to the walls with anchor bolts set in the concrete or in solid-filled horizontal courses of the masonry. As in other situations, the combined stresses on these connections must be carefully investigated to determine the critical load conditions.

Another shear transfer problem is that which occurs at the base of a shear wall in terms of a sliding effect. For a wood-framed wall some attachment of the wall sill member to its support must be made. If the support is wood, the attachment is usually achieved by using nails or lag screws. If the support is concrete or masonry, the sill is

usually attached to preset anchor bolts. The lateral load capacity of the bolts is determined by their shear capacity in the concrete or the single shear limit in the wood sill. The *UBC* Sec. 1806.6 gives some minimum requirements for sill bolting, which should be used as a starting point in developing this type of connection.

For walls of concrete and masonry, in which there is often considerable dead load at the base of the wall, sliding resistance may be adequately developed by friction. Doweling provided for the vertical wall reinforcing also offers some lateral shear resistance. If a more positive anchor is desired, or if the calculated load requires

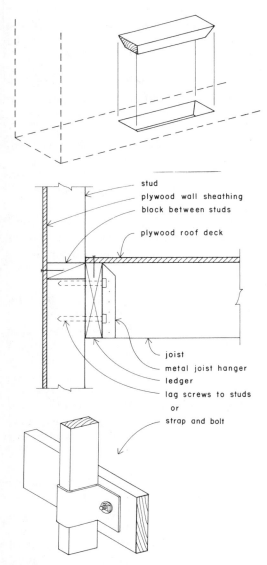

FIGURE 6.18 Transfer of lateral loads between horizontal and vertical elements.

stud
plywood wall sheathing
block between studs
plywood roof deck

joist
metal joist hanger
ledger
lag screws to studs
or
strap and bolt

lateral resistive system can be quite complex in some cases. Figure 6.18 shows a joint between a horizontal plywood diaphragm and a vertical plywood shear wall. For reasons other than lateral load resistance, it is desired that the studs in the wall run continuously past the level of the roof deck. This necessitates the use of a continuous edge-framing member, called a *ledger*, that serves as the vertical support for the deck as well as the chord and edge collector for the lateral forces. This ledger is shown to be attached to the faces of the studs with two lag screws at each stud. The functioning of this joint involves the following:

1. The vertical gravity load is transferred from the ledger to the studs directly through lateral load on the lag screws.
2. The lateral shear stress in the roof deck is transferred to the ledger through lateral load on the edge nails of the deck. This stress is in turn transferred from the ledger to the studs by horizontal lateral load on the lag screws. The horizontal blocking is fit between the studs to provide for the transfer of the load to the wall plywood which is nailed to the blocking.
3. Outward loading on the wall is resisted by the lag screws in withdrawal. This is generally not considered to be a good positive connection, although the load magnitude should be considered in making this evaluation. A more positive connection is achieved by using the bolts and straps shown in the lower sketch in Fig. 6.18.

it, shear keys may be provided by inserting wood blocks in the concrete, as shown in Fig. 6.18.

Transfer of Forces

The complete transfer of force from the horizontal to the vertical elements of the

6.5 SEPARATION JOINTS

During the swaying motions induced by earthquakes, different parts of a building tend to move independently because of the differences in their masses, their fundamental periods, and variations in damping, support constraint, and so on. With regard to the building structure, it is usually desirable to tie it together so that it moves as a whole as much as possible. Sometimes, however, it is better to separate parts from one another in a manner that permits them a reasonable freedom of motion with respect to one another.

Figure 6.19 shows some building forms in which the extreme difference of period of adjacent masses of the building makes it preferable to cause a separation. Designing the building connection at these intersections must be done with regard to the spe-

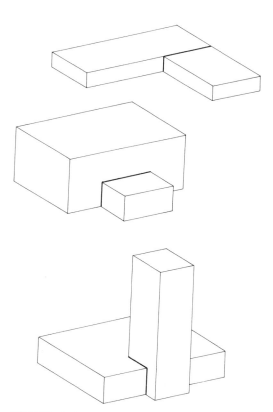

FIGURE 6.19 Situations requiring consideration for seismic separation joints.

cific situation in each case. Some of the considerations to be made in this are the following.

Specific Direction of Movements. In generally rectangular building forms, such as those shown in the examples in Fig. 6.19, the primary movements are in the direction of the major axes of the adjacent masses. Thus the joint between the masses has two principal forms of motion: a shear effect parallel to the joint, and a together–apart motion perpendicular to the joint. In building forms of greater geometric complexity, the motions of the respective masses are more random and the joint action is much more complex.

Actual Dimensions of Movement at the Joint. If the joint is to be truly effective and if the adjacent parts are not allowed to pound each other, the actual dimension of the movements must be safely tolerated by the separation. The more complex the motions of the separate masses, the more difficult it is to predict these dimensions accurately, calling for some conservative margin in the dimension of the separation provided.

Detailing the Joint for Effective Separation. Because the idea of the joint is that structural separation is to be provided while still achieving the general connection of the adjacent parts, it is necessary to make a joint that performs both of these seemingly contradictory functions. Various techniques are possible using connections that employ sliding, rolling, rotating, swinging, or flexible elements that permit one type of connection while having a freedom of movement for certain directions or types of motion. The possibilities are endless, and the specific situation must be carefully analyzed in order to develop an effective and logical joint detail. In some cases the complexity of the motions, the extreme dimension of movement to be facilitated, or other considerations may make it necessary to have complete separation—that is, literally to build two separate buildings very close together.

Facilitating Other Functions of the Joint.   It is often necessary for the separation joint to provide for functions other than those of the seismic motions. Gravity load transfer may be required through the joint. Nonstructural functions, such as weather sealing, waterproofing, and the passage of wiring, piping, or ductwork through the joint, may be required. Figure 6.20 shows two typical cases in which the joint achieves structural separation while providing for a closing of the joint. The upper drawing shows a flexible flashing or sealing strip used to achieve weather or water tightness of the joint. The lower drawing symbolizes

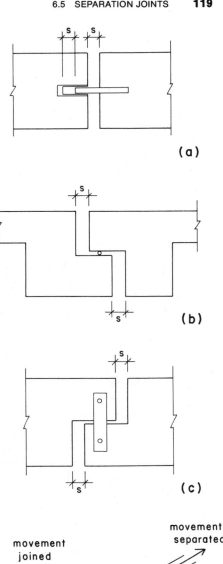

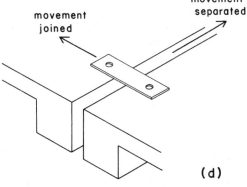

FIGURE 6.21   Means of achieving partial separation.

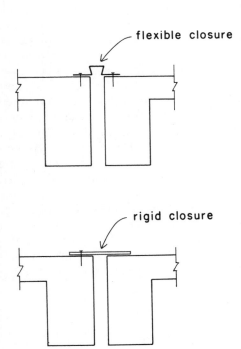

FIGURE 6.20   Closure for horizontal separation joints.

the usual solution for a floor in which a flat element is attached to one side of the joint and is allowed to slip on the other side.

Figure 6.21 shows a number of situations in which partial structural separation is achieved. The details of such joints are often quite similar to those used for joints designed to provide separation for thermal expansion. Figure 6.21*a* shows a key slot, which is the type of connection usually used in walls where the separation is required only in a direction parallel to the wall plane. Figure 6.21*b* and *c* show means for achieving the transfer of vertical gravity forces through the joint while permitting movement in a horizontal direction. Figure 6.21*d* shows a means for achieving a connection in one horizontal direction while permitting movement in the perpendicular direction.

6.6   MITIGATION OF LATERAL FORCES

Most of our efforts to date in designing for earthquake resistance have been aimed at strengthening structures or controlling their behavior in some way. A different approach is that of trying to reduce the impact of the earthquake itself by mitigation (reduction) of the seismic effects. One way to achieve this is to use what amounts to a form of shock absorber between the ground and a building, generally called *base isolation.*

Since seismic force is induced in a building by movement of the supporting ground, the isolation consists of absorbing some of this motion in the building-to-ground connection. Techniques for achieving this have been developed experimentally and are used increasingly where the method is appropriate. At the present, the methods available work best for a limited form and size of building. As these limitations have some restrictions on the planning of buildings, they have not so far been widely used.

When achieved effectively, base isolation permits a reduction in the required strength of the lateral bracing system. An application of some significance is that developed as a retrofit effort on an existing building with construction that does not yield easily to being strengthened. Old masonry buildings are one such example, and several old buildings with cultural value have been preserved by this method. The buildings are made less vulnerable without strengthening of the existing construction.

Another mitigating technique that has been used for wind forces on tall buildings is the use of a device that delivers a counterpunch to the wind force, literally pushing the building in the opposite direction. This is done with a large inertia mass of some form, mounted on a track, and launched when a sensor notes the presence of the wind force. The method is also being developed experimentally for seismic forces.

In recent years a monumental international effort has been undertaken to study and develop methods for designing structures for wind and earthquakes. As new methods are developed and examples are tested in real windstorms and earthquakes, there will certainly be an extension of the available means to make buildings safe against these disastrous natural forces. For now, however, the vast majority of buildings are still designed and built with well-known, time-tested means for lateral bracing.

7

SPECIAL PROBLEMS FOR LATERAL RESISTIVE SYSTEMS

Development of a structural system for resistance to wind or earthquakes must be done in the context of the entire building design. Many problems relating to this are discussed in Chapter 4 and various situations are illustrated in the example design cases in Part III. In this chapter we present some general considerations for the overall planning of lateral resistive systems, with respect to their structural as well as their architectural design problems.

7.1 SYSTEMS PLANNING CONSIDERATIONS

With respect to its singular function of resisting the effects of wind or earthquakes, some basic considerations for a structure are the following.

1. *System Type.* Basic choice for a shear wall, trussed frame, or rigid frame system—

or for some other special type—depends partly on the type and magnitude of loads or other structural demands. If tolerance for movements is low, the basic choice of system may be affected; possibly a very stiff shear wall system versus a flexible rigid frame. In general, every type of system, achieved with every variation of materials and details, has some range of behaviors and overall feasibility. Thus the precise definition of the conditions will produce some identification with the most logical system choice. This is not a simple task, but many data are available for such a study.

2. *General Form.* Within a basic system choice—for example, shear walls—there is considerable range of selection. Should one use a perimeter-only system, or a core-braced system, or a mixed perimeter- and core-system, and so on? Besides the feasibility or general correctness of a system

type, the most appropriate form may also be optimized. Needs for symmetry, dimensional limits, proportions (height-to-width ratios, etc.), or specific details may be part of determining the real effectiveness of any system.

3. *Appropriate Combinations.* Most whole systems for building structures consist of mixtures rather than a single system type using a single material. Thus a building may have a concrete foundation, masonry walls, wood frame floor construction, and a steel truss roof system. Similarly, its lateral resistive system may be a collection of components: a steel roof deck diaphragm, wood or concrete floor diaphragms, masonry shear walls, and a rigid frame concrete base. The behavior of each of these components, their interactions for various load transfers, and their collective performance as a whole system all affect the assessment of the appropriateness of the design. There is such a thing as right or wrong, but the potential combinations make for a very large number of possibilities to consider.

4. *Building Code Requirements.* At any one time, in any one place, a specific set of requirements will be established by the enforceable building code. The common practice is to design strictly for conformance to those requirements—no less, no more. Economic pressures make any other design objectives very difficult to defend. The fact that every professional designer knows full well that the code purpose is to define minimum requirements, resulting in a marginally acceptable design, does not much affect this situation. It usually takes a major event—earthquake or windstorm— to change things; and even then only very slowly. In any case, the precise requirements of the building code are a critical reference, wherever the designer decides to go from that point.

7.2 EFFECTS OF LACK OF SYMMETRY

The significance of symmetry with regard to the behavior of lateral force resistive systems is discussed extensively in Chapter 4. Torsional, twisting effects on buildings are somewhat more critical for earthquakes, but can also cause problems for relatively flexible structures in windstorms.

The critical issue of concern here is a close relationship between the center of the acting forces (total effect of the wind or earthquake) and the center (centroid, etc.) of the total resistive system (see Figs. 4.5 and 6.2). Of course, this will be achieved most simply by having a building form and its lateral resistive system arranged in perfect symmetry of construction. If that arrangement works for the general building design, it will usually result in a simpler task for all concerned: architect, structural designer, and builder.

However, it is *structural symmetry* that is at issue here; getting the location of the resultant of the acting forces to coincide with the center of stiffness of the resisting system. In theory, that can be achieved with any force system and any resisting system if the situation is carefully managed and investigated. But the more complex the loading (read: building form) and the more complex the resisting system (unsymmetrical, multielement, etc.), the more the design effort involved for careful analysis and final detailing of the construction. If the time, effort, and expense for that structural design work is affordable, the complex systems may be accommodated with reasonable safety.

7.3 NONSTRUCTURAL ELEMENTS

Typically, the majority of building construction is not considered to perform the task of providing lateral bracing. It is there to anchor the building for wind effects and to add to the seismic force by mass but does not participate as part of the lateral force resisting system. This simplistic view is increasingly coming into conflict with reality.

A stiff wall is a stiff wall, whether or not it is labeled a shear wall. Thus many curtain walls and interior partitions provide major shear wall effects, even though there is theoretically a whole separate system assumed to provide independent bracing. This situation results in two potential problems, both of which are the source of much damage, as demonstrated in recent events.

1. Due to its stiffness, the nonstructural construction may receive a share of the lateral force for which it does not have sufficient resistive strength. This is a primary source of cracks in exterior plaster (stucco). It is also a major reason for breakage of window glass in earthquakes. The possible remedies for this are to isolate the stiff nonstructural elements from the acting force, stiffen the true bracing system, or use control joints or other means to affect the behavior of the nonstructural construction and its interactions with the bracing structure.

2. Coincident structural action of nonstructural elements may modify the behavior of the lateral bracing structure. A common situation of this type is the production of a soft-story effect due to the existence of nonstructural bracing in some stories and not others in a multistory building. Another situation is the inadvertent stiffening of columns in a rigid frame system that was designed on the basis of flexible columns. The remedies here include those mentioned above, but mostly careful development of the structural attachments between the nonstructural elements and the bracing structure.

Nonstructural construction also includes all the building service systems for electric power, water, ventilating, and so on. It is important that these remain in use after an event for emergency service and that they not require costly repair or replacement. They must be developed to have their own independent bracing as well as to be isolated from actions of the structural bracing system. Basic planning of these systems should be coordinated with the general planning of the lateral bracing system to avoid interference and recognition of the basic needs of all systems.

8

SITE AND FOUNDATION CONCERNS

Building sites provide support for buildings but are also sources for potential concerns for lateral forces. For wind the site represents the ultimate resisting element that must absorb the wind forces by anchoring of the building. For earthquakes, this function is the same for conceptual purposes, but in truth the ground is the *source* for the earthquake forces. In this chapter we present some of the problems of dealing with sites for buildings with regard to wind and earthquake effects.

8.1 BUILDING/SITE RELATIONS

For buildings that sit on the ground, an objective for design is to support the building so that it does not move after its initial construction. In a technical sense, this is not possible, since any stress developed in bearing pressure will produce some strain and deformation, even with bearing on solid rock. It therefore becomes a matter of relative magnitude of movement. Dropping vertically by a tiny fraction of an inch must be accepted as insignificant; sinking out of site below the ground surface is obviously not acceptable.

In reality what is required is some definition of acceptable safety in terms of both strength and deformation, just as it is with almost any structural design work. Foundation design begins with the establishment of the limits for strength and movements and proceeds with determination of the interface structure (foundations) required to achieve the building-to-ground connection.

In a broad consideration of site problems it is necessary to consider two additional problems, besides that of the building foundations.

Site Surface Development. What is necessary to achieve the desired site profile,

124

to achieve any site construction (pavements, walls, etc.), and to support any site plantings.

Macro-site Considerations. Although legally defined by boundary lines, the site is a continuous part of a larger geological formation.

For earthquakes there are some additional potential concerns, including the following:

Proximity to Existing Faults. This will be a major factor in the definition of the risk zone but may also relate to the form of the particular faults.

Potential for Soil Liquefaction or Other Site Response Behavior. This is the basis for the building–site interaction factor for seismic base shear.

8.2   SITE PROBLEMS

Each building presents a unique situation in regard to the combination of true site conditions, design requirements for wind and earthquakes, and the specific form of the building. However, most cases fall into a few commonly encountered situations. In this section we consider some of these common situations.

General Case, Basic Considerations

Foundation design problems have considerable variety because of the wide range of possible soil conditions, load magnitudes, building size and shape, and type of structural system. For any building it is wise to have a subsoil investigation, lab tests on representative soil samples, and a recommendation from an engineer with experience in soil behaviors and foundation problems. We illustrate some of the ordinary and simple design problems in the

examples in Part III, but do not attempt to deal with all the special problems that can occur in the design of foundations for lateral loads.

For seismic load the foundations have a dual role. Initially, they are the origin point for load on the building, being directly attached to the ground and used by the ground to shake the building. In our analysis procedures, however, the seismic load is considered to be a result of the inertial effect of the moving building mass. From this viewpoint the loading condition becomes similar to that for the wind load. Thus we visualize the horizontal inertial force as being transmitted through the building structure into the foundation, and finally into the ground, which is now considered to offer passive resistance, as it does actually to the wind force.

For seismic loads it is usually desirable that the entire building foundation act essentially as a single rigid unit. If elements of the foundation are isolated from one another, as in the case of individual column footings, it may be necessary to provide struts or grade walls to tie the structure into a unit. Where they exist, of course, the ordinary elements such as basement walls, grade walls, wall footings, and grade-level framing members may be used for this tying function.

For wind loads the foundations function essentially as an anchor for the building superstructure, resisting the overturn, uplift, and horizontal sliding effects. The predominant effects on the foundations depend on the magnitude of the wind load, the type of foundation system, the ground conditions, and the relative magnitude of the weight of the building construction. For very light construction, for example, the principal problems may be the resistance of uplift and overturn. For a building on deep foundations (piles or piers) with no basement, the principal problem may be the simple resistance to horizontal movement. For a building of heavy masonry or concrete construction, the wind load on the foundations may be negligible and the predominant concern be that for vertical soil pressure.

In many cases the existence of basement walls, grade walls, wall footings, or other parts of the building substructure makes the addition of ties unnecessary. When individual footings are truly isolated, the tie member is designed as both a tension and a compression member. The required size of the concrete cross section and the main reinforcing is usually determined by the column action in compression. The tension tie is provided by extending the tie reinforcing into the footings in dowel action.

When calculated lateral loads exist for the footings, they may be used as the design loads on the ties. When no calculated load exists, it is common practice to design for a minimum lateral load of 0.10 times the vertical load. For isolated footings under structural elements that are not part of the lateral resistive structural system, there is no quantifiable basis for the tie design. Such footings are usually designed using the minimum requirements for concrete compressive members.

Design Criteria and Data. Chapter 18 of the *UBC* provides some data and recommendations for use in designing for sliding or lateral passive resistance of soil. The *UBC* material is quite conservative in most cases, and a complete soils investigation and report may provide more rational data and permit a less conservative design. In many cases soil conditions are general local phenomena, and local building codes often have special requirements and procedures for foundation design. Local design and construction procedures are sometimes based more on histories of successes and failures than on scientific analysis or engineering judgment.

Recontoured Sites

Most sites receive some trimming up to achieve a final surface definition. As this redevelopment of the ground surface profile extends in magnitude, various problems may be created.

For buildings with exceptional uplift or overturn effects, the foundation must usually provide an anchor through its own sheer dead weight. For a large building with shallow foundations and no basement, the necessary dead load may not exist in the foundations designed for gravity loads alone. This situation may require some additional mass in the foundation itself or the use of soil weight in the form of backfill over the foundation.

Another problem with shallow foundations is that they often bear on relatively compressible soil. Figure 8.1 illustrates the effect that can occur with a large lateral load and a strong overturning moment. With repeated applications and reversals of such a load, the soil beneath the foundation edges becomes compressed, resulting in an increasing tendency for the structure to rock, thus producing a loss of stability or a significant change in the dynamic load behavior of the structure. When the bearing strata of soil are very compressible, it is generally advisable to avoid the extreme condition of soil stress shown in the illustration.

Figure 8.2 illustrates the typical problem of tying isolated footings, which is done primarily for the purpose of ensuring that the building structure move as a single mass with respect to the ground. It is also sometimes done to allow for the sharing of the lateral load when the load on a single footing cannot be resisted by that footing alone.

Many seismic load failures of building structures are precipitated or aggravated by soil movements in fill or other highly compressible soil deposits. An especially hazardous situation is that which occurs when

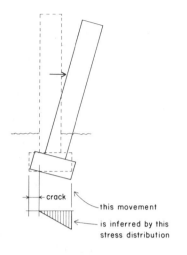

crack

this movement

is inferred by this
stress distribution

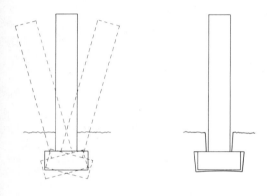

repeated actions can result in some
loss of resistance to rocking effects

FIGURE 8.1 Potential effect of repeated reversal of
lateral forces.

different parts of the foundation are placed
on soil of significantly different com-
pressibility, as may occur because of a hilly
site, variations in the level of the foundations, extensive regrading, or non-hori-
zontal soil strata. Subsidence of the fill,
major differences in settlements, or lateral
movement of the soil mass can produce
various critical situations. Figure 8.3 illustrates some of the possibilities that can produce structural failures for both gravity
and lateral load conditions.

Poor Soil at Desired Bearing Level

Whenever possible a shallow bearing foundation is used, simply because of its easier
construction and lower cost. However, when
the necessary criteria for safety and lack of
settlement cannot be assured, the usual
solution is a deep foundation, that is, one
that reaches down farther into the ground
mass. The three general types of deep foundations are shown in Fig. 8.4.

Although the building foundation in
this case is essentially bypassing the sur-
face-level soils, the upper soil levels still
constitute the site surface and must be dealt
with in the full site development. They also
provide lateral support for the tall elements
of the deep foundation system, as discussed in Sec. 8.5.

With a deep foundation there are actually three separate elements that interact:
the supported building, the deep foundation supports, and the in-between mass of

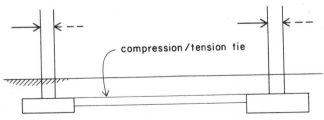

compression/tension tie

FIGURE 8.2 Tying of isolated footings.

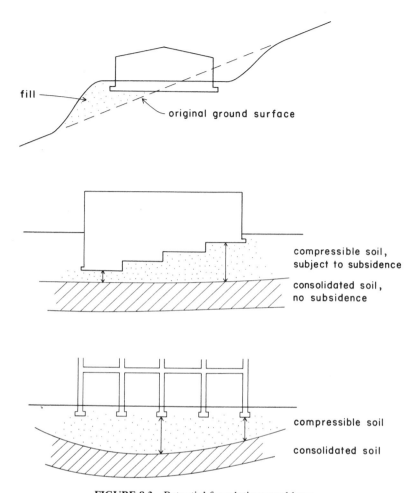

FIGURE 8.3   Potential foundations problems.

the upper soil levels. The actions of all three of these must be considered and their interactions studied carefully.

The tying of the pole structure is usually performed above the level of the ground by the rest of the structure or by added ties and struts. The use of piles and caissons usually involves two considerations: the actual lateral load and the need for lateral stability under the vertical load. Tying is required for both of these reasons, and the previously cited *UBC* requirement is based on the latter problem—that of stability under

the vertical load. Any calculated lateral loads delivered to the tops of the pile and caisson foundations should be carried away to other parts of the construction for development of the actual lateral resistance, which is essentially similar to the procedure used for isolated footings.

Sloping Sites

Sloping sites present various problems, depending considerably on the angle of the slope. At low angles of slope the main prob-

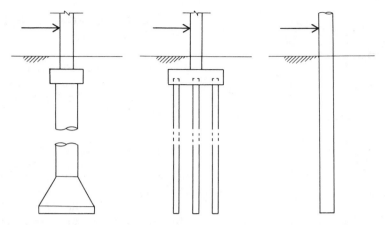

FIGURE 8.4 Types of deep foundations: cast piers (caissons), driven piles, and buried posts.

lem may simply be one of control of surface water flow. As the angle increases, other problems may emerge, including erosion of site surface materials and general retaining of the surface soils.

Figure 8.5 illustrates the situation of a footing in a hillside location that could be a wall footing or an isolated column footing. The dimension A in the figure, called the *daylight dimension*, must be sufficient to provide resistance to the failure of the soil as shown. Lateral load on the footing will further aggravate this type of failure if A is too small. The preferred solution, if the

construction permits it, is to use a tension tie to transfer the lateral effect to some other part of the structure. Otherwise, the level of the footing should be lowered until a conservative distance is developed for the daylight dimension.

Site with Major Level Changes

For various reasons it is often necessary to achieve abrupt changes in the site surface level. The building construction itself (particularly basement walls) can sometimes be used to facilitate these changes. However, some site construction remote from the building may also be necessary.

For major changes in site elevation a retaining wall is generally used. Special soil problems occur when there is a seismic load on retaining walls. As shown in Fig. 8.6a, when designing for gravity loads, a lateral soil pressure is usually assumed on the basis of an equivalent horizontal fluid pressure. If a surcharge exists, or if the ground slopes to the wall as shown in the illustration, the level of the equivalent fluid is assumed to be somewhere above the true ground level at the back of the wall.

When lateral load has been caused by seismic movement, the earth mass behind

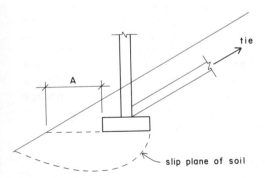

FIGURE 8.5 Critical dimension for bearing footings in hillside conditions (daylight dimension).

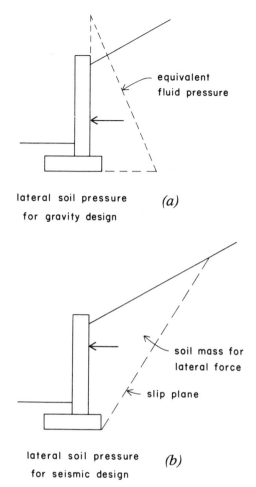

lateral soil pressure   *(a)*
for gravity design

lateral soil pressure   *(b)*
for seismic design

FIGURE 8.6   Lateral load conditions for tall retaining walls with gravity plus seismic loads.

8.3   LATERAL FORCE RESISTANCE IN SOILS

The horizontal movement of an object buried in soil is usually resisted by some combination of fricton and passive soil pressure.

Passive Soil Pressure

Passive soil pressure is visualized by considering the effect of pushing some object through the soil mass. If this is done in relation to a vertical cut, as shown in Fig. 8.7a, the soil mass will tend to move inward and upward, causing a bulging of the ground surface behind the cut. If the slip-plane type of movement is assumed, the action is similar to that of active soil pressure, with the directions of the soil forces simply reversed. Since the gravity load of the upper soil mass is a useful force in this case, passive soil resistance will generally exceed active pressure for the same conditions.

If the analogy is made to the equivalent fluid pressure, the magnitude of the passive pressure is assumed to vary with depth below the ground surface. Thus for structures whose tops are at ground level, the pressure variation is the usual simple triangular form as shown in the left-hand illustration in Fig. 8.7b. If the structure is buried below the ground surface, as is the typical case with footings, the surcharge effect is assumed and the passive pressures are correspondingly increased.

As with active soil pressure, the type of soil and the water content will have some bearing on development of stresses. This is usually accounted for by giving values for specific soils to be used in the equivalent fluid pressure analysis.

Soil Friction

The potential force in resisting the slipping between some object and the soil depends

the wall has the potential of delivering a horizontal force while simultaneously offering gravity resistance to the overturn of the wall. Figure 8.6b shows the usual assumption for the soil failure mechanism that defines the potential mass whose weight develops the lateral seismic force. In a conservative design for the analysis of the wall, this force should be added to the lateral seismic forces of the wall and footing weights.

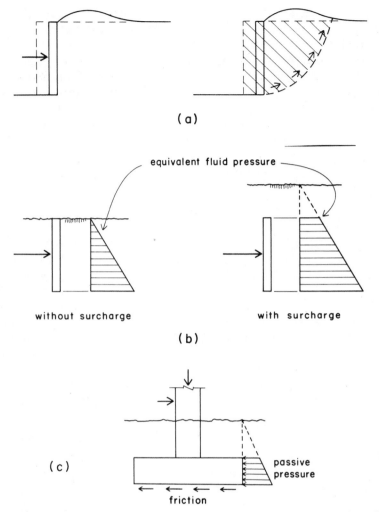

FIGURE 8.7 Aspects of the development of lateral soil pressure: (*a*) development of passive resistance; (*b*) equivalent fluid pressure, with and without surcharge; (*c*) resistance to horizontal movement of a footing.

on a number of factors, including the following principal ones:

Form of the Contact Surface. If a smooth object is placed on the soil, there will be a considerable tendency for it to slip. Our usual concern is for a contact surface created by pouring concrete directly onto the soil, which tends to create a very nonsmooth, intimately bonded surface.

Type of Soil. The grain size, grain shape, relative density, and water content of the soil are all factors that will affect the development of soil friction. Well-graded dense, angular sands and gravels will develop considerable friction. Loose, rounded, saturated, fine sand

and soft clays will have relatively low friction resistance. For sand and gravel the friction stress will be reasonably proportional to the compressive pressure on the surface, up to a considerable force. For clays, the friction tends to be independent of the normal pressure, except for the minimum pressure required to develop any friction force.

Pressure Distribution on the Contact Surface. When the normal surface pressure is not constant, the friction will also tend to be nonuniform over the surface. Thus, instead of an actual stress calculation, the friction is usually evaluated as a total force in relation to the total load generating the normal stress.

Friction seldom exists alone as a horizontal resistive force. Foundations are ordinarily buried with their bottoms some distance below the ground surface. Thus pushing the foundation horizontally will also usually result in the development of some passive soil pressure, as shown in Fig. 8.7c. Since these are two totally different stress mechanisms, they will actually not develop simultaneously. Nevertheless, the usual practice is to assume both forces to be developed in opposition to the total horizontal force on the structure.

In situations where simple sliding friction is not reliable or the total resistance offered by the combination of sliding and passive pressure is not adequate for total force resistance, a device called a *shear key* is used. The shear key is produced by trenching a short distance below the footing to produce a short, cantilevered stub beneath the footing. The enhancement of force resistance offered by a shear key is particularly desirable when the soil at the footing bottom is quite slippery (wet clay, etc.) or the footing bottom is a very short distance below grade.

8.4 LATERAL FORCES ON BEARING FOUNDATIONS

Wind and earthquake effects, combined with gravity effects, produce various requirements for foundations, typically involving some combined resistance to vertical pressure, horizontal sliding, and moment. Figure 8.8 shows a situation in

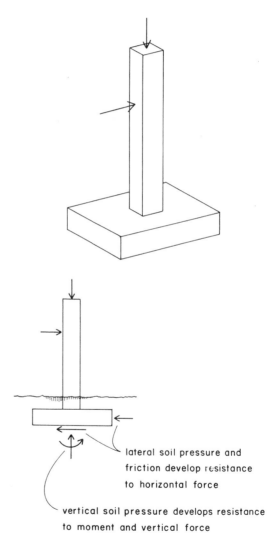

lateral soil pressure and friction develop resistance to horizontal force

vertical soil pressure develops resistance to moment and vertical force

FIGURE 8.8   Typical force development in moment-resistive footings.

which a simple rectangular footing is subjected to forces that require the development of resistance to vertical force, horizontal sliding, and an overturning moment. We will first consider the combined effects of the vertical force and the moment.

Figure 8.9 illustrates our usual approach to the combined direct force and moment on a cross section. In this case the "cross section" is the contact face of the footing with the soil. However the combined force and moment may originate, we make a transformation into an equivalent eccentric force that produces the same effects on the cross section. The direction and magnitude of this mythical equivalent e are related to properties of the cross section in order to qualify the nature of the stress combination. The value of e is established by simply dividing the moment by the force normal to the cross section, as shown in the figure. The net, or combined, stress distribution on the section is visualized as the sum of the separate stresses due to the normal force and the moment. For the stresses on the two extreme edges of the footing the general formula for the combined stress is

$$P = \frac{N}{A} \pm \frac{Nec}{I}$$

We observe three cases for the stress combination obtained from this formula, as shown in the figure. The first case occurs when e is small, resulting in very little bending stress. The section is thus subjected to all compressive stress, varying from a maximum value on one edge to a minimum on the opposite edge.

The second case occurs when the two stress components are equal, so that the minimum stress becomes zero. This is the boundary condition between the first and third cases, since any increase in the eccentricity will tend to produce some tension

stress on the section. This is a significant limit for the footing since tension stress is not possible for the soil-to-footing contact face. Thus case 3 is possible only in a beam or column where tension stress can be developed. The value of e that corresponds to case 2 can be derived by equating the two components of the stress formula as follows:

$$\frac{N}{A} = \frac{Nec}{I} \qquad e = \frac{I}{Ac}$$

This value for e establishes what is called the *kern limit* of the section. The kern is a zone around the centroid of the section within which an eccentric force will not cause tension on the section. The form of this zone may be established for any shape of cross section by application of the formula derived for the kern limit. The forms of the kern zones for three common shapes of section are shown in Fig. 8.10.

When tension stress is not possible, eccentricities beyond the kern limit will produce a so-called *cracked section*, which is shown as case 4 in Fig. 8.9. In this situation some portion of the section becomes unstressed, or cracked, and the compressive stress on the remainder of the section must develop the entire resistance to the force and moment.

Figure 8.11 shows a technique for the analysis of the cracked section, called the *pressure wedge method*. The pressure wedge represents the total compressive force developed by the soil pressure. Analysis of the static equilibrium of this wedge and the force and moment on the section produces two relationships that may be utilized to establish the dimensions of the stress wedge. These relationships are:

1. The total volume of the wedge is equal to the vertical force on the section. (Sum of the vertical forces equals zero.)

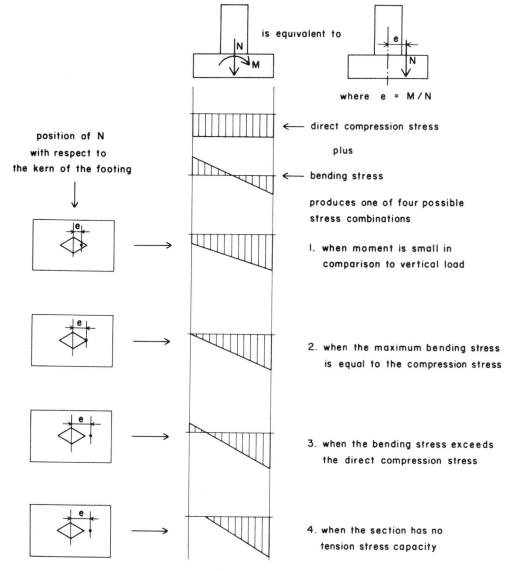

FIGURE 8.9   Investigation for vertical pressure due to combined compression and moment.

2. The centroid of the wedge is located on a vertical line with the force on the section. (Sum of the moments on the section equals zero.)

Referring to Fig. 8.11, the three dimensions of the stress wedge are w, the width of the footing; p, the maximum soil pressure; and x, the limit of the uncracked portion of the section. With w known, the solution of the wedge analysis consists of determining values for p and x. For the rectangular footing, the simple triangular stress wedge will have its centroid at the third point of the

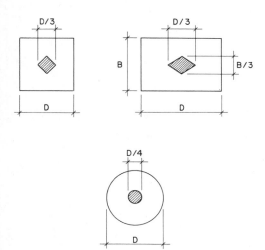

FIGURE 8.10 Kern limits for common shapes.

triangle. As shown in the figure, this means that x will be three times the dimension a. With value for e determined, a may be found and the value of x established.

The volume of the stress wedge may be expressed in terms of its three dimensions as follows:

$$V = \tfrac{1}{2}wpx$$

Using the static equilibrium relationship stated previously, this volume may be equated to the force on the section. Then, with the values of w and x established, the value for p may be found as follows:

$$N = V = \tfrac{1}{2}wpx$$

$$p = \frac{2N}{wx}$$

All four cases of combined stress shown in Fig. 8.9 will cause rotation of the footing due to deformation of the soil. The extent of this rotation and the concern for its effect on the supported structure must be considered carefully in the design of the footing. It is generally desirable that long-term

loads (such as dead loads) not develop uneven stress on the footing. This is especially true when the soil is highly deformable or is subject to long-term continued deformation, as is the case with soft, wet clay. Thus it is preferred that stress conditions as shown for case 2 or 4 in Fig. 8.9 be developed only with short-term live loads.

When foundations have significant depth below the ground surface other forces will develop to resist moment, in addition to the vertical pressure on the bottom of the footing. Figure 8.12a shows the general case for such a foundation. The moment effect of the horizontal force is assumed to develop a rotation of the foundation at some point between the ground surface and the bottom of the footing. The position of the rotated structure is shown by the dashed outline. Resistance to this movement is visualized in terms of the three major soil pressure effects plus the friction on the bottom of the footing.

When the foundation is quite shallow, as shown in Fig. 8.12b, the rotation point for the foundation moves down and toward the toe of the footing. It is common in this case to assume the rotation point to be at the toe, and the overturning effect to be resisted only by the weights of the structure, the foundation, and the soil on top of the footing. Resisting force A in this case is considered to function only in assisting the friction to develop resistance to the horizontal force in direct force action.

When a foundation is very deep and is essentially without a footing, as in the case of a pole, resistance to moment must be developed entirely by the forces A and B, as shown in Fig. 8.13. If the structure is quite flexible, its bending will cause the two forces to develop quite close to the ground surface, making the extension of the element into the ground beyond this point of little use in developing resistance to moment. Behavior of pole foundations is discussed more fully in Sec. 8.7.

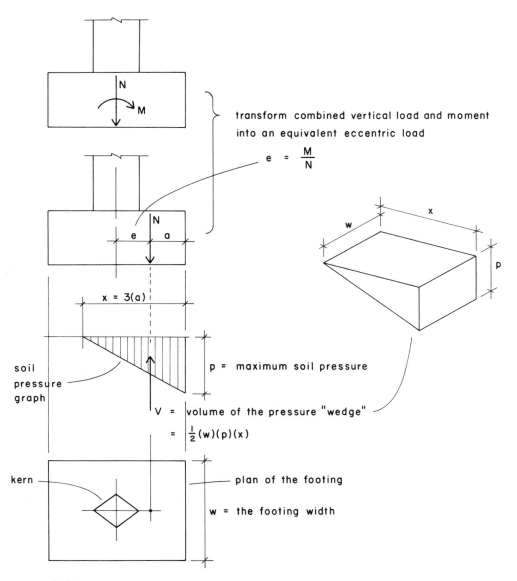

transform combined vertical load and moment into an equivalent eccentric load

$$e = \frac{M}{N}$$

x = 3(a)

p = maximum soil pressure

V = volume of the pressure "wedge"

$$= \frac{1}{2}(w)(p)(x)$$

soil pressure graph

kern

plan of the footing

w = the footing width

FIGURE 8.11 Investigation of the cracked section by the pressure wedge method.

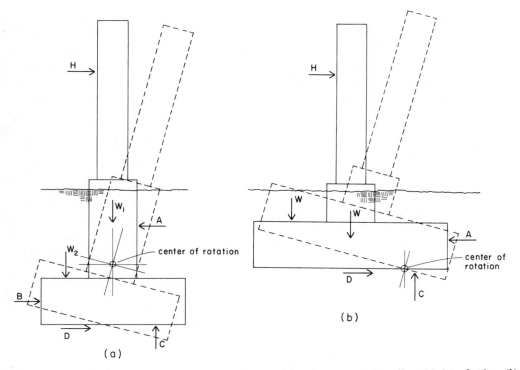

FIGURE 8.12 Force actions and movement of footings subjected to overturning effect: (*a*) deep footing; (*b*) shallow footing.

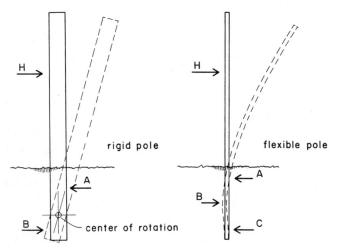

FIGURE 8.13 Force actions and movements of pole-type structures with lateral loading.

8.5 LATERAL AND UPLIFT FORCES ON DEEP FOUNDATIONS

The resistance to horizontal forces, vertically directed upward forces, and moments presents special problems for deep foundation elements. Whereas a bearing footing has no potential for the development of tension between the structure and the ground, both piles and piers have considerable capacity for uplift resistance. On the other hand, the sliding friction that constitutes a major resistance to horizontal force by a bearing footing is absent with deep foundation elements. The following discussion deals with some of the problems of designing deep foundations for force effects other than the primary one of vertically directed downward load.

Lateral Force Resistance

Resistance to horizontal force at the top of both piles and piers is very poor in most cases. The relatively narrow profile offers little contact surface for the development of passive soil pressure. In addition, the process of installation generally causes considerable disturbance of the soil around the top of the foundation elements. Although there are procedures for determination of the resistance that can be developed by the passive pressure, this resistance is seldom utilized as the major force-resolving effect in building design. In addition to the low magnitude of the force that can be developed, there is the problem of considerable movement due to the soil deformation, which constitutes a dimensional distortion that few buildings can tolerate.

The usual method for resolving horizontal forces with deep foundation systems is to transfer the forces from the piles or piers to other parts of the building construction. In most cases this means the use of ties and struts to transfer the forces to grade walls or basement walls that offer considerable surface area for the development of passive soil pressures. This procedure is also used with bearing footings when the footings themselves are not capable of the total force resistance required.

For large freestanding structures without grade walls, the horizontal force resistance of pile groups is usually developed through the use of some piles driven at an angle. These so-called *battered piles* are capable of considerable horizontal force resistance, in both compression and tension. This practice is common in the construction of foundations for large towers, bridge abutments, and so on, but is rarely utilized in building construction, where the load sharing described previously is usually the more feasible design option.

Uplift Resistance

Friction piles and large piers have considerable resistance to upward forces. If skin friction is truly the main resistive force that constitutes the pile capacity to sustain downward load, then it should also resist force in the opposite direction. An exception is the pile with a tapered form, which will have slightly higher resistance in one direction. Another exception is the unreinforced concrete pile, which has considerably more resistance to compressive stresses than it has to tensile stresses on the pile shaft.

The combined weight of the shaft and bell of a drilled pier offers a considerable potential force for resistance to upward loads. In addition, if the bell is large in diameter, considerable soil pressure can be developed against the upward withdrawal of the pier from the ground. Finally, if the shaft is long and the hole is without a permanent steel casing, some skin friction will be developed. These effects can be combined into a major resistance to upward force. However, the concrete shaft must be heavily reinforced, or prestressed, in order to develop the full potential resistance of the pier.

End-bearing piles usually have much less resistance to uplift in comparison to their compressive load capacities. However, despite the existence of relatively weak upper soil strata, there is usually some potential for skin friction resistance to withdrawal of the pile.

Uplift resistance of piles must usually be established by field load tests if they are to be relied on for major design loads. For large piers, and also for very large individual piles, this may not be feasible, because of the load magnitudes involved. However, in many design situations the uplift forces required are less than the limiting capacity of the deep elements, in which case it is sometimes acceptable to rely on a very conservative calculation of the uplift capacity.

The potential for uplift resistance on the part of deep foundation elements is an element of design that is not present with bearing foundations. Thus design for moments or for actual tension anchorage may be approached differently with deep foundations. In the case of the shear wall examples illustrated in Sec. 8.6, for example, the concerns for overturn and maximum soil pressure must be resolved without any reliance on tensile resistance between the footing and the soil. Thus the footing length and the length of the large grade beam must both be increased until the necessary relationships are developed through the use of weight and compression stress on the soil. With deep foundations this structure could possibly be reduced bin shortening the grade beam and relying on the tensile capacity of the deep foundation elements.

Moment Resistance

When used for direct support of vertical loads, piles and piers are generally designed for axial compression only. When moments are required to be resisted—such as those due to overturning effects—the resolution to the supporting piles or piers may be in the form of axial forces distributed to the group rather than to single elements. Figure 8.14 shows the case of a moment on a pile group, represented by an eccentric vertical load. Just as for a bearing footing, the group has a kern limit. If the load is

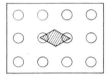

kern limit for the group

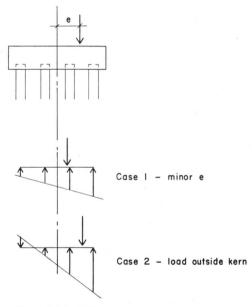

FIGURE 8.14 Effects of an eccentric load on a pile group.

inside this limit, the loads on individual piles will all be compressive, although not equal in magnitude.

By contrast with a footing, it may be possible for the pile group to develop some tension when the vertical load falls outside the kern limit. If uplift resistance of the piles can be reliably determined, this may allow for increased moment resistance by the group.

Individual piers are usually designed for some minor moment at their tops, in order that they be able to sustain some actual eccentricity of vertical loads. However, major moments must usually be developed by groups of piers, as in the case of piles. Uplift resistance can be somewhat more reliably predicted for the piers if their simple dead weight is sufficient.

A special case for piles or piers occurs when they are used in the manner of poles. This is discussed in general in Sec. 8.7.

8.6 FOOTINGS FOR SHEAR WALLS

Shear walls typically function as vertical cantilevers, with the fixed end being represented by the shear wall's foundation. When shear walls rest on bearing foundations the situation is usually one of the following:

1. The shear wall is part of a continuous wall and is supported by a foundation that extends beyond the shear wall ends.
2. The shear wall is a separate wall and is supported by its own foundation, in the manner of a freestanding tower.

We will consider the second of these two situations first. The basic problems to be solved in the design of such a foundation are the following:

Anchorage of the Shear Wall. The shear wall anchorage consists of the attachment of the shear wall to the foundation to resist the sliding and the overturning effects due to the lateral loads on the wall. This involves a considerable range of possible situations, depending on the construction of the wall and the magnitude of forces.

Overturning Effect. The overturning effect is taken into consideration by performing the usual analysis for the overturning moment due to the lateral loads and the determination of the safety factor resulting from the resistance offered by the dead loads and the passive soil pressure.

Horizontal Sliding. Horizontal sliding is the direct, horizontal force resistance in opposition to the lateral loads. It may be developed by some combination of soil friction and passive soil pressure or may be transferred to other parts of the building structure.

Maximum Soil Pressure and Its Distribution. The magnitude and form of distribution of the vertical soil pressure on the foundation caused by the combination of vertical load and moment must be compared with the established design limits.

We will illustrate some of the issues involved in dealing with the last three of these problems in the two examples that follow.

Example 1: Independent Shear Wall Footing—Minor Load

The wall and proposed foundation are shown in Fig. 8.15a. The wall is assumed to function as a bearing wall as well as a shear wall, and the vertical loads applied to the top of the foundation are the sum of the wall weight and the support loads on the wall.

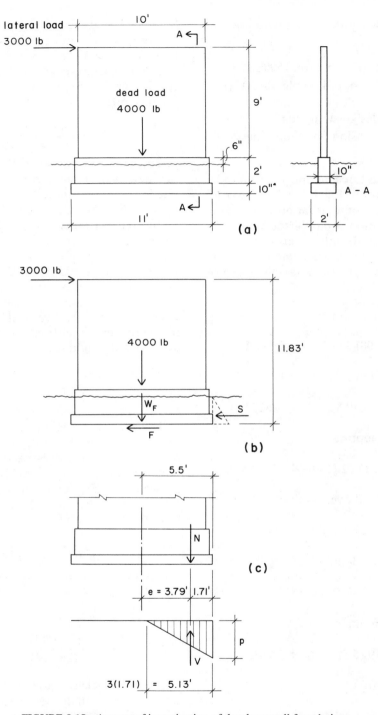

FIGURE 8.15 Aspects of investigation of the shear wall foundation.

The following are design data and criteria:

Allowable soil pressure: 1500 psf

Soil type: Group 4, Table 18-I-A, (Ref. 1)

Concrete design strength: $f'_c = 2000$ psi

Allowable tension on reinforcing: 20,000 psi

The various forces acting on the foundation are shown in Fig. 8.15*b*. For the overturning analysis the usual procedure is to assume a rotation about the toe of the footing and to include only the gravity loads in determining the resistive moment. With these assumptions the analysis is as follows:

Overturning moment:

$$M = (3000)(11.83) = 35,490 \text{ lb/ft}$$

Weight of foundation wall:

$$(2)(10/12)(10.5)(150) = 2625 \text{ lb}$$

Weight of footing:

$$(10/12)(11)(2)(150) = 2750 \text{ lb}$$

Weight of soil over footing:

$$(1.17)(1.5)(11)(80) = 1544 \text{ lb}$$

Total vertical load:

$$4000 + 2625 + 2750 + 1544 = 10,919 \text{ lb}$$

Resisting moment:

$$(10,919)(5.5) = 60,055 \text{ lb-ft}$$

Safety factor:

$$\text{SF} = \frac{60,055}{35,490} = 1.69$$

Since this safety factor is greater than 1.5, the foundation is not critical for overturning effect due to wind. (See Fig. 6.9.)

For the soil group given the soil friction coefficient is 0.25, and the total sliding resistance offered by friction is thus

$$F = (0.25)(10,919) = 2730 \text{ lb}$$

Since we assume that the lateral load on the shear wall is due to either wind or seismic force, this resistance may be increased by one-third. Thus, although there is some additional resistance developed by the passive soil pressure on the face of the foundation wall and the footing, it is not necessary to consider it in this example.

For the soil stress analysis we combine the overturning moment as calculated previously with the total vertical load to find the equivalent eccentricity as follows, deducting soil weight for *N:*

$$e = \frac{M}{N} = \frac{35,490}{9375} = 3.79 \text{ ft}$$

This eccentricity is considerably outside the kern limit for the 11-ft-long footing ($\frac{11}{6}$, or 1.83 ft) so that the stress analysis must be done by the pressure wedge method. As illustrated in Fig. 8.15*c*, the analysis is as follows: Distance of the eccentric load from the footing end is

$$5.5 - 3.79 = 1.71 \text{ ft}$$

Therefore,

$$x = (3)(1.71) = 5.13 \text{ ft}$$

$$p = \frac{2N}{wx} = \frac{(2)(9375)}{(2)(5.13)} = 1827 \text{ psf}$$

Since this is less than the allowable design pressure with the permissible increase of one third [$p = (1.33)(1500) = 2000$], the condition is not critical, as long as this type of soil pressure distribution is acceptable. This acceptance is a matter of judgment, based on concern for the rocking effect, as discussed in Sec. 8.1 and illustrated in Fig. 8.1. In this case, with the wall relatively short with respect to the footing length, we would judge the concern to be minor and would probably accept the foundation as adequate.

The design considerations remaining for this example have to do with the structural adequacy of the foundation wall and footing. The short wall in this case is probably adequate without any vertical reinforcing, although it would be advisable to provide at least one vertical dowel at each end of the wall, extended with a hook into the footing. Both the wall and footing should be provided with some minimal longitudinal reinforcing for shrinkage and temperature stresses.

Example 2: Independent Shear Wall Footing—Major Load

The wall and proposed foundation for this example are shown in Fig. 8.16a. Additional design data and criteria are as follows:

Allowable soil pressure: 3000 psf

Soil type: Group 4, Table 18-I-A, *UBC* (Ref. 1)

Concrete design strength: $f'_c = 3000$ psi

Allowable tension on reinforcing: 20,000 psi

In this case the supporting foundation wall and footing are extended some distance past the end of the shear wall to increase the stability and reduce the soil pressures. The forces acting on the structure are shown in Fig. 8.16. Following the usual procedure, we assume the overturning to be resisted only by the gravity forces

and the rotation point for overturn to be at the toe of the footing. With these assumptions, the analysis is as follows:

Overturning moment:

$$M = (24)(46) + (40)(34) + (40)(22)$$
$$= 1104 + 1360 + 880 = 3344 \text{ kip-ft}$$

Weight of foundation wall:

$$(1.5)(6)(28)(0.150) = 37.8 \text{ kips}$$

Weight of footing:

$$(2)(6)(30)(0.150) = 54 \text{ kips}$$

Weight of soil over footing:

$$(4.5)(5.5)(30)(0.08) = 59.4 \text{ kips}$$

Total vertical load:

$$240 + 37.8 + 54 + 59.4 = 391.2 \text{ kips}$$

Resisting moment:

$$M = (391.2)(15) = 5868 \text{ kip-ft}$$

Safety factor:

$$SF = \frac{5868}{3344} = 1.75$$

Since this is greater than the required factor of 1.5, the foundation is not critical for wind overturning effect.

For the soil group given, the soil friction coefficient is 0.25, and the total sliding resistance offered by friction is thus

$$F = (0.25)(391.2) = 97.8 \text{ kips}$$

Since this is slightly less than the total horizontal load of 104 kips, we will proceed with a determination of the additional re-

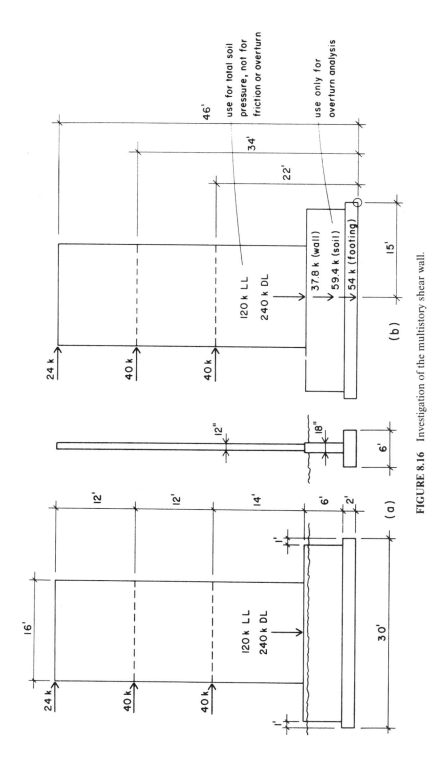

FIGURE 8.16 Investigation of the multistory shear wall.

sistance offered by the passive soil pressure on the end of the footing and foundation wall. Using the value for passive soil resistance for the group 4 soil as given in Table 18-I-A of the *UBC* (Ref. 1), the pressures as shown in Fig. 8.17*a* and are calculated as follows:

Table value for pressure/ft of depth: 150 psf

Maximum pressure at bottom of wall: (5.5)(150) = 825 psf

Pressure at bottom of footing: (7.5) (150) = 1125 psf

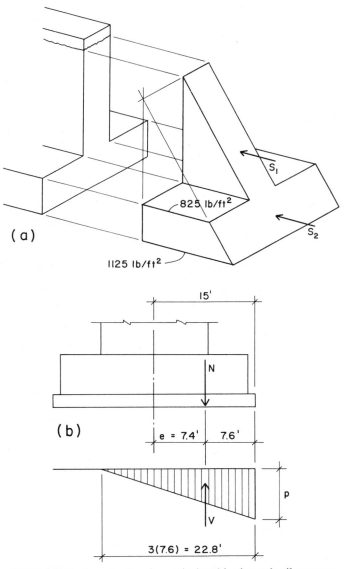

FIGURE 8.17 Investigation for vertical and horizontal soil pressure.

Total resistive forces:

On the end of the wall:

$$S_1 = \tfrac{1}{2}(1.5)(5.5)(0.825) = 3.4 \text{ kips}$$

On the end of the footing:

$$S_2 = (2)(6) \frac{0.825 + 1.125}{2}$$

$$= 11.7 \text{ kips}$$

total force $= S_1 + S_2 = 15.1$ kips

This increases the total resistive force due to the combination of sliding friction plus passive soil pressure to 112.9 kips, which exceeds the total horizontal load.

For the vertical soil pressure on the bottom of the footing we consider the combined effect of the vertical load and the overturning moment. Although the passive soil pressure offers some resistance to the moment, it is relatively minor in this case and we will ignore it. The vertical load in this case should not include the weight of the soil over the footing, but must include the design live load. The loads and the resulting eccentricity are thus as follows:

Moment:

$$M = 3344 \text{ kips-ft}$$

Vertical load:

$$N = 391.2 + 120 - 59.4 = 451.8 \text{ kips}$$

Equivalent eccentricity:

$$e = \frac{M}{N} = \frac{3344}{451.8} = 7.40 \text{ ft}$$

This is considerably in excess of the kern limit of 5 ft for the footing and makes the design questionable. However, we will proceed with an analysis for the maximum soil pressure by the pressure wedge method. Referring to Fig. 8.17b, the analysis is as follows.

The distance from the load to the edge of the footing is

$$15 - 7.4 = 7.6 \text{ ft}$$

Then

$$x = (3)(7.6) = 22.8 \text{ ft}$$

$$p = \frac{2N}{wx} = \frac{(2)(451,800)}{(6)(22.8)} = 6605 \text{ psf}$$

With the increased in allowable stress due to wind or seismic force, this would require a basic allowable soil pressure of

$$p = (\tfrac{3}{4})(6605) = 4954 \text{ psf}$$

which is greater than the given limit of 3000 psf in this example.

Reduction of the soil pressure requires an increase in the size of the footing. If this increase consists entirely of adding width, the gain is only a linear function of the increase. Increase in length is similar to adding depth to a beam section, which is considerably more effective in increasing bending resistance. However, in this situation increasing the footing length produces an increase in the cantilever distance for the foundation wall. We will therefore compromise with increases in both the width and length, as shown in Fig. 8.18. These changes result in added weight of the foundation as follows:

New wall weight:

$$(1.5)(6)(32)(0.150) = 43.2 \text{ kips}$$

New footing weight:

$$(2)(8)(34)(0.150) = 81.6 \text{ kips}$$

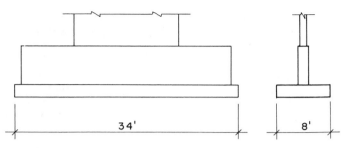

FIGURE 8.18 Modification of the foundation.

New vertical load:

$$N = 360 + 43.2 + 81.6 = 484.8 \text{ kips}$$

The new combined load analysis is thus as follows:

Eccentricity:

$$e = \frac{M}{N} = \frac{3344}{484.8} = 6.90 \text{ ft}$$

Distance from end of footing:

$$17 - 6.90 = 10.10 \text{ ft}$$

For the pressure wedge:

$$x = (3)(10.1) = 30.3 \text{ ft}$$

Maximum soil pressure:

$$p = \frac{2N}{wx} = \frac{(2)(484,800)}{(8)(30.3)} = 4000 \text{ psf}$$

If the wedge type of soil stress distribution is acceptable, this is within the limit for the given soil with the permissible increase for wind and seismic loads. For this example the rocking phenomenon, as discussed in Sec. 8.1 and illustrated in Fig. 8.1, may be marginally critical. However, the overall height of the wall above the bottom of the footing is only 1.35 times the length of the footing, so the problem should only be a critical one if the soil is highly compressible or the building structure is highly sensitive to lateral deflections.

In Example 2 we have assumed the shear wall and its foundation to be completely independent of the building structure, and have dealt with it as a freestanding tower. This is sometimes virtually the true situation, and the design approach that we have used is a valid one for such cases. However, various relations may occur between the shear wall structure and the rest of the building. One of these possibilities is shown in Fig. 8.19. Here the shear wall and its foundation extend some distance below a point at which the building structure offers a bracing force in terms of horizontal constraint to the shear wall. This situation may occur when there is a basement and the first floor structure is a heavy rigid concrete system. If the floor structure is capable of transferring the necessary horizontal force directly to the outside basement walls, the shear wall foundation may be relieved of the usual sliding resistance function.

As shown in Fig. 8.19, when the upper-level constraint is present, the rotation point for overturn moves to this point. The forces that contribute to the resisting moment become the gravity load W, the sliding friction F, and the passive soil pressure S. The following example illustrates the analysis for such a structure.

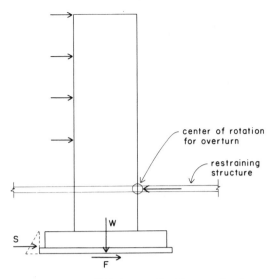

FIGURE 8.19   Tall shear wall with upper-level constraint.

Example 3: Shear Wall Footing— Upper-Level Constraint

As shown in Fig. 8.20, this structure is a modification of the one in Example 2. We assume the construction to be the same as that shown in Fig. 8.16a, except for the added height of the wall and the constraint at the first-floor level. Data and criteria for design remain the same as in Example 2.

The only modification of the vertical loads from those determined for Example 2 is the additional basement wall. This added load is

$$w = (12 \text{ ft})(16 \text{ ft})(0.150 \text{ kip/ft}^2)$$

$$= 28.8 \text{ kips}$$

Added to the total dead load calculated previously, this results in a new total dead load of

$$W = 391.2 + 28.8 = 420 \text{ kips}$$

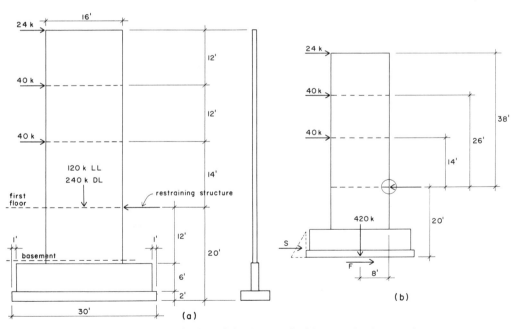

FIGURE 8.20   Investigation of the shear wall with upper-level constraint.

The overturning analysis in this case begins with a comparison of the overturning moment and the resisting moment due to the dead load. If this does not result in the necessary safety factor of 1.5, we proceed to investigate the added forces that are necessary.

Overturning moment:

$$M = (24)(38) + (40)(26) + (40)(14)$$
$$= 912 + 1040 + 560 = 2512 \text{ kip-ft}$$

Required dead load moment:

$$(2512)(\text{SF of } 1.5) = 3768 \text{ kip-ft}$$

Actual dead load moment:

$$(420)(8) = 3360 \text{ kip-ft}$$

Required additional resisting moment:

$$3768 - 3360 = 480 \text{ kip-ft}$$

If we rely on the development of sliding friction for this moment, the necessary friction force is

$$F = \frac{408}{20} = 20.4 \text{ kips}$$

which is quite a nominal force in view of the footing size and the magnitude of the dead load.

Since the friction is easily capable of the necessary added moment in this case, we do not need to consider the potential capability of added moment due to passive soil pressure. Were it necessary to do so, we would determine this potential force as was done for Example 2 and is illustrated in Fig. 8.17a.

Considering the equilibrium of the structure as shown with the forces in Fig. 8.20b, we can now determine the required force that must be developed by the constraining structure at the first-floor level. This will consist of the sum of the horizontal loads and the required friction force. Thus

$$R = H + F = 104 + 20.4 = 124.4 \text{ kips}$$

If we consider the rotational stability of the wall to be maintained in the manner assumed in the preceding calculations, the vertical soil pressure on the footing is relieved of any moment effect. Thus the pressure is simply that due to the vertical loads and is determined as follows:

Total vertical load:

$$420 \text{ kips (dead load)} + 120 \text{ kips (live load)}$$
$$= 540 \text{ kips}$$

Maximum soil pressure:

$$p = \frac{540}{(6)(30)} = 3 \text{ kips/ft}^2$$

Since this is precisely the limit given, the footing is adequate in this example.

Another relationship that may occur between the shear wall structure and the rest of the building is that of some connection between the shear wall footing and other adjacent foundations. This occurs commonly in buildings designed for high seismic risk, since it is usually desirable to assure that the foundation system moves in unison during seismic shocks. This may be a useful relationship for the shear wall, in that additional horizontal resistance may be developed to add to that produced by the friction and passive soil pressure on the shear wall foundation itself. Thus, if the elements to which the shear wall foundation is tied do not have lateral load requirements, their potential friction and passive pressures may be enlisted to share the loads on the shear wall.

Shear walls on the building exterior often occur as individual wall segments, consisting of solid portions of the wall between openings or other discontinuities in the wall construction. In these situations the foundation often consists of a continuous wall and footing or a grade beam that extends along the entire wall. The effect of the overturning moment on such a foundation is shown in Fig. 8.21. The loading tends to develop a shear force and moment in the foundation wall, both of which are one half of the forces in the wall. If the foundation wall is capable of developing this shear and bending, it functions as a distributing member, spreading the overturning effect along an extended length of the foundation.

The overturning effect just described must be added to other loadings on the wall for a complete investigation of the foundation wall and footing stresses. It is likely that the continuous foundation wall also functions as a distributing member for the gravity loads.

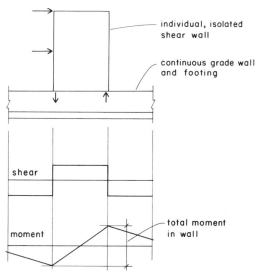

FIGURE 8.21 Isolated shear wall on a continuous foundation.

8.7 POLE STRUCTURES

A type of construction used extensively in ancient times and still used in some regions is that which employs wood poles as vertical structural members. Although processed poles, cut to have a constant diameter, are possible, most poles are simply tree trunks with the branches and soft outer layers of material stripped away. There are generally three ways such poles may be used: as timber piles, driven into the ground; as vertical building columns in a framed structure; or as buried poles, partly below the ground and partly extending above it. The following discussion is limited to consideration of the buried-end pole.

As a foundation element, the buried-end pole is typically used to raise a building above the ground. Although timber piles may also be used this way, especially in waterfront locations, they are used more often as buried foundations. With regard to the building construction, the two chief means of using poles are for pole-frame buildings and pole-platform buildings. For a pole-frame building, the poles are extended above ground to become building columns. For the pole-platform, the poles are cut off at some level above the ground and a flat structure (platform) is built on top of them, providing support for some conventional wood frame construction.

Pole foundations must usually provide both vertical and lateral support for a building. For vertical loads, the pole end simply transfers vertical load primarily by direct bearing. The three common forms for buried-end pole foundations are shown in Fig. 8.22. In Fig. 8.22a, the bottom of the dug hole is filled with concrete to provide a footing, a preferred method when the soil at the bottom of the hole is quite compressible. In Fig. 8.22b the pole bears directly on the bottom of the hole, the hole is partly backfilled, and a concrete collar is poured around the pole before backfilling is com-

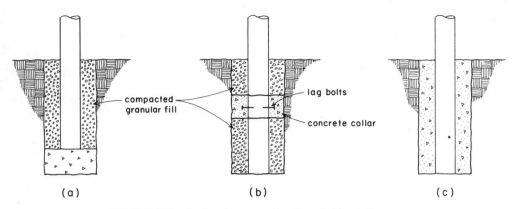

FIGURE 8.22 Optional construction for pole foundations.

pleted. To anchor the pole to the concrete collar, lag bolts are installed in the pole at this location. The collar helps to increase the vertical load capacity, but mostly aids in the lateral stability of the pole. In Fig. 8.22c the hole is backfilled completely with concrete, a choice made mostly in poorer soils.

Pole building construction is quite regionally limited, although it is used widely for service buildings. Local codes and practices often determine details of the construction as well as load capacities for the pole foundations. For vertical loads, capacities of 6 to 10 kips per pole are common, depending on soil conditions and the wood species of the poles.

For lateral loads, the situation is usually quite different for pole-frame and pole-platform buildings. For the pole-frame building, the poles are normally held in considerable constraint above the ground, as shown in Fig. 8.23a. This makes lateral movement of the pole in the hole of somewhat less concern, as the lower (usually stiffer) end of the pole works as a short cantilever. For the pole-platform building, especially when the platform is quite close to ground level, there is less possibility to restrain the poles at their tops, and they thus may function primarily as cantilevers from the holes. Because of the typical greater concern for lateral movements, poles for the platform structure building are usually

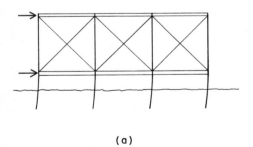

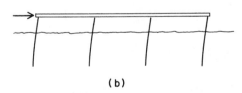

FIGURE 8.23 Response to lateral forces in pole-type structures: (a) pole–frame construction, columns cantilever downward; (b) pole–platform construction, columns (poles) cantilever upward.

buried slightly deeper in the ground than those for pole-frame construction—other factors not withstanding.

Backfilling of holes, if not done with concrete, should be done with some material that can be compacted to a dense, stable condition. A well-graded sand or sandy gravel is usually the first choice.

For lateral forces building code requirements generally deal with the pole structure as being one of two situations, as illustrated in Fig. 8.24. If construction exists at grade level, the lateral movement of the pole may be constrained at this location, so that a rotation of the pole occurs at grade level and the development of resistance to lateral load is as shown in Fig. 8.24*a*. If grade-level constraint is minor or nonexistent, lateral resistance will be developed by opposing soil pressures on the buried portion of the pole, as shown in Fig. 8.24*b*.

The following example illustrates the use of design criteria for the unconstrained pole taken from the *Uniform Building Code* (see Appendix B) and the *City of Los Angeles Building Code.*

Example 4: Lateral Force on a Pole

A 12-in.-diameter round wood pole is used as shown in Fig. 8.25. The soil around the buried pole is generally a medium compacted silty sand. Investigate the adequacy of the 10-ft embedment.

Using criteria from the *City of Los Angeles Building Code*, a determination is made of the two critical soil stresses f_1 and f_2, as shown in Fig. 8.24. These computed stresses are then compared to the allowable pressures. Using the formulas from the code, we find

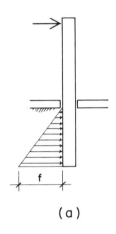

(a)

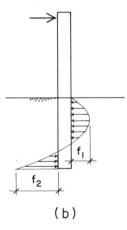

(b)

FIGURE 8.24 Development of lateral resistance in buried poles: (*a*) with grade-level constraint; (*b*) with no grade-level constraint.

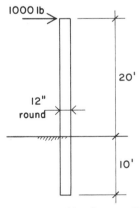

FIGURE 8.25 Form and loading condition for the example.

$$f_2 = \frac{7.62P(2h + d)}{bd^2} = \frac{7.62(1000)(50)}{(1)(10)^2}$$

$$= 3810 \text{ psf}$$

$$f_1 = \frac{2.85P}{bd} = \frac{f_2}{4} = \frac{(2.85)(1000)}{(1)(10)} + \frac{3810}{4}$$

$$= 285 + 953 = 1238 \text{ psf}$$

From the Los Angeles code the allowable lateral-bearing pressure for the compact silty sand is 233 psf per foot of depth. For f_1 the depth is taken as one-third the total; thus

allowable $p = \frac{10}{3} \times 233 \times \frac{4}{3} = 1036$ psf

(assuming that the lateral force is due to wind or seismic load and the increase of one-third is permitted).

For f_2 the allowable pressure is

$$p = 10 \times 233 \times \frac{4}{3} = 3107 \text{ psf}$$

As both of the computed pressures exceed the allowable values, it is observed that the embedment is not adequate.

Solving for the required depth of embedment is not very direct with the formulas from the Los Angeles code. On the other hand, the *UBC* provides a formula for the direct determination of the depth of embedment as follows.

From *UBC*, Sec. 1806.7.2.1,

$$d = \frac{A}{2} \left(1 + \sqrt{1 + \frac{4.36h}{A}} \right)$$

$$A = \frac{2.34P}{S_1 b}$$

From *UBC* Table 18-I-A, p for silty sand = 150 psf per foot of depth; thus

$$S_1 = 150 \times \frac{10}{3} \times \frac{4}{3} = 667 \text{ psf}$$

$$A = \frac{(2.34)(1000)}{(667)(1)} = 3.51$$

$$d = \frac{3.51}{2} \left[1 + \sqrt{1 + \frac{(4.36)(20)}{3.51}} \right]$$

$$= 14.24 \text{ ft}$$

which also indicates that the proposed embedment is not adequate.

8.8 GROUND TENSION ANCHORS

Tension-resistive foundations are a special, although not unique, problem. Some of the situations that require this type of foundation are the following:

Anchorage of very lightweight structures, such as tents, air-inflated structures, light metal buildings, and so on

Anchorage of cables for tension structures or for guyed towers

Anchorage for uplift resistance as part of the development of overturn resistance for the lateral bracing system for a building

Figure 8.26 illustrates a number of elements that may be used for tension anchorage. The simple tent stake is probably the most widely used temporary tension anchor. It has been used in sizes ranging from large nails up to the huge stakes used for large circus tents. Also commonly used is the screw-ended stake, which offers the advantages of being somewhat more easily inserted and withdrawn, and having less tendency to loosen.

Ordinary concrete bearing foundations offer resistance to tension in the form of their own dead weight. The so-called deadman anchor consists simply of a buried block of concrete, similar to a simple foot-

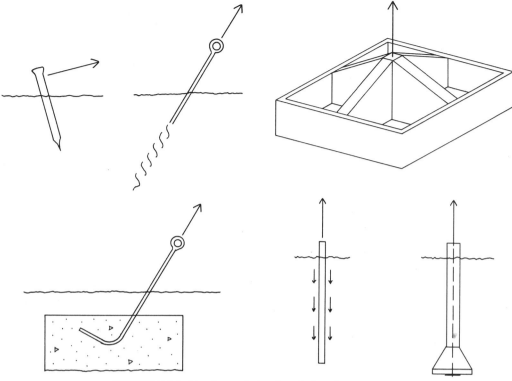

FIGURE 8.26   Tension-resistive foundation elements.

ing. Column and wall footings, foundation walls, concrete and masonry piers, and other such heavy elements may be utilized for this type of anchorage. Many lightweight building structures are essentially anchored by being fastened to their heavy foundations.

Where resistance to exceptionally high uplift force is required, special anchoring foundations may be used that develop resistance through a combination of their own dead weight plus the ballast effect of earth fill placed in or on them. Friction piles and piers may also be used for major uplift resistance, although if their shafts are of concrete, care should be taken to reinforce them adequately for the tension force.

A special technique is to use a belled pier to resist force by its own dead weight plus that of the soil above it, since the soil must be pushed up by the bell in order to extract the pier. One method for the development of the tension force through the bell end is to anchor a cable to a large plate, which is cast into the bottom of the bell, as shown in Fig. 8.26.

The nature of tension forces must be considered as well as their magnitude. Forces caused by wind or seismic shock will have a jarring effect that can loosen or progressively weaken anchorage elements. If the surrounding soil is soft and easily compressed, the effectiveness of the anchor may be reduced.

8.9 SPECIAL SITE CONDITIONS

Unusual site conditions may require special design for lateral forces. The following are some common conditions that present this requirement.

Sloping Site with Deep Foundations

The general problem of providing lateral stability for deep foundations is discussed in Sec. 8.2. The usual basic source for the bracing at the top of piles or piers is the upper soil mass. A special problem occurs when the site has a significant slope and there is a potential for a downslope movement of the soil mass at the surface. This is a problem for many hillside and waterfront locations.

A special structure sometimes used for this situation is that shown in Fig. 8.27, sometimes called a *downhill frame*. This consists of a grade-level concrete founda-

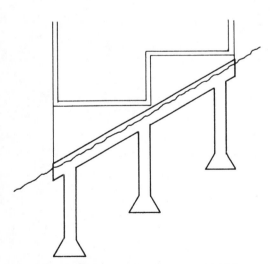

FIGURE 8.27 Hillside foundation with rigid frame (downhill frame).

tion beam structure used in combination with a set of deep foundation elements. The horizontal beams and vertical piles or piers are made to form a rigid frame. With sufficient burying of the lower ends of the deep elements in a stable soil mass, the upper part of the structure is made to resist the lateral movement of the soil at the surface level—as opposed to the usual case of relying on it for bracing. This is essentially similar to the action of the pole-frame system described in Sec. 8.7.

Soil Liquefaction

A special phenomenon that sometimes occurs during an earthquake is that of soil liquefaction. Due to a combination of ground configuration, special soil properties, and a high groundwater level, the ground-level soil on a site may experience greatly exaggerated back-and-forth movement. This excessive motion may be applied to buildings that sit on or otherwise rely on the soil mass for lateral stability. The condition exists most frequently in areas with extensive fill—either human-made or natural but of recent origin.

Major concentrations of damaged buildings from the Loma Prieta and Northridge earthquakes were located in areas of high potential for soil liquefaction, even at considerable distances from the epicenter. Locations of old waterways, swamps, lakes, or reclaimed waterfront sites are at high risk for this action. There is no real "design" solution for this situation, other than a recognition of the potential for considerable magnification of ordinary seismic effects. The only defense against disaster is a thorough knowledge of the geology of the region in the vicinity of the site and the history of damage from previous earthquakes in the region.

PART III

DESIGN EXAMPLES

This part contains a number of examples of the design of building structures for the effects of wind and seismic actions. The examples have been chosen to illustrate relatively common situations with regard to size and shape of buildings and type of construction. A few example buildings have been used repetitively, resulting in the presentation of alternative solutions for the lateral bracing system for the same basic building form.

The examples are mostly limited to low-rise buildings with relatively simple forms. For this category of buildings, it has generally been acceptable to use the simplified procedures of the equivalent static load method (basically, the requirements of the *Uniform Building Code*). Recent experiences, however, have called this procedure into question, so that in the future the use of more stringent dynamic analyses and deformation-limit, rather than strength-limit, design procedures may be required for most buildings in high-risk zones. For now, the *UBC* remains the chief guide for simple buildings and is, at the least, a point for beginning of the design development of a reasonable lateral bracing system.

9

DESIGN PROCESS AND METHODS

Before presenting the examples of design, we consider briefly in this chapter some aspects of the basic process of design and some aspects of making it happen in the context of current practices of building design and construction.

9.1 DESIGN MANAGEMENT

Design of complex objects, involving many people and having interrelations with other design developments occurring simultaneously, requires careful management if progress is to be made toward a reasonable conclusion. Many designs bog down in the process due to bad management, despite the work of skilled and experienced designers. It behooves everyone involved—and most notably those in charge—to fully understand the process and some of the problems of the management of design work.

Design is essentially a continuing task of inquiry and decision. The inquiry continues as long as potential questions can be brought up; decisions must be made with much judgment and with the weighing of the importance of many factors, some of which are usually in opposition.

Final design solutions for complex systems, such as those for building structures, often contain many compromises and many relatively arbitrary choices. The most economical, most fire-resistive, most quickly erected, most handsome, or most architecturally accommodating structure is only accidentally likely to be the optimal choice for the resistance of lateral forces.

We cannot pretend to show the complete process of structural design in these examples, although consideration of factors other than lateral force actions is frequently mentioned. The design solutions developed are thus unavoidably somewhat simplistic and myopic in their concentration on the lateral force problem.

In general, with regard to the lateral force resistive system, the design process incorporates the following:

1. *Determination of the Basic Scheme.* This includes the choice of type and layout of the basic elements of the lateral resistive system.

2. *Determination of Loads.* This involves the establishment of criteria and the choice of investigative methods.

3. *Determination of the Load Propagation.* This consists of the tracing of the load through the structure, from element-to-element of the system, until it is finally resolved into the supporting ground.

4. *Design of Individual Elements.* Based on their load-sharing roles, each separate element of the system must be investigated and designed.

5. *Design for Interactions.* Connections between elements of the structure, and between structural and nonstructural parts of the building, must be investigated and designed.

6. *Design Documentation.* Because the design as such is essentially only an imagined idea, all the information necessary to clearly and unequivocally communicate the idea must be documented.

In the discussions of the examples that follow, all of these aspects of the design process are given some treatment.

9.2 METHODS AND AIDS

Methods used to achieve design work depend greatly on the working context of the designer as well as the nature of the design work. For the design of building structures, the work includes some amount of mathematical, engineering-based, analytical work, but also includes a lot of visualization, planning, and the development of construction plans and details. It also typically involves the efforts of more than a single person, and frequently even more than a single design organization.

The most practical method used to achieve a single task—for example, design of a single reinforced concrete column—may be determined relatively easily. The best method for achieving a whole system design, however, is subject to so many special considerations for each system that it is quite elusive. This is the essence of design management and requires major management skills as well as basic design skills.

Whatever methods are used, they must account for some fundamental aspects of the work, as follows:

1. Complete investigation of the definable loading demands. What can happen that should be considered for an assurance of safety?

2. Reasonable consideration of all the feasible alternatives for solution of the problem. Final choices should be based on some effort to assess all the possibilities.

3. Integration of the subsystem design with all the other subsystems with which it interacts and within the predominant concerns for the whole system of which it is a part.

4. Final definition of the design in a communicable form for those who will perform the construction. In the end, that typically means written specifications and annotated graphics (working drawings).

Whatever works to get that done in a responsible, reliable manner in reasonable time at minimum cost is a good method.

All design work utilizes various aids in the execution of the work. The chief aids are typically any existing records of previous designs of a similar nature that proved to be reasonably successful. As most design work is highly repetitive, and nothing shines as bright as demonstrated success, this is usually the first place to look for help.

Textbooks, handbooks, industry-supplied data, research reports, models, and computer-aided processes are all used when the occasion indicates their usefulness. It is primarily the familiarity of the designer with the aids that is critical for their effective use. That starts with knowing they exist, how to access them, and how to use them. The various industry organizations and professional organizations that relate to a specific area are good sources for determination of the availability of design aids.

9.3 REFERENCES FOR CRITERIA AND STANDARDS

Every designer approaches a particular design task armed with their individual collection of design experience, developed skills, and personal store of knowledge. Routine, familiar tasks may be performed out of hand, with no other references. However, for most design work, considerable use will be made of reference sources for assistance. Some of these references may provide direct information for individual design tasks. Other sources will provide basic information that establishes fundamental design criteria and standards for design requirements, without which measurement of design acceptability cannot be made.

Criteria and standards apply to all phases of the design work. Referring to the itemized list in Sec. 9.1, the following are some of the potential sources that can be used for the various stages of design.

1. *Determination of the Basic Scheme.* This is largely a matter of the designer's judgment. However, various sources of information may assist in the comparison of basic schemes in terms of their appropriateness for different situations. Thus the degree of risk of windstorms or earthquakes, the size and form of a building, any special site conditions, and other factors will define a particular situation. For that particular situation, some evaluation of the feasibility or effectiveness of various basic schemes may be made.

Evaluations of this kind may be largely judgmental, or may be made with some reliable data from observed performances, research, or analysis of basic criteria and standards. Many studies of this kind have been made and are part of the technical literature. For common situations these studies will probably be of some practical value. For unique situations, they will provide a starting point for a more precise design study for the situation at hand.

In the end, it may be necessary to carry design work for more than one scheme to some level of detail in order to make an evaluation for a particular design. Since the basic scheme for the lateral bracing system may have considerable relation to the basic architectural design, this kind of study may be required by others as well as the structural designer.

The reality here is that there is really no identifiable source with singular reliability for support of this phase of the design work. Any borrowed information used here is subject to someone else's judgments or opinions.

2. *Determination of Loads.* The enforceable building code for a particular project is obviously a starting point for this and must be satisfied for a minimal design. The

question is, is a minimal design acceptable for this work? If real optimization is desired, a much deeper investigation must be made for design criteria and eventually for design performance evaluation in later stages of the design work. This means either some educated judgments by the designer or some research into what the building codes themselves used as reference sources. A really *safer* design may simply begin with some increase in the loading as defined by the codes. If the rest of the design work is performed reliably, that will certainly provide a stronger structure.

Required loads provided in the building codes are not developed arbitrarily. If you want to challenge them, you will have to find out how they were determined and pursue your own evaluations in the same way that those who wrote the codes did.

3. *Determination of the Load Propagation.* This consists of the structural investigation for the internal forces and deformations produced by the loads. Performing this reliably means using whatever analytical methods are available to the designer. For simple structural systems (such as most of those illustrated in this book) the work may be performed relatively easily. For a true dynamic load analysis of a highly indeterminate structure using strength methods and deformation-limit criteria, it may strain the capabilities of the most sophisticated computer-aided programs available.

A fundamental question here is what is acceptable as an investigation. Building codes may sometimes stipulate the required method of investigation, but mostly it is a matter of judgment—by the designer and by anyone who has to review the acceptability of the design work. At any given time there is an existing body of knowledge that is expected to be known by professional designers, and some degree of skill in its application to design work. That

situation changes over time, advancing primarily with new research, theories, and experience. The best sources at any given time are usually the latest popular professional text and reference books on the topic.

For review of the design work as part of the general review for acceptability of the proposed construction, some documentation of the structural investigation will usually be required for anything other than very simple buildings. This means that the form of the investigation should be one that is generally familiar, or the reviewers will not be able to follow the work.

4. *Design of Individual Elements.* Unless a totally unique invention is being contrived by the designer, this is probably the easiest task in terms of available references. Promotional work by the manufacturers of materials and products generally include reference data and recommended designs for just about anything that could conceivably be done with the materials and products. Industry organizations such as the ACI (American Concrete Institute), AISC (American Institute for Steel Construction), and BIA (Brick Institute of America) provide both standards that are widely accepted by building codes and entire libraries of design guides. Name a structural material or product and there probably is one or more organizations that provide some reference materials of this kind.

For the most widely used materials and the most widely used basic elements and systems, there are also basic reference textbooks, handbooks, and information from individual manufacturers or suppliers in competition with each other as well as with all other materials and products.

5. *Design for Interactions.* There are two levels to this. First is the interaction of the separate parts of the structural system, necessary to assure load transfers within the

system and the overall performance of the system. For performance analysis this is an extension of items 3 and 4. For a final design development, it involves thorough study of the necessary construction details to assure the necessary structural actions. This requires that the structural designer follow through with a review of the execution of the construction—all the way to the actual erection process, if necessary. It assumes, of course, that the designer has considerable knowledge of construction work and can pass judgment on both its specification and graphic detailing as well as its actual execution by workers.

For various reasons, this is an area where a lot of problems occur. First, not all designers are knowledgeable about specifications, graphics, and construction work. It therefore means that someone else (if anyone) must follow through beyond the stage of basic structural computations. The more people added to the chain—from structural investigation to final construction—the more opportunities for lack of follow-through and for the possibility that the design falls through the cracks somewhere along the line.

A problem here is that reference sources do not often themselves follow through from basic structural investigation to final construction. It therefore behooves *somebody*—hopefully the structural designer—to do so.

6. *Design Documentation.* The principal purpose of the design documentation is for communication of the design to others. For structures, the basic components of the documentation are the structural computations, the structural specifications, and the parts of the building construction drawings that deal with the structure. That package is the definition of the final design and the total communication from designer to builder.

However individual the design or whatever the designer's style, the communication must be in common language, understandable by the builder as well as many others. Standard references in use by the construction industry must be used so as not to confuse the communication process. This means using whatever exists in the form of accepted common language, notation, symbols, and even writing and drafting styles. The single best source for this common language is the CSI (Construction Specifications Institute), which is both the author and principal user of the standard materials.

The detailed illustrations of construction in this part (framing plans and construction section details) are done in a general form similar to the standards in use, but are developed here principally for illustration and are not complete in most cases.

A critical concern for the separate elements of the design documentation is that all the parts say the same thing. This is not so easy to assure when each element (computation, specification, drawing) is produced by different people—which is often the case in large design offices. This coordination needs to be assigned to someone in the design process, preferably the structural designer.

The Bibliography for this book contains several useful references for the design of lateral bracing systems, but a complete list of all such references would possibly be the size of this entire book. A few very general references may be used by most designers, but for a specific project a relatively short, specialized list of useful references must be custom-assembled.

9.4 COMPUTER-AIDED DESIGN

Computers can be used to aid the work in all phases of design as described earlier.

Their use is more critically necessary for some tasks, particularly the investigation of very complex systems for multiple loading conditions. Increasingly, however, a significant application is in using the steadily accumulating data regarding evaluations of performances of previously designed structures, which holds the promise of providing a much more intelligent basis for the work in early stages of design. Programs for the investigation and design of most ordinary structures are readily available from various commercial sources as well as from the various industry and professional organizations.

9.5 INSPECTION OF CONSTRUCTION

Recent studies, including those of damaged buildings following major windstorms and earthquakes, indicate that a considerable amount of what is designed and specified in the way of special construction for lateral bracing does not actually get installed during the building construction work. This applies especially to special anchors and connectors for wood structures and to reinforcement for concrete and masonry structures, items not in view once the construction is completed.

Poor construction is often attributable to the lack of responsible inspection by others of the builder's work at critical points during construction. The proper time to perform this inspection is often a short interval, and if the inspector is not present at the appropriate time, the work continues. Or, of course, the inspector may be incompetent, irresponsible, or in collusion with the builder to defraud the building owner.

In any event, someone should ascertain that the work of building a structure is properly done, and no one knows what should be done better than the designer of the structure. If the services of the designer do not include that work—as it increasingly does not in practice—some gap between design intentions and actual construction is more likely to occur.

Most responsible professional structural design firms provide some inspection of the construction at critical points as part of their full consulting service. This is simply good business, both for the full service to the client but also for the assurance of the firm that the work for which they accept responsibility will not fail.

10

SHEAR WALL SYSTEMS

Probably because they occur in most buildings, vertical planes of solid wall construction are the most frequently used form of vertical lateral bracing element. This is true primarily for modest-sized buildings and for low-rise construction, and is not the case for high-rise or long-span structures, or for buildings in general with very open plans and extensively glazed exterior walls. In this chapter we present some design examples for shear wall systems of the three most common forms of construction: surfaced wood frames, reinforced masonry, and reinforced concrete.

10.1 WOOD-FRAMED SHEAR WALLS

Because of its widespread use, the light wood frame (stud construction) with applied paneling is the most common form of shear wall. In this section we present some examples of design of this type of wall.

Example 5, Building A

The upper part of Fig. 10.1 shows the plan, partial elevation, and partial section for a one-story building. We will refer to this building as Building A, and in this example will design it with a light wood frame with plywood wall sheathing and roof deck. The following criteria will be used:

For wind: basic wind speed = 80 mph (see *UBC* map, Fig. 16-1, in Appendix B). Assume exposure condition *C*.

For seismic: *UBC* map zone 3 (see *UBC* map, Fig. 16-2, in Appendix B).

Building construction:

Roof: with ceiling, suspended items, DL = 20 psf.

Walls: 2X frame, plywood + stucco on exterior, gypsum drywall on interior, DL = 20 psf for exterior, 10 psf for interior.

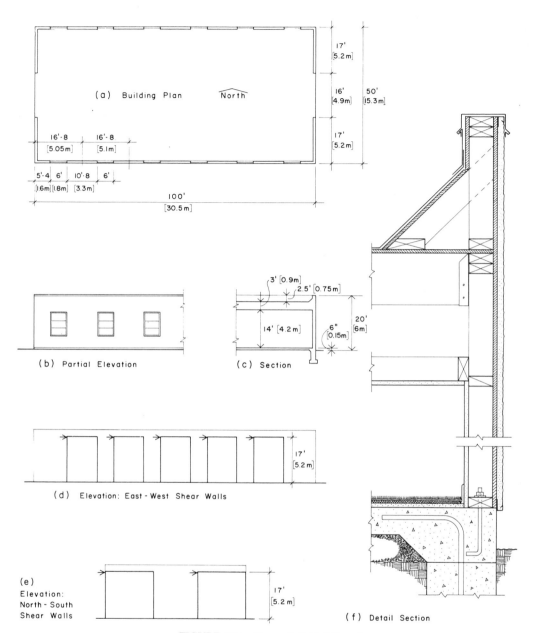

FIGURE 10.1 Example 5: Building A.

Materials for design: Framing is Douglas fir–larch, No. 1; plywood is fir, structural grade.

Assume that equipment for an air exchange system used for heating and cooling will be installed on the roof with a total equipment weight of 5 kips.

It is quite common, when designing for both wind and seismic forces, to have some parts of the structure designed for wind and others for seismic effects. In fact, what is necessary is to investigate for both effects and to design each element of the structure for the condition that produces the greater effect.

Design for Wind. The *UBC* Sec. 1618 defines the *design wind pressure* as

$$p = C_e C_q q_s I_w \ [UBC \ \text{Formula (18-1)}]$$

In this formula the factor C_e combines concerns for the height above ground level, exposure conditions, and gust effects. From *UBC* Table 16-G (see Appendix B), assuming exposure condition C, $C_e = 1.2$ for the height zone from zero to 20 ft above the ground surface.

The quantity q_s is the *wind stagnation pressure* at the standard measuring height of 10 meters (approximately 30 ft) above ground. From *UBC* Table 16-F (see Appendix B) the q_s value for a wind speed of 80 mph is 17 psf.

The importance factor I is given in *UBC* Table 16-K (see Appendix B). We will assume for this example that the building use does not qualify for the heightened concerns indicated in the table and will therefore use a value of 1.0 for I.

The factor C_q is the pressure coefficient for the structure or portion of structure under consideration, as given in *UBC* Table 16-H (see Appendix B). Values are given for individual building elements, such as the

exterior walls and roof surfaces, as well as for items such as parapets, eaves, and canopies. For design of the building structural system (called the *primary frame and system* in the code), two methods are given. The first, called method 1 or the normal force method, consists of applying individual forces to the various components of the building surface. This method is required for gable frames and optional for other cases. The second method, called method 2 or the projected area method, may be used for any building less than 200 ft high, except those with gabled frames. Method 2 is applied by considering the projected building profile as a single vertical or horizontal surface acted on by direct pressure (Fig. 10.2). We will demonstrate the use of method 2 in this example.

The wind pressures and total wind forces must be considered for a number of situations, including the following:

Direct pressure effects on walls and roof surfaces, affecting design of rafters and studs as well as the consideration of window glazing, attachment of cladding, and so on

Uplift effect of wind on the roof that is possibly critical for lifting of the roof structure or even the entire building if construction is very light

Horizontal sliding of the building off the foundation, or sliding of the foundation where depth of penetration below grade is shallow

Overturning of the entire building

Horizontal force effects on the various elements of the lateral bracing system

All of these, plus other possible concerns, must be investigated for a complete building design. Because our concern here is with the design of the lateral bracing system, we will limit our involvement to con-

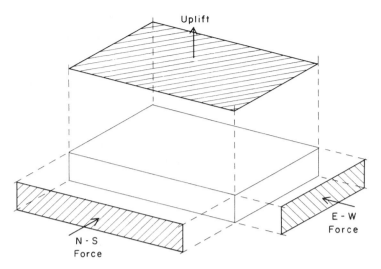

FIGURE 10.2 Generation of wind forces: Building A.

siderations of the major elements of that system. Figure 10.3 shows the basic elements of the lateral bracing system and the manner in which wind forces are applied as lateral effects. We observe that the exterior walls function in spanning vertically between the floor slab and the edge of the roof structure. There are two possibilities for the function of the parapet structure, as described in the illustration. We will assume the parapet to be cantilevered from the roof structure and the wall studs to be simple spans, as shown for case 2 in Fig. 10.3a. For this situation, the critical design consideration will most likely be that occurring under the combined wind and gravity loads, with the columns subjected to combined axial compression and bending.

We will assume that the roof functions and the wall construction produce a design gravity load of 200 lb/ft of the wall length. With the studs at 16-in. centers, the load on a single stud is thus $1.33 \times 200 = 267$ lb. For the wind loading, the critical condition is that of direct inward pressure for which *UBC* Table 16-H yields a value of $C_q = 1.2$. Thus, using the other data established pre-

viously, the wind pressure on the wall is determined as

$$p = C_e C_q q_s I_w$$
$$= (1.2)(1.2)(17)(1.0)$$
$$= 24.48 \text{ psf} \qquad \text{say, 25 psf}$$

and the load on a single stud will be $1.33 \times 25 = 33.25$ lb/ft.

This lateral loading will produce bending in the studs in a form related to the stud wall construction. The bending must be combined with the appropriate gravity load effects for a full design of the studs. In many cases the usual increase of allowable stress for the wind or seismic loads will result in no increased requirements for the studs over that determined for gravity load alone.

If the investigation demonstrates that an overstress condition exists, calling for some remedy, possibilities for this include:

A check of the data for accuracy, if the overstress is quite low. If some dimensions are not accurate, or loads are

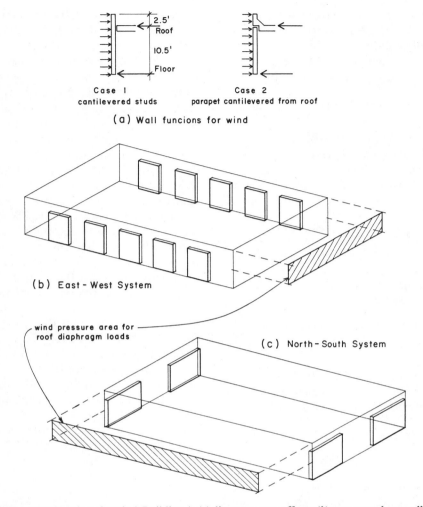

FIGURE 10.3 Wall functions for wind: Building A: (*a*) direct pressure effects; (*b*) east–west shear wall system; (*c*) north–south shear wall system.

overestimated, the overstress may not be real.

Increase the stress grade of the wood to obtain some higher design values for F_c, F_b, and E.

Place the studs on 12-in. centers to lower the loading on individual studs.

Increase the stud size to 3 × 6 or 2 × 8.

Install kick braces from the bottom of the roof structure to the stud just above the ceiling. This will reduce the L/d ratio and the span for bending.

Referring again to Fig. 10.3, we note that the lateral bracing system for wind in a north–south direction consists of the roof-deck acting as a horizontal diaphragm, and the four end shear walls acting as vertical diaphragms and providing the reactions for the simple span of the roof diaphragm. On the basis of the functioning of the wall as shown for case 2 in Fig. 10.3a, the wind load delivered to the edge of the roof diaphragm is thus (see Fig. 10.3b)

$$H_{ns} = p \times 100 \text{ ft} \times 11 \text{ ft} = 1100p$$

in which p is the design wind pressure.

From UBC Table 16-H (see Appendix B) the value of C_q to be used finding p in this case is 1.3. Thus, using the other data established previously, the design pressure is

$$p = C_e C_q q_s I = (1.2)(1.3)(17)(1.0)$$
$$= 26.52 \text{ psf} \qquad \text{say, 27 psf}$$

and the north–south load to the roof diaphragm is thus

$$H_{ns} = 1100p = (1100)(27)$$
$$= 29{,}700 \text{ lb}$$

Similarly, the force applied to the roof diaphragm by wind in the east–west direction is determined as (see Fig. 10.3c)

$$H_{ew} = 27 \times 50 \times 11$$
$$= 14{,}850 \text{ lb}$$

The action of the roof deck as a simple beam subjected to the uniformly distributed loading is illustrated in Fig. 10.4. This indicates the condition for north–south loading, with the reactions representing the loads to the end shear walls, and the shear and moment representing actions of the roof diaphragm. Before proceeding with the wind investigation we will determine the seismic loading to see which loading is critical for the various elements of the system.

Design for Lateral Seismic Force.   For the seismic load we consider the building

mass as a horizontally impelled force. Thus the weight of the roof and any items on top of it or hung directly from it will constitute load to the roof diaphragm. Added to this will be the portions of the walls that depend on the roof for lateral support.

As discussed in Sec. 3.1 and illustrated in Fig. 3.2, the wall function for seismic load depends on the direction of the load with respect to the plane of the wall surface. As with wind load, the parapet on the exterior wall has two possibilities, and we will again assume case 2, as shown in Fig. 10.3a. With these considerations, the building weight used for the determination of the loads to the roof diaphragm are tabulated in Table 10.1. Note that the wall loads are included only when the load direction is perpendicular to the wall. For the tabulation we have assumed a nominal amount of interior partition wall using a light-framed wall weight of 10 psf.

For the one-story building, the total base shear would be used only to investigate the horizontal effect on the foundations. With the roof diaphragm and shear walls, the design of the lateral bracing elements will use only a portion of the total base shear. We will nevertheless use the UBC formula for base shear to find these forces. Thus

$$V = \frac{ZIC}{R_W} W$$

where

$Z = 0.30$ (for zone 3, UBC Table 16-I)

$I = 1$ (UBC Table 16-L, no special qualification)

$C = 2.75$ [maximum required from UBC Sec. 1628.2.1, with S not determined]

$R_W = 8$ (UBC Table 16-N, plywood shear wall)

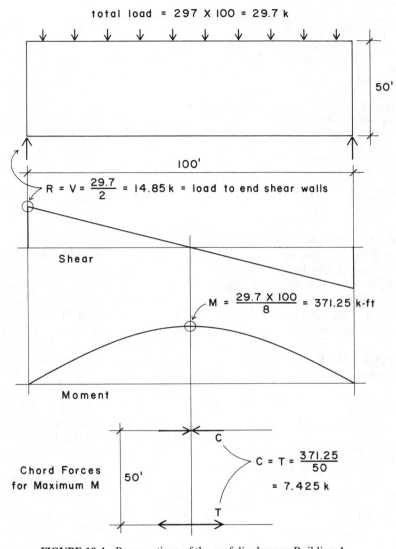

total load = 297 X 100 = 29.7 k

50'

100'

R = V = $\frac{29.7}{2}$ = 14.85 k = load to end shear walls

Shear

M = $\frac{29.7 \times 100}{8}$ = 371.25 k-ft

Moment

Chord Forces
for Maximum M

50'

C

C = T = $\frac{371.25}{50}$

= 7.425 k

T

FIGURE 10.4 Beam actions of the roof diaphragm: Building A.

Using these values for the factors

$$V = \frac{(0.30)(1)(2.75)}{8} W = 0.103125 \, W$$

and using the values for W from Table 10.1,

$V = (0.103125)(156) = 16.09$ kips
(in the north–south direction)

$V = (0.103125)(134) = 13.82$ kips
(in the east–west direction)

In both directions these loads are less than those determined previously for wind. We

TABLE 10.1   BUILDING WEIGHTS FOR THE ROOF DIAPHRAGM LOADS

Load Source and Calculations	North–South Load (kips)	East–West Load (kips)
Roof dead load		
20 psf × 50 ft × 100 ft	100	100
East and west exterior walls		
20 psf × 11 ft × 50 ft × 2		22
North and south exterior walls		
20 psf × 11 ft × 100 ft × 2	44	
Interior walls		
10 psf × 7 ft × 100 ft (estimate)	7	7
Rooftop HVAC unit	5	5
Total weights	156	134

thus proceed with the design of the bracing system using the wind loadings.

Figure 10.4 shows the function of the roof diaphragm as a simple span beam with uniformly distributed load. The maximum shear is one half the total load and results in a unit shear in the roof deck of

$$v = \frac{\text{maximum shear}}{\text{deck width}} = \frac{29,700/2}{50}$$
$$= 297 \text{ lb/ft}$$

and the maximum chord force is

$$C = T = \frac{\text{maximum moment}}{\text{building width}}$$
$$= \frac{(29.7 \times 100)/8}{50}$$
$$= 7.425 \text{ kips}$$

For the roof deck, referring to *UBC* Table 23-I-J-1 (see Appendix B), one option is

Nominal-$\frac{1}{2}$-in. (actually $\frac{15}{32}$-in.) plywood, structural II grade, blocked diaphragm

8d nails at 4 in. at diaphragm boundary and any continuous panel edges parallel to the load, and at 6 in. at other panel edges

Table load value: 320 lb/ft with 2× framing

For the chord we will assume a Douglas fir–larch No. 2 grade to be used for the double 2 × 6 top plate. With an allowable tension stress of 625 psi, the capacity of a single 2 × 6 is thus 1.33 × 625 × 8.75 = 6858 lb. This indicates that the single 2 × 6 is not enough for continuity at the splices. Options are to design a splice joint with bolts or steel straps to develop the full double plate as a continuous member, or to use a larger member or higher stress grade wood to obtain enough capacity from a single member. If a single member is adequate, and the joints in the two plates are staggered a sufficient distance, a splice joint is probably not necessary.

Depending on the details of the construction at the joint between the roof and wall, it is also possible that some other part of the framing may be used for the chord. It must, of course, be something to which the roof deck is nailed to develop the shear transfer at the boundary edge.

Although the roof must also function as a diaphragm for the load in the east–west direction, it is obvious that the unit shear in the deck will be much lower in that direction and the chord force will be quite small. Thus the deck and chord design are of primary concern for the north–south loading.

The wind loading condition for the end

shear wall is shown in Fig.10.5a. The end force in the roof deck is divided between the pair of walls at one end. The total shear force to the wall is thus 7.425 kips and the unit shear in the wall is computed as

$$v = \frac{\text{shear force}}{\text{wall length}} = \frac{7425}{17} = 437 \text{ lb/ft}$$

Referring to *UBC* Table 23-I-K-1 (see Appendix B) options for the wall are:

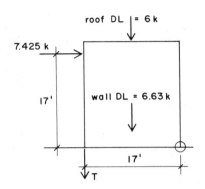

(a) east and west shear walls — N-S loading

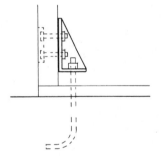

(b) typical tie down
      anchor

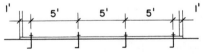

(c) minimum
      code-required
      sill bolting

(d) north and south shear walls –
      E-W loading

FIGURE 10.5 Design considerations for the shear walls: Building A.

A $\frac{3}{8}$-in. structural I plywood with 8d nails at 3 in. at all panel edges. Capacity: 1.20 x 460 = 552 lb/ft. Note that the table footnote allows a 20% increase in the table values for $\frac{3}{8}$-in. plywood for certain conditions.

Nominal-$\frac{1}{2}$-in. (actually $\frac{15}{32}$-in.) structural II plywood with 10d nails at 4 in. Capacity: 460 lb/ft.

For overturning of the wall, we consider the moment of the lateral force about the wall base as opposed by the moment of the wall dead weight plus any other dead load carried by the wall. The code requires that the restoring moment (due to dead load) be at least one and one half times the overturning moment; if it is not, anchorage must be provided. Referring to Fig. 10.5*a*, the investigation is as follows:

Overturning moment:

$$7.425 \times 17 \times 1.5 = 189.3 \text{ kip-ft}$$

Restoring moment:

$$12.63 \times 8.5 = \underline{-107.4}$$

Net moment for anchor force:

$$= 81.9 \text{ kip-ft}$$

Anchor force required:

$$T = \frac{81.9}{17}$$

$$= 4.82 \text{ kips}$$

For the wind load investigation, the overturning moment is multiplied by the safety factor of 1.5; if the restoring moment equals or exceeds this, no anchorage is required. In the example, the net force is what must be supplied by the anchorage for assis-tance of the dead load moment. Because the safety factor is already included, this force is really not a service load, although most designers treat it as such.

Note that some roof dead load is assumed to be carried by the wall. Note also that the determination of this effect must include consideration for the wind uplift force on the roof (see Fig. 10.2).

If anchorage must be provided, it may be possible to use a device such as that shown in Fig. 10.5*b*. This device is bolted to the member that forms the end framing of the wall and is secured by an anchor bolt in the foundation. It is also possible that anchorage is not really required. At the building corner, for example, the end wall is attached to the wall on the north or south side. If both are adequately attached to the end framing at the corner, anchorage is probably redundant. At the other end of the wall the wall may be constituted as a post for support of a header over the opening or for a beam from the roof-framing system. If the dead load delivered by such framing is sufficient, anchorage at this point may also be redundant.

For the design of the end-framing members in the wall any investigation should include the force generated by the chord function due to the overturning moment. For this consideration the uplift effect should probably not be included.

Another consideration for the wall is that of the sliding at the base due to the lateral force. From *UBC* Sec. 1806.6 the minimum bolting of the wall sill to the foundation must be done with $\frac{1}{2}$-in. bolts a maximum of 6 ft on center, with one bolt not more than 12 in. from each end of the wall. In this case, this minimum bolting could be provided by four bolts, as shown in Fig. 10.5*c*. With a Douglas fir–larch No. 2 2 $\times$ 6 sill and $\frac{1}{2}$-in. bolts, *UBC* Table 23-III-J gives a value for one bolt in single shear of 610 lb (see *UBC* Sec. 2336.2.3). The four bolts will thus provide a total resistance of $1.33 \times 4 \times 610 = 3245$ lb, which is considerably less than the lateral force of 7425 lb. If we use a $\frac{3}{4}$-in. bolt, the number of bolts required is

$$N = \frac{7425}{1.33 \times 1190} = 4.7 \text{ or } 5$$

This is a reasonable solution, although others are also possible. With five bolts the spacing will be 3 ft 9 in., which is quite reasonable. However, the concrete contractor may prefer to set fewer bolts, and a larger bolt allowing the 5-ft spacing could be used.

In the east–west direction the wind load is half of that in the north–south direction and there are five shear walls on each side. The load on each wall is thus quite low, although the shorter wall will result in some increased overturn effect due to the shorter moment arm for the restoring moment. The loading condition for this wall is as shown in Fig. 10.5d. The unit shear in the wall is

$$v = \frac{1485}{10.67} = 139 \text{ lb/ft}$$

This is quite a low stress, and if plywood is used, the thinnest, lowest-grade plywood with minimum nailing is more than adequate (see UBC Table 23-I-K-1 in Appendix B). Other wall treatments are possible, including the following: From UBC Table 23-I-K-2 (see Appendix B); $\frac{5}{16}$-in. particleboard with 6d nails at 6 in. Capacity = 180 lb/ft.

A final consideration for the shear walls is the wall foundation. If there is a net overturning moment, it must be resisted by the foundation, and the foundation design should include this together with the gravity loads. The general problem of shear wall foundations is discussed in Sec 8.6. We will not deal with the design here, as we are not designing the complete building structure.

Building A, as described in Fig. 10.1, has a symmetrical plan layout of its shear wall system. If the disposition of the building weight affecting the roof diaphragm is also symmetrical, there is in theory no torsional effect on the building during seismic actions. However, UBC Sec. 1628.5 requires that the seismic force be considered to have an accidental torsion with an eccentricity equal to 5% of the maximum building dimension. This provision is intended mostly for multistory buildings and rigid horizontal diaphragms, and not for the single-story building with a light wood frame and a plywood diaphragm. If rigidly interpreted, however, the procedure is to compute the torsional moment as the seismic force times the eccentricity and to determine the torsional rigidity of the shear wall system. The added stress to the shear walls is then determined as for a cross section in torsion. This procedure is demonstrated in Sec. 10.2, Example 3 for Building D, which has an unsymmetrical shear wall layout—a case that clearly calls for such an investigation.

Parapet Design The detailed wall section in Fig. 10.1 shows the parapet (the wall extension above the roof surface) to be constructed as a short stud wall on top of the roof deck. Connected to this wall is some diagonal framing that serves the dual purposes of lateral bracing for the wall and forming of the cant (45° transition) between the roof and wall. Many variations of these details are possible, with variables including the height and construction of the parapet, type of roofing, and arrangement of the roof framing. The following discussion relates to the construction as shown in Fig. 10.1 and the considerations for the latereral wind and seismic effects.

Wind pressure for this case is the same as was determined previously for the design of the wall studs: 23 psf. Assuming a maximum height for the parapet of 4 ft, the total wind force on the parapet is thus

25 psf × 4 ft = 100 lb/ft of wall length

For the lateral seismic force, the *UBC* gives the following formula for determination of force on building parts:

$$F_p = ZIC_pW_p$$

where

F_p = lateral force on the part
C_p = factor from *UBC* Table 16-O
W_p = dead load of the part

For a parapet the C_p factor is 2. Assuming an average weight of 30 psf for the parapet construction, the lateral force is thus

$$F_p = (0.3)(1)(2)(30 \text{ psf} \times 4)$$
$$= 7 \text{ lb/ft of wall length}$$

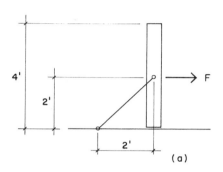

(a)

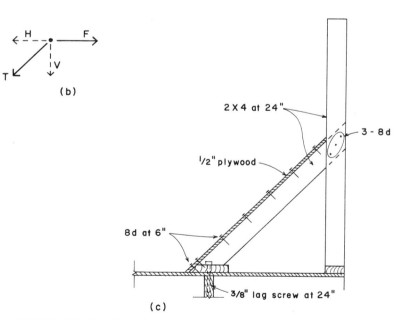

(b)

(c)

FIGURE 10.6 Considerations for lateral force on the parapet: Building A.

We thus observe that the wind load is critical and the load for design of the parapet is as shown in Fig. 10.6a. With the brace arranged as shown in Fig. 10.6a, the resolution of the forces at the wall-to-brace connection are thus as shown in Fig. 10.6b, and the force T is

$$T = (1.4142)(100)$$

$$= 142 \text{ lb/ft of wall length}$$

and with studs at 24 in. on center, the force for one brace is $142 \times 2 = 284$ lb.

To develop the construction as shown in Fig. 10.1, a plywood decking is placed on the braces so that the roofing can be run up the cant. The connection between the brace and the stud is by nailing the brace to the stud. Connection to the roof is achieved by nailing the decking to the braces, then nailing the edge of the decking to the mitered sill, and then bolting the sill to a rafter or to continuous blocking between rafters. As shown in Fig. 10.6c, the construction is adequate for the computed forces in this case. We show this example to illustrate the process of design of individual parts for lateral force, which includes the necessity of following the force transfers through the construction.

Example 6, Building B

This building is similar to Building A, except for the plan dimensions, which result in the very long, narrow plan form shown in Fig. 10.7. We will assume a light wood-framed construction with exterior shear walls of plywood and will design for seismic load using the following data:

Seismic zone 4 (considered critical, $Z = 0.4$)

Roof, ceiling, and supported items = 16 psf

Exterior walls = 20 psf

Canopy added to wall = 225 lb/ft
Rooftop HVAC units = 15 kips total

Obviously the critical concern is for the north–south load; thus we limit our investigation to this condition and to the resulting effects on the roof deck and the two end shear walls. Table 10.2 presents the summary of the building weights for determination of the north–south seismic load to the edge of the roof diaphragm. As in Example 1, we determine the design load as

$$V = \frac{ZIC}{R_W} W = \frac{(0.4)(1)(2.75)}{8} (166.4)$$

$$= 22.88 \text{ kips}$$

maximum shear in the roof deck

$$V = \frac{22.88}{2}$$

$$= 11.44 \text{ kips}$$

maximum unit shear

$$v = \frac{11,440}{30}$$

$$= 301 \text{ lb/ft}$$

To compound matters, we may consider the addition of stress due to the code-required 5% eccentricity (*UBC* Sec. 1628.5). For this computation the main portion of J will be due to the end shear walls; considering these only,

$$J = 2 \times 18 \times (60)^2 = 129,600 \text{ ft}^3$$

and the added shear is

$$v = \frac{Fec}{J}$$

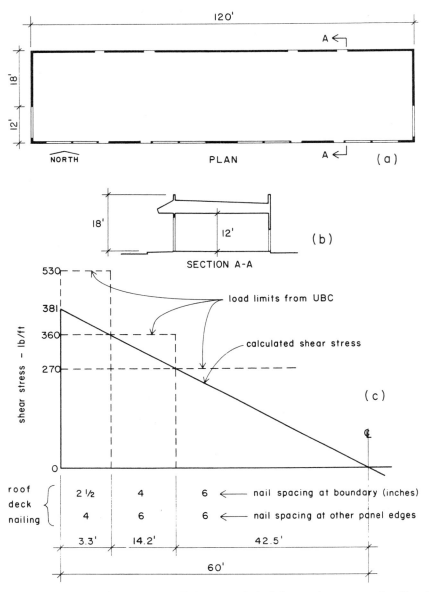

FIGURE 10.7   Example 6, Building B: (*a*) building plan; (*b*) building section; (*c*) zoned nailing for the roof diaphragm; stability of end shear wall.

$$= \frac{22{,}880 \times (0.05 \times 120) \times 60}{129{,}600}$$

$$= 64 \text{ lb/ft}$$

This should be added to the end shear wall

shear stress. Without the torsion the end shear wall stress is

$$v = \frac{11{,}440}{18} = 636 \text{ lb/ft}$$

TABLE 10.2 BUILDING WEIGHTS FOR THE ROOF DIAPHRAGM LOADS: NORTH–SOUTH DIRECTION

Load Source and Calculations	Load (kips)
Roof and ceiling	
16 psf × 120 ft × 30 ft	57.6
East and west exterior walls	
20 psf × 240 ft × 11 ft	52.8
Interior walls	
10 psf × 200 ft × 7 ft	14.0
Canopy	
225 lb/ft × 120 ft	27.0
Rooftop HVAC units	15.0
Total	166.4

and the maximum stress with accidental torsion is

$$v = 636 + 64 = 700 \text{ lb/ft}$$

Using Douglas fir plywood for the blocked roof deck (see Appendix B for *UBC* Table 23-I-J-1), some options are as follows:

- $\frac{15}{32}$-in. structural II with 2× framing, blocking, and 8d nails at $2\frac{1}{2}$-in. at boundaries and 4 in. at other edges.
- $\frac{19}{32}$-in. structural II with 3× framing, blocking, and 10d nails at 4 in. at boundaries and at 6 in. at other edges
- $\frac{15}{32}$-in. structural I with 3× framing, blocking, and 10d nails at 4 in. at boundaries and at 6 in. at other edges

Because the high stresses occur only near the ends of the building, it is reasonable to consider the possibility of zoning the deck nailing in this case. By using the $\frac{15}{32}$-in. structural II plywood, it is possible to use two fewer nail spacings, as given in the *UBC* table. Thus the range of nail spacings and corresponding load ratings are the following:

8d at $2\frac{1}{2}$ in. at boundaries, 4 in. at other edges: load = 530 lb/ft

8d at 4 in. at boundaries, 6 in. at other edges: load = 360 lb/ft

at 6 in. at all edges: load = 270 lb/ft

In Fig. 10.7*c* these allowable loads are plotted on the graph of stress variation in the deck to permit the determination of the areas in which the various nailing spacings are usable. The maximum nailing is seen to be required on only a very small area at each end of the roof. The actual dimensions of the specified nailing zones may be adjusted slightly to correspond to modules of the roof framing and plywood sheet layouts as long as the limits of the calculated zone boundaries are not exceeded.

Using Douglas fir plywood for the wall sheathing (see Appendix B for *UBC* Table 23-I-K-1), possible options are:

- $\frac{15}{32}$-in. structural II with 3× boundary framing and 10d nails at 2 in. at all edges: load = 770 lb/ft
- $\frac{3}{8}$-in. structural I with 8d nails at 2 in. at all edges: load = (610)(1.2) = 732 lb/ft (see footnote 3 in the *UBC* table)
- $\frac{15}{32}$-in. structural I with 3× boundary framing and 10d nails at 2 in. at all edges: load = 770 lb/ft

And if it is possible to have plywood on both sides of the wall, use:

- $\frac{3}{8}$-in. structural II, both sides, with 8d nails at 4 in. at all edges: load = (2)(320)(1.2) = 768 lb/ft

Although technically permitted by the code, none of the foregoing is desirable. The closely spaced heavy nailing is likely to cause some splitting of framing members. Placing plywood on both sides of the wall can entail some difficulty in the framing details and in the installation of wiring or other elements within the wall cavity. The heavy shear load on the wall is also likely to be a problem in terms of stress on the wall boundary framing.

In addition to the shear stress problem, the required tie-down force is considerable and requires a very heavy anchorage device, heavy end framing in the wall, and a strong foundation. Development of sliding resistance for the total load requires heavy bolting of the wall sill to the foundation. Although the chord force can be developed with ordinary framing members, the length of the building will result in several splices, each of which needs considerable bolting to maintain the continuity of the chord.

A final consideration is the horizontal deflection of the roof diaphragm at the center of the 120-ft span. The diaphragm depth-to-span ratio of 4 to 1 is just at the limit permitted by the *UBC* (see *UBC* Table 23-I-I in Appendix B). Even though the diaphragm is within the code limit, the actual dimension of the deflection should be determined and its possible effect on interior partitions considered.

We abandon this example without attempting to solve all its problems because we actually consider it to be a poor solution. It was presented to show what bad planning can produce in the way of difficult situations and not as an illustration of proper design.

This example presents a situation in which a strong argument can be made for the use of at least one interior shear wall. Figure 10.8 shows a modification of the plan with the introduction of a central north–south wall extending for the full width of the building. Using the same loading as in the previous computation, the analysis for the north–south seismic loading of the roof diaphragm is as shown in Fig. 10.8*b*. In this analysis the roof deck is assumed to be sufficiently flexible to justify an assumption of distribution of load to the vertical elements on a peripheral basis. This is consistent with the usual practice for a wood diaphragm with the depth-to-span ratio in the example. With this assumption half of the total load is taken by the center wall, and the load on the end walls is reduced to one half that in the preceding computations.

Ignoring torsional effects, the maximum stress in the roof diaphragm drops to 190 lb/ft, occurring at the ends and at both sides of the interior wall. This is less than the lowest rated capacity for $\frac{15}{32}$-in. plywood with edge blocking and minimum nailing as given in *UBC* Table 23-I-J-1 (see Appendix B). Nail zoning is therefore not a consideration, unless it is acceptable to use an unblocked diaphragm for some portions of the roof.

The shear stress in the end wall drops to 318 lb/ft. This is still a significant stress, but it can be achieved with $\frac{3}{8}$-in. structural II plywood with 8d nails at 4 in. at all panel edges, which is quite reasonable. The overturn and tiedown requirement are also considerably reduced.

The center shear wall must carry a considerable force—the same as that on the end shear walls in the previous example. However, the wall is longer in plan, which results in a lower shear stress and less overturn effect. The design of this wall must include the full consideration of its use in the building. Possible issues are its use for fire separation, acoustic separation, and load bearing for the roof structure. For our analysis we consider it to be a single stud wall and to serve as a bearing wall. If it were not a bearing wall, the net overturning effect would be higher. It happens in this example that the same plywood and nailing can be used for the end walls and the interior wall. From the *UBC* table, the capacity is 384 lb/ft, which includes the allowable 20% increase noted in the footnote to the table.

If the interior wall is not a bearing wall, there may be some tiedown requirement. However, the connection of this wall to the

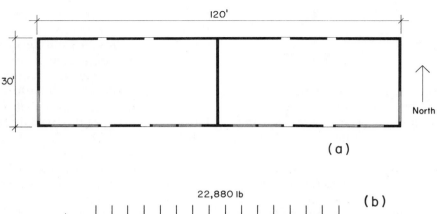

(a)

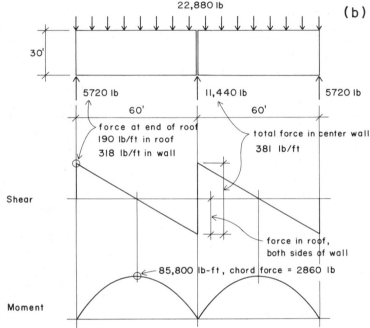

(b)

FIGURE 10.8 Modification of Building B with interior wall: (*a*) building plan; (*b*) load investigation of the roof diaphragm.

walls at the front and the rear of the building should be adequate for this function.

If sliding friction is ignored and the entire horizontal force is taken on the sill bolts, the choice is whether to use a lot of small bolts or a few large ones. This is a matter of individual preference by designers and builders. If the maximum code spacing of 6 ft is used, the two walls will have a minimum of four bolts in the end wall and six in the center wall. Thus the minimum required capacity for this minimum number of bolts is as follows:

$$\text{End wall: load} = \frac{5720}{4}$$

$$= 1430 \text{ lb/bolt}$$

Center wall: load $= \dfrac{11,440}{6}$

$= 1907 \text{ lb/bolt}$

If the bolt size required for these loads is not excessive, the minimum number of bolts may be used—which is the usual preference of both the concrete and framing contractors.

A construction detail that must be developed for this example is the connection between the roof diaphragm and the center shear wall. This connection must transfer the total force from the roof to the wall. Because there are several options for both the roof and the wall construction, the potential variations for this connection become quite numerous. Figure 10.9 illustrates some of the possibilities based on the assumption that the wall is a single stud wall and that the roof framing consists of joists perpendicular to and supported by the wall. There are three basic structural functions to be considered for this situation: the vertical gravity load transfer, the hold-down against wind uplift, and the transfer of diaphragm force parallel to the wall.

Detail A in Fig. 10.9 shows the ordinary construction used if the gravity load alone is considered. The joists are either butted end to end, or they are lapped on top of the wall and are toe nailed to the top plate. Vertical blocking is usually provided between the joists for their stability as well as to provide a nailer for the plywood edges perpendicular to the joists. This joint provides only minor resistance to uplift (relying on withdrawal of the toenails) and virtually no capacity for transfer of the shear.

In detail B of Fig. 10.9, a second horizontal block is added to facilitate the transfer of diaphragm shear. The vertical block is nailed to the horizontal block, and the horizontal block is nailed to the top plate. With the roof deck nailed to the top of the vertical block, the shear transfer is achieved

from the roof deck to the top plate of the wall. With the wall surfacing nailed to the top plate, the transfer from horizontal to vertical diaphragms is then complete.

It should be noted that, although the stress in the roof deck is only 190 lb/ft at this location (see Fig. 10.8b), the total load transfer to the wall is twice this, resulting from a delivery of load to both sides of the wall by the deck. Thus all the nailing shown in detail B of Fig. 10.9—the deck to the vertical block, the blocks to one another, and the block to the top plate—must be designed for the load transfer of 381 lb/ft. If a roof plywood panel edge occurs at this point, the nailing will be adequate because there will actually be two edge nailings at the joint. If this is an interior support point for a plywood panel, the usual minimum nailing with nails at 12-in. centers will not be sufficient, and a nailing must be specified that is capable of the load transfer. It is questionable whether the nailing of the blocks for this load magnitude could be achieved without splitting the blocks, so this option is probably not the best for the example.

A variation on detail B is shown in detail C of Fig. 10.9. A second vertical block is added, and the horizontal block is bolted to the top plate. This results in an extended range of load capacity because there are now two rows of nails and the bolts are much stronger than nails in lateral load resistance. However, with fairly closely spaced joists, this detail would require a considerable number of bolts through the top plates.

Another approach to this connection is shown in details D and E of Fig. 10.9 in which metal framing anchors are used for the attachment to the top plates. In detail D the anchors are attached to the joists, and in detail E they are attached to the blocks. Assuming that the anchor devices used have a rated load capacity adequate for the load transfer, either of these options is ac-

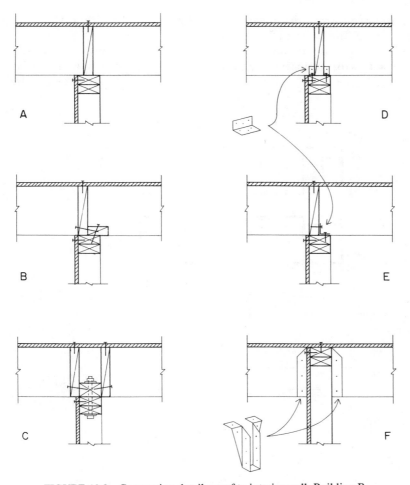

FIGURE 10.9 Connection details; roof to interior wall: Building B.

ceptable for the lateral shear function. Detail D is slightly better for wind uplift because there is more attachment between the roof deck and the joists than there is between the deck and the blocks.

A third technique for this connection is shown in detail F of Fig. 10.9 in which the top plate of the wall is raised to the level of the roof deck. This procedure allows for the simplest transfer of the wind shear because both the roof deck and the wall sheathing are directly nailed to the top plate. The joists are supported by saddle-type metal

hangers hung from the top plate. One problem with this detail is that the upper panels of the wall sheathing must be installed before the roof framing can be placed, which is not the usual sequence of the construction.

We do not attempt to judge which of these, or of other possible options, is the best solution for this connection. From the viewpoint of structural design, anything that "works" is all right. In real situations there are many issues to consider in addition to the necessary structural functions.

Thus the influence of roof drainage, wall surface finishes, ceiling construction, ductwork installation, and so on, may provide the deciding factors for choice between viable alternatives.

Example 7, Building C

Figure 10.10 shows a building that consists essentially of stacking the plan for Building A to produce a two-story building. The

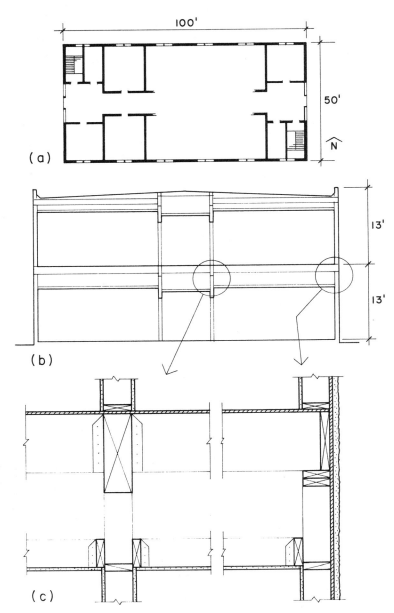

FIGURE 10.10 Building C, general form.

profile section of the building shows that the structure for the second floor is developed essentially the same as the roof structure for Building A. For the roof, however, a clear span structure could be provided by 50-ft span trusses.

The two-story building will sustain a greater total wind force, although the shear walls in the second story will essentially be the same as those for Building A. The major effect in this building will be the forces in the first-story shear walls.

Design for Wind. The general design for wind includes the considerations enumerated at the beginning of this section for Building A. Investigation of the second-story studs in the exterior walls would be similar to that made for Building A. At the first story in Building C the studs carry considerably more axial compression, but the bending due to wind is approximately the same as at the second story. The 2 × 6 studs at 16-in. centers are probably adequate at the first story. If an investigation similar to that made for Building A shows an overstress condition, the stud spacing can be reduced to 12 in. or a higher-grade wood can be used.

For lateral load the roof deck in Building C is basically the same as that in Building A. With the trusses it may be more practical to use an unblocked deck, and the investigation should consider this factor. The footnotes to *UBC* Table 23-I-J-1 (see Appendix B) should be studied with regard to the patern of the layout of the plywood panels, especially for unblocked decks. Various special deck panels with tongue-and-groove edges are available in thicknesses greater than in $\frac{1}{2}$ in., thus permitting truss spacings up to 4 ft. Special data for lateral load resistance may be available from the manufacturers for these products, although local code approval must be determined.

The wind loading condition for the two-story building is shown in Fig. 10.11a. This indicates a loading to the second floor diaphragm of 235 lb/ft. With a $\frac{15}{32}$-in.-thick deck as a minimum, the shear in the 50-ft-wide deck will not be critical. However, the stair wells at the east and west ends reduce the actual diaphragm width at the ends to only 35 ft. Figure 10.11b shows the loading for the second floor deck and the critical shear and moment values for the diaphragm actions. At the ends the critical unit shear in the deck is

$$v = \frac{11,750}{35} = 336 \text{ lb/ft}$$

From *UBC* Table 23-I-J-1 (see Appendix B) it may be determined that this requires a bit more than the minimum nailing for the deck. Options at this location include:

1. Using a $\frac{15}{32}$-in. structural II deck with 8d nails at 4 in. at the diaphragm boundary and other critical edges
2. Using $\frac{15}{32}$-in. structural I deck with 10d nails at 6 in. throughout and 3× framing
3. Using $\frac{19}{32}$-in. structural II deck with 10d nails at 6 in. throughout and 3× framing

At 8 ft from the building ends the deck resumes its full 50-ft width, and the unit shear at this point drops to

$$v = \frac{9870}{50} = 197 \text{ lb/ft}$$

Since this value is well below the capacity of the $\frac{15}{32}$-in. structural II deck with minimum nailing, it may be the most practical to elect option 1, which involves only the use of 4-in. nailing in approximately 12% of the total second-floor deck.

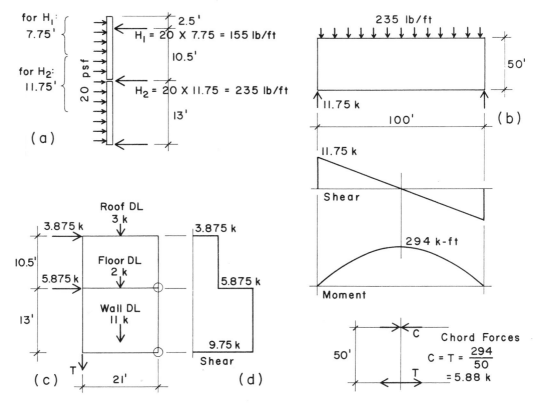

FIGURE 10.11 Building C, development of lateral force due to wind: (*a*) determination of the wind loads to the horizontal diaphragms; (*b*) spanning functions of the second floor diaphragm; (*c*) loading of the two-story end shear wall; (*d*) shear diagram for the shear wall.

The diaphragm chord force for the second-floor deck is approximately 6 kips, and must be developed in the framing at the wall, as shown in Fig. 10.10c. The most likely member to use for this is the continuous edge member at the face of the joists. The only real design consideration for this situation is developing the splicing of the member that will be made up of several pieces in the 100-ft length. Splicing may be achieved in a number of ways, the details must be developed to work within the construction of the wall and floor at this location. A joint using a steel strap with wood screws or one with bolts and steel plates will likely cause the least intrusion in the construction.

The loading for the two-story end shear wall is shown in Fig. 10.11c and the shear diagram for this load is shown in Fig. 10.11d. The second-story wall is essentially similar to the end wall in Building A for which an investigation was made in Example 5. Because minimal construction is adequate here, and no anchorage for overturn is required, the only problem for concern is the development of sliding resistance to the lateral force of 3875 lb. Since this wall does not sit on a concrete foundation, other means of anchorage than steel anchor bolts must be considered.

The lateral force in the second-story wall must be transferred to the lower (first-story) wall. Essentially this occurs directly if the plywood is continuous past the construction at the level of the second-floor, as it indeed is as shown in Fig. 10.10c. A critical location for stress transfer in this location is at the top of the second-floor joists. At this point the lateral force from the second-floor deck is transferred to the wall through the continuous edge member. Therefore, the nailing and plywood requirements for the first-story wall begin at this location. The last point for the nailing and plywood requirements for the second-story wall are at the location of the sill for the second-story wall (on top of the second-floor deck, as shown in Fig. 10.10c). The plywood for the wall and its nailing from this point down must satisfy the requirements for the first-story wall.

In the first-story wall the total shear force is 9750 lb and the unit shear is

$$v = \frac{9750}{21} = 464 \text{ lb/ft}$$

If the $\frac{3}{8}$-in. structural II plywood selected for Building A is used for the second floor (see Example 5), it may be practical to use the same plywood for the entire two-story wall and simply to increase the nail size and/or reduce the nail spacing at the first story. *UBC* Table 23-I-K-1 yields a value of 410 lb/ft for $\frac{3}{8}$-in. structural II plywood with 8d nails at 3 in. If the conditions of table footnote 3 are met, this value can be increased by 20% to 492 lb/ft. There are other options and many other design considerations for the choice of the wall construction, but this is an adequate choice for the lateral design criteria.

At the first-floor level, the investigation for overturn of the end shear wall is as follows (see Fig. 10.11c):

overturning moment =
$$(3.875)(23.5)(1.5) = 136.6 \text{ kip-ft}$$
$$+ (5.875)(13)(1.5) = \underline{114.6 \text{ kip-ft}}$$
$$\text{total} = 251.2 \text{ kip-ft}$$

restoring moment =
$$(3 + 2 + 11)(21/2) = 168 \text{ kip-ft}$$

net overturning moment = 83.2 kip-ft

This requires an anchorage force at the wall ends of

$$T = \frac{83.2}{21} = 3.96 \text{ kips}$$

Since the safety factor of 1.5 for the overturn has already been used in the computation, it is reasonable to consider reducing this anchorage requirement to 3.96/1.5 = 2.64 kips if it is used in the form of a service load. In addition, the wind loading permits an increase of one-third in allowable stress, which may also be used to reduce the requirement. Finally, there are added dead load resistances at both ends of the wall. At the corridor the beam sits on the end of the wall and at the building corner this wall is reasonably firmly attached to the wall around the corner. Thus the real need for an anchorage device is questionable. However, most structural designers would probably prefer the positive reassurance of such a device.

10.2 MASONRY SHEAR WALLS

Although many forms of masonry may produce walls with sufficient strength for use as shear walls, the construction most widely used in areas with severe windstorms or high risk of earthquakes is that using units of precast concrete (CMU construction). CMU construction can be used with only minor reinforcement (technically qualified as *unreinforced*) but is usually developed as reinforced masonry for structural purposes. The examples in this section use reinforced CMU construction that qualifies as reinforced masonry by *UBC* standards.

Use of Stiffness Factors

A common situation that occurs in shear wall systems is that in which a number of individual shear walls (also called *piers*) share some lateral force. When this occurs the amount of force in the individual piers must be determined. For plywood walls the stiffness of individual walls is assumed to be proportionate to their plan length, assuming similar construction for all the walls. For walls of masonry or concrete construction, a more precise investigation is usually performed, based on the stiffness of the piers.

Stiffness of masonry or concrete piers is based on the heights and plan lengths (also called the pier width) of the piers or on the ratios of these dimensions. It is also af-

fected by the degree of fixity at the top or bottom of the piers. Figure 10.12 shows some common situations in which piers occur in masonry and concrete walls.

In Fig. 10.12a a wall is formed by a series of separated, but linked, masonry walls, with lighter construction forming the wall portions between the masonry. If the individual masonry piers are all of the same size and similarly constructed, the total lateral force in the wall will be simply divided equally between the piers. If they have different dimensions, however, their relative stiffnesses must be used to apportion the load to the piers.

For the wall in Fig. 10.12b a similar situation occurs if the distribution of shear on a horizontal plane through the window openings is considered. In this case the indi-

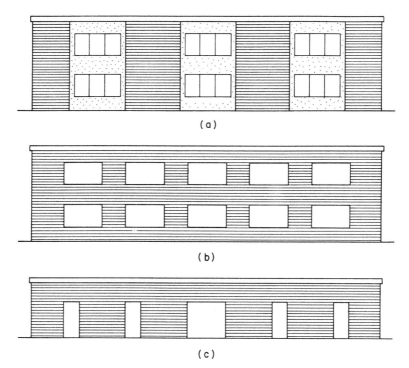

(a)

(b)

(c)

FIGURE 10.12   Form variations for masonry shear walls: (a) individual, isolated, linked piers (vertical cantilevers with fixed bases); (b) continuous wall with fully fixed, individual piers; (c) continuous wall with individual cantilever piers, fixed at their tops.

vidual piers of masonry between the window openings act as fully fixed elements, as shown in Fig. 6.10a.

For the piers between the door openings in Fig. 10.12c, the condition may be one of full fixity (Fig. 10.12b) or simple cantilever (Figs. 10.12a or 6.10c), depending on the nature of the anchorage and support at the base of the piers.

Assuming that the piers are all similarly constructed, the basic factor that distinguishes them from each other in any of the walls in Fig. 10.12 is their aspect ratio (vertical dimension to horizontal dimension). Strictly on the basis of this ratio, plus their qualification of single fixity (simple cantilever) or double fixity, their relative stiffnesses can be established and used as a basis for distribution of lateral loads. Factors for this purpose are given in the tables in Appendix C, and their use is demonstrated in the following example.

Example 8, Pierced Masonry Wall

A lateral force is delivered to the wall shown in Fig. 10.13. Find the percentage of the total load (H) resisted by each of the individual piers at the level of the window openings.

In this case the piers are considered to be fixed at their tops and bottoms. Stiffness factors, R_c, for the individual piers are thus obtained from Table C.2, on the basis of the h/d ratios for the piers. The load distribution to each individual pier is then determined by multiplying the total load by a distribution factor, DF:

$$DF = \frac{\text{factor for the individual pier}}{\text{sum of the factors for all the piers}}$$

The computations for the distribution are summarized in Table 10.3.

TABLE 10.3   LOAD DISTRIBUTION TO THE MASONRY PIERS

Pier	h (ft)	d (ft)	h/d	R_c	DF[a]	Share of Lateral Load (%)
1	8	4	2.0	0.1786	0.087	8.7
2	8	8	1.0	0.6250	0.304	30.4
3	8	10	0.8	0.8585	0.417	41.7
4	8	6	1.33	0.3942	0.192	19.2
				Sum = 2.0563		

[a] $DF = \dfrac{R_c \text{ for pier}}{\text{sum of } R_c}.$

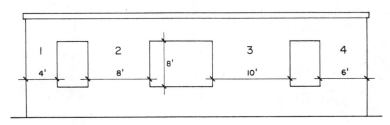

FIGURE 10.13   Multiple-pier wall for the example problem.

Example 9, Building A

This example uses the same building form as in Example 5 in Sec. 10.1, with a difference consisting of the use of masonry shear walls and a steel deck roof in place of the all wood structure. The slightly modified plan and the typical construction for this example are shown in Fig. 10.14. A pilaster is used on the inside of the wall piers on the north and south sides to create a column for the roof beams. In addition, this pilaster provides a brace, permitting the use of a thinner wall, for the approximately 16-ft unbraced height would otherwise probably require a thick masonry wall with hollow concrete units.

With the building form the same, the wind loads for this example are the same as those for Example 5 in Sec. 10.1. However, the seismic forces are different for two reasons. First, the dead loads of the exterior walls are much greater; we assume an average of 60 psf for the unpierced walls on the east and west and an average of 50 psf for the walls and windows on the north and south. A second consideration is that the R_W factor for the masonry shear walls is 6, while for the plywood walls it was 8. By comparison with Example 5 in Sec 10.1, the lateral shear is thus determined as

$$V = \frac{ZIC}{R_W} W = \frac{(0.3)(1)(2.75)}{6} W$$

$$= 0.1375W$$

Using these data, the computation of the lateral seismic loads for the roof diaphragm is summarized in Table 10.4. For these computations, we have also made the following assumptions:

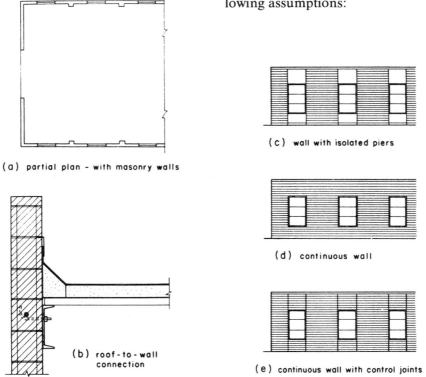

(a) partial plan - with masonry walls

(b) roof-to-wall connection

(c) wall with isolated piers

(d) continuous wall

(e) continuous wall with control joints

FIGURE 10.14   Masonry wall details: Example 9. Building A.

1. The dead load of the roof, interior walls, and HVAC unit are the same as in Example 5 in Sec. 10.1.
2. The center portion of the east and west walls has an average dead load of 20 psf.

From these computations it may be observed that the seismic forces in both directions are greater than the wind loads as found for Example 5 in Sec. 10.1. An additional observation is that the design criteria for this example (see Example 1 in Sec. 10.1) specifies seismic zone 3, meaning that the forces would be an additional one-third higher for seismic zone 4.

The steel deck diaphragm is similar in many ways to the plywood diaphragm. Individual panels of decking must be attached to each other and to the supports to achieve the continuous diaphragm surface structure. A wide variety of deck units is marketed, although a few basic types are quite widely used. The details in the illustrations for this example show a common unit in which ribs are formed on 6-in. centers in a panel that usually comes in 2-ft-wide units and can be obtained in lengths up to about 30 ft. Edges of the panels have a slip-fit interlock and ends are simply overlapped. Units are typically welded to steel supports by placing a thick washer in the bottom of the corrugation and welding the inside hole of the washer to the support by burning through the thin deck material.

Diaphragm shear capacity and relative stiffness of steel decks depend on a number of considerations, including the following:

1. *Thickness (Gauge) of the Deck Sheet.* Decks can be obtained in a range of weights, but the thinner gauges are most often used for cost reduction.
2. *Form of the Deck Corrugation, Depth of the Ribs, and Spacing of the Ribs.*
3. *Spacing of the Welded Connections to Supports.*
4. *Enhancement of the Edge Connection Between Units.* This is equivalent to providing blocking for a plywood deck. The interlocking joint can be crimped (pinched and twisted) to make a mechanical fastening.
5. *Effects of Concrete Fill.* For floors, the presence of structural-grade concrete fill will result in added strength and stiffness for the deck. For roofs, lightweight, insulating concrete or other materials on the deck will have some effect, but are usually not considered

TABLE 10.4   COMPUTATION OF SEISMIC FORCES: EXAMPLE 9

Load Source and Calculations	North–South Load (kips)	East–West Load (kips)
Roof (assumed as for Example 5 in Sec. 10.1)	100	100
East and west walls		
20 psf × 16 × 11 × 2	—	7
60 psf × 34 × 11 × 2	—	45
North and south walls		
50 psf × 100 × 11 × 2	110	—
Interior walls (same as Example 5 in Sec. 10.1)	7	7
Rooftop HVAC	5	5
Total dead loads (W)	222	164
Horizontal force to roof edge ($V = 0.1375W$)	30.53	22.55

for structural response. Of course, concrete fill will add significantly to the mass for seismic load.

Rating of decks for shear capacity and relative stiffness is the responsibility of individual producers, and information must be obtained from the manufacturer or marketer for an individual product. Design criteria are developed by the Steel Deck Institute, but building structural designers usually rely on the manufacturers to obtain code approval of design values for their products.

As with the plywood diaphragm, a complete design also includes the development of diaphragm chords, the provision for transfer of forces to vertical bracing elements of the lateral resistive system, and the development of any collectors, drag struts, and so on. Because steel deck units are most often used with a steel-framing system, the various components of the framing system will usually serve to form the chords and help in the transfer of forces.

For seismic resistance masonry shear walls consist most often of walls made from units of precast concrete (concrete blocks). Other forms of masonry are possible, but the most widely used construction is that with hollow units of concrete with both horizontal and vertical reinforcing. The reinforcing usually consists of small size steel reinforcing rods installed in continuous voids in the wall which are filled with concrete. The filled voids and reinforcing rods literally form a reinforced concrete rigid frame within the hollow block wall (see Fig. 10.15). Although unreinforced construction is also possible, reinforced masonry is the only type permitted by codes in the higher risk seismic zones.

The code requires minimum reinforcing and grouting of voids and has various special requirements for doweling of vertical reinforcing, added reinforcing at ends, tops, and around openings in walls; the attachment of walls to roofs and floors result in a "minimum" construction with a capacity usually already above that of the heaviest plywood shear wall. Unlike the plywood wall, however, the masonry wall is most often also a bearing wall for gravity loads and its complete design must consider the full range of load combinations. Choice of the units, the type of mortar joints, the pattern of the unit arrangement, and other considerations are also of some concern when the wall is exposed to view.

In this example we limit our concern to that for the shear walls on the north and south sides of the buildings. Three possibilities for the form of these walls are shown in Fig. 10.14. In Fig. 10.14c walls consist of individual masonry panels separated by the window units that are formed by light frame construction. The masonry shear walls thus function as isolated piers that resist force independently, although they are linked together by the rest of the wall and roof structure and thus have the same deflection under lateral load. This form of behavior is illustrated in Fig. 10.16a and is the same as that which was assumed for the plywood walls in Example 5 in Sec. 10.1. Shear stress is constant throughout the height of the wall and overturn, sliding, and any critical chord force effects are investigated as for the plywood walls.

Another possibility for the masonry wall is to build it as a continuous wall with openings for the windows as shown in Fig. 10.14d. In this case, the wall functions under lateral loading as shown in Fig. 10.16b. The portions of wall between openings function as fixed end columns, inflecting at mid-height, with end restraint provided by the upper and lower portions of the wall. There are two values of unit shear to consider in the wall: that in the continuous

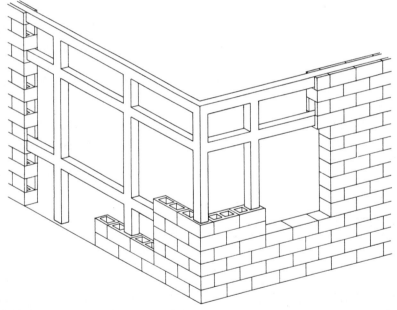

FIGURE 10.15   Reinforced masonry wall construction with hollow concrete blocks.

upper and lower portions and that in the portions of wall between openings.

From the loadings previously determined, total east–west seismic load = 22.55 kips. Ignoring torsion, the total load on the north or south wall is

$$\frac{22.55}{2} = 11.28 \text{ kips}$$

The load in the wall is then divided between the various elements in the wall plane. For the cantilevered, isolated piers

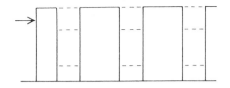

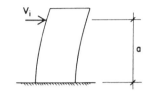

(a) wall as linked isolated piers

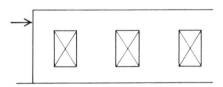

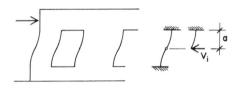

(b) continuous wall

FIGURE 10.16 Functioning of the masonry shear walls.

(Fig. 10.16a), considering only the larger intermediate piers, the load per pier is thus

$$\frac{11.28}{5} = 2.26 \text{ kips}$$

and the unit shear force in the 10-ft 8-in.-long pier is

$$\frac{2260}{10.67} = 212 \text{ lb/ft}$$

This is a very low stress for the reinforced masonry wall and will most likely be developed with the minimum construction required by codes. Stresses in the individual piers between window openings with the continuous wall (Fig. 10.16b) will be the same—again, if the narrow end piers are ignored. Stress will be even lower in the continuous wall above and below the windows. The low stress condition does not make the construction a poor choice, but it leaves little concern for further consideration of this example. We will therefore use some later examples to discuss additional concerns for masonry walls.

As with the plywood walls, a major concern may be the development of the foundations, especially for the isolated piers. Anchorage is usually adequately provided for the wall by the doweling of the vertical reinforcing, particularly that installed at the ends of the walls. What the cantilevered piers are cantilevered from, however, is the foundation. If this is a shallow-bearing-grade beam-type footing, it must be designed for the shears and moments induced by the cantilevered walls as well as the effects of gravity loading. As we are not designing the whole building structure, we will again dodge this problem, although some discussion is given in Sec. 8.6 on shear wall foundations.

Transfer of loads from the roof deck to the walls is usually achieved through the connections of the steel framing to the masonry. With the parapet wall, as shown in Fig. 10.14a, a steel ledger (braced angle or channel) is usually bolted to the wall using anchor bolts set in the masonry. This member and its connections to the wall must be designed for the loading combinations possible with these three different loads:

1. The transfer of the roof diaphragm shear from wind or seismic force
2. The bracing of the wall against out-

ward force normal to the wall plane—wind suction effect or seismic force due to the weight of the wall

3. Vertical load transfer of the roof gravity loads

For seismic anchorage, especially item 2 above, the bolts should be set in a concrete-filled horizontal course of the block wall and preferably hooked around a reinforcing rod, as shown in Fig. 10.14*b*.

If the wall construction is as shown in Fig. 10.16*b*, the construction between the isolated masonry piers must be adequate to connect the piers for combined action. Framing at the roof will most likely be developed for the roof diaphragm shear collection and serve as the connector between piers.

Example 10, Building D

Building D is a simple, rectangular, one-story building that is similar to Building A, except that the walls on the front and rear are different, resulting in a lack of symmetry in the lateral resistive system. This applies only to the east-west loading and we will limit our concern in this example to the problems of the north and south walls. The general building plan and details are shown in Figs. 10.17 and 10.18.

As shown in the roof-framing plan and the wall sections, the roof structure consists of large steel beams supporting 6 ft on center steel purlins, which in turn support a steel-formed deck. The walls consist of reinforced masonry with hollow concrete units. The walls provide both vertical and lateral support. We consider the design of the walls for direct wind load plus vertical gravity load and for their actions as shear walls resisting the east–west seismic force on the building.

For lateral forces we consider the following conditions:

Wind: basic wind speed = 70 mph

Seismic: zone 4

We consider the concrete units to be medium weight, grade N, ASTM C90 with f'_m = 1350 psi laid with type S mortar. Reinforcing is Grade 40, with f_y = 40 ksi.

We assume that the walls will consist of reinforced, hollow concrete blocks with finishes of stucco (cement plaster) on the exterior and gypsum drywall on wood furring strips on the interior. We assume this construction to weigh approximately 70 psf of wall surface.

The exterior walls must be designed for the following combinations of vertical gravity and lateral wind or seismic forces (see Fig. 10.19):

1. Gravity dead plus live loads
2. Gravity vertical load plus bending due to lateral load
3. Horizontal shear and overturn due to shear wall actions

We first consider the long expanses of wall at the building ends and rear. For the end wall the laterally unsupported height varies because of roof slope. We assume it to be a maximum of 15 ft at the end of the solid wall portion nearest the front of the building. With an 8-in. block thickness, the maximum h/t of the wall is thus (15 × 12)/ 7.625 = 23.6, which is just short of the usual limit of 25.

Assuming that code-required inspection is not provided during construction, the maximum stress for vertical compression is as indicated below. (*Note*: The following structural computations for masonry are based on the working stress method as provided for in earlier editions of the *UBC*. The *UBC* does not permit these methods for seismic design in high-seismic-risk zones. However, the methods shown here are still

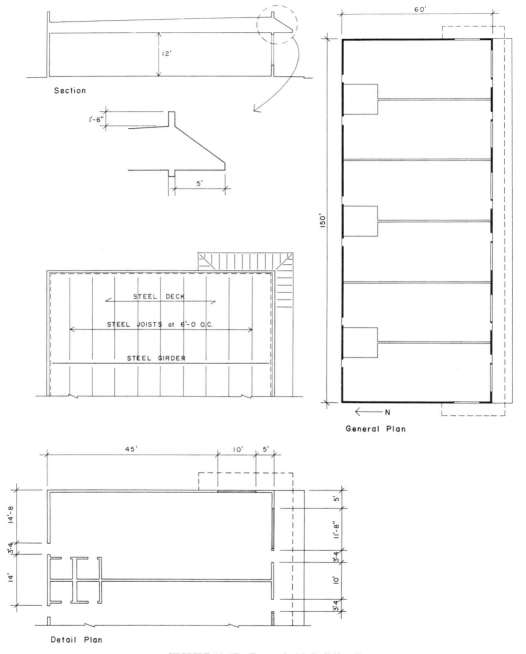

FIGURE 10.17 Example 10, Building D.

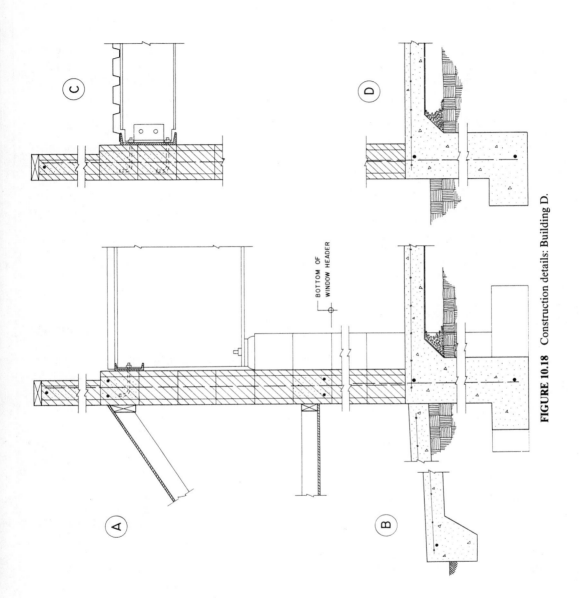

BOTTOM OF
WINDOW HEADER

FIGURE 10.18 Construction details: Building D.

197

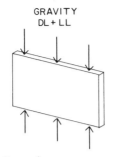

GRAVITY
DL + LL

Case I

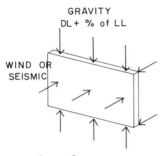

GRAVITY
DL + % of LL

WIND OR
SEISMIC

Case 2

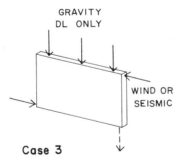

GRAVITY
DL ONLY

WIND OR
SEISMIC

Case 3

FIGURE 10.19 Loading cases for the exterior walls.

adequate for preliminary design work, and are still acceptable for many local building codes.)

$$F_a = 0.10f'_m \left[1 - \left(\frac{h}{42t} \right)^3 \right]$$

$$= (0.10)(1350) \left[1 - \left(\frac{180}{42 \times 7.625} \right)^3 \right]$$

$$= 111 \text{ psi}$$

and the maximum allowable bending stress is

$$F_b = 0.166f'_m = 224 \text{ psi}$$

For a total wall height of 18 ft the wall dead load is $18 \times 70 = 1260$ lb/ft. Assuming the clear purlin span to be 24 ft, the loads from the purlins are

$$\text{dead load} = 12 \times 25 \text{ psf} = 300 \text{ lb/ft}$$

$$\text{live load} = 12 \times 20 \text{ psf} = 240 \text{ lb/ft}$$

The total gravity vertical load on the wall is thus 1800 lb/ft and the average net compression stress, assuming the wall to be 65% solid, is

$$f_a = \frac{P}{A} = \frac{1800}{(0.65)(12 \times 7.625)}$$

$$= 30.3 \text{ psi}$$

Assuming the purlins to be supported by a ledger that is bolted to the wall surface, the roof loading will cause a bending moment equal to the load times one-half the wall thickness; thus

$$M = 540 \times \frac{7.625}{2}$$

$$= 2059 \text{ in.-lb per foot of wall length}$$

Using Fig. C.1, Appendix C, we find an approximate bending stress as follows.

Assume an average reinforcing with No. 5 bars at 40-in. centers. Thus

$$p = \frac{(0.31)(12/40)}{(12)(7.625)} = 0.001$$

$$np = 44 \times 0.001 = 0.044$$

$$K = \frac{M}{bd^2} = \frac{2059}{(12)(3.813)^2} = 11.8$$

From the graph, $f_m = \pm 90$ psi $= f_b$. Then

$$\frac{f_a}{F_a} + \frac{f_b}{F_b} = \frac{30.3}{111} + \frac{90}{224}$$

$$= 0.27 + 0.40 = 0.67$$

Because this is less than 1.0, the wall is adequate for the vertical gravity load alone.

For the case of gravity load plus lateral bending we must determine the maximum bending moment due to wind or seismic load. We thus determine the following:

For wind:

$$p = C_e C_q q_s I$$

where

C_e = 1.2 (exposure C, *UBC* Table 16-G)
C_q = 1.2 (wall, inward pressure, *UBC* Table 16-H)
q_s = 13 psf (70-mph wind, *UBC* Table 16-F)
I = 1 (as for wind)

Then

$$p = (1.2)(1.2)(13)(1) = 18.72 \text{ psf}$$

For seismic:

$$F_p = ZI_p C_p W_p$$

where

Z = 0.4 (for zone 4, *UBC* Table 16-I)
I_p = 1 (as for wind)
C_p = 0.75 (wall, *UBC* Table 16-O)
W_p = 70 psf (assumed wall weight)

Then

$$F_p = (0.4)(1)(0.75)(70) = 21 \text{ psf}$$

This indicates that the seismic effect is critical, so we will use it for the combined load investigation. The wall spans the vertical distance of 15 ft from the floor to the roof. The doweling of the reinforcing at the base plus the cantilever effect of the wall above the roof will reduce the positive moment at the wall midheight. We thus use an approximate moment for design of

$$M = \frac{qL^2}{10} = \frac{(21)(15)^2}{10}$$

$$= 473 \text{ ft-lb}$$

To this we add the moment due to the eccentricity of the roof dead load; thus

$$M = 300 \times \frac{7.625}{2 \times 12}$$

$$= 95 \text{ ft-lb}$$

and we now design for a total moment of 568 ft-lb. Assuming an approximate value of $j = 0.85$, we find that the required reinforcing is

$$A_s = \frac{M}{f_s jd} = \frac{0.568 \times 12}{(1.33 \times 20)(0.85)(3.813)}$$

$$= 0.079 \text{ in.}^2/\text{ft}$$

We try No. 5 at 32 in.

$$A_s = (0.31)\left(\frac{12}{32}\right) = 0.116 \text{ in.}^2/\text{ft}$$

Then

$$p = \frac{A_s}{bd} = \frac{0.116}{12 \times 3.813} = 0.0025$$

$$np = 44 \times 0.0025 = 0.112$$

$$K = \frac{M}{bd^2} = \frac{568 \times 12}{(12)(3.813)^2} = 39$$

From Fig. C.1, $f_m = \pm 240$ psi. For axial compression due to dead load only,

$$f_a = \frac{1560}{0.65(12 \times 7.625)}$$

$$= 26.2 \text{ psi}$$

and

$$\frac{f_a}{F_a} + \frac{f_b}{F_b} = \frac{26.2}{111} + \frac{240}{224}$$

$$= 0.24 + 1.07 = 1.31$$

This indicates a combination close to the limit of 1.33. However, the analysis is conservative because the axial stress used is actually that at the bottom of the wall and the tension resistance of the masonry is ignored.

The rear walls have less load from the roof and a slightly shorter unsupported height. For these walls it is possible that the minimum reinforcing required by the code is adequate. The code requirements are:

1. Minimum of 0.002 times the gross wall area in both directions (sum of the vertical and horizontal bars)
2. Minimum of 0.0007 times the gross area in either direction
3. Maximum spacing of 48 in.
4. Minimum bar size of No. 3
5. Minimum of one No. 4 or two No. 3 bars on all sides of openings

With the No. 5 bars at 32 in., the gross percentage of vertical reinforcing is

$$p_g = \frac{0.116}{12 \times 7.625} = 0.00127$$

To satisfy the requirement for total reinforcing, it is thus necessary to have a minimum gross percentage for the horizontal reinforcing of

$$p_g = 0.002 - 0.00127 = 0.00073$$

which requires an area of

$$A_s = p_g \times A_g = (0.00073)(12 \times 7.625)$$

$$= 0.067 \text{ in.}^2/\text{ft}$$

This can be provided by

No. 4 at 32 in.:

$$A_s = 0.20 \times \frac{12}{32}$$

$$= 0.075 \text{ in.}^2/\text{ft.}$$

No. 5 at 48 in.:

$$A_s = 0.20 \times \frac{12}{48}$$

$$= 0.0775 \text{ in.}^2/\text{ft.}$$

Choice of this reinforcing must also satisfy the requirements for shear wall functions.

At the large wall openings the headers will transfer both vertical and horizontal loads to the ends of the supporting walls. The ends of these walls will be designed as reinforced masonry columns for this condition. Figure 10.20 shows the loading condition for the header columns. In addition to this loading the columns are part of the wall and must carry some of the axial load and bending as previously determined for the typical wall.

Figure 10.21 shows a plan layout for the entire solid front wall section between the window openings. A pilaster column is provided for the support of the girder. Because of the stiffness of the column, it will tend to take a large share of the lateral load. We will thus assume the end column to take only a 2-ft strip of the wall lateral load. As shown in Fig. 10.21, the end column is a doubly reinforced beam for the direct lateral load.

LOADS ON HEADER

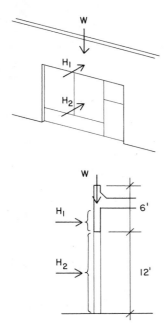

The gravity dead load on the header is:

$$\text{Roof} = 100 \text{ lb/ft}$$

$$\text{Wall} = (70 \text{ psf})(6 \text{ ft})$$

$$= 420 \text{ lb/ft}$$

$$\text{Canopy} = 100 \text{ lb/ft (assumed)}$$

$$\text{Total load} = 620 \text{ lb/ft}$$

For the lateral load we will use a design pressure of 20 psf because the weight of the window wall will produce a low seismic force. Assuming the window mullions span vertically, the wind loads are as shown in Fig. 10.20.

$$H_1 = (20 \text{ psf})(2 \text{ ft} \times 15 \text{ ft})$$

$$= 600 \text{ lb}$$

$$H_2 = (20 \text{ psf})(12 \text{ ft} \times 15 \text{ ft})$$

$$= 3600 \text{ lb}$$

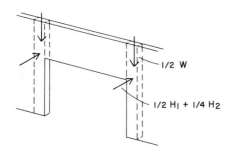

LOADS ON HEADER COLUMNS

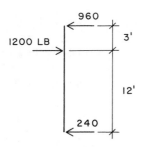

Thus the column loads from the header are:

$$\text{vertical load} = (620 \text{ plf})(15/2)$$

$$= 4650 \text{ lb}$$

$$\text{horizontal load} = \tfrac{1}{2}H_1 + \tfrac{1}{4}H_2$$

$$= 300 + 900$$

$$= 1200 \text{ lb}$$

$$\text{moment} = (960)(3)$$

$$= 2880 \text{ ft-lb}$$

(see Fig. 10.20)

For the direct wind load on the wall we assume a 15-ft vertical span and a 2-ft-wide strip of wall loading. Thus

$$M = \frac{wL^2}{8} = \frac{(20 \text{ psf})(2)(15)^2}{8}$$

$$= 1125 \text{ ft-lb}$$

FIGURE 10.20 Loads on the headers and columns.

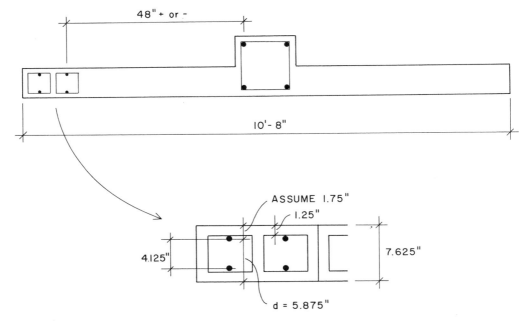

FIGURE 10.21 Details of the front wall.

These two moments do not peak at the same point; thus without doing a more exact analysis we assume a maximum combined moment of 3800 ft-lb. Then, for the moment alone, assuming a j of 0.85,

$$\text{required } A_s = \frac{M}{f_s j d} = \frac{(3.8)(12)}{(26.7)(0.85)(5.9)}$$

$$= 0.34 \text{ in.}^2$$

$$\text{approximate } f_m = \frac{M}{bd^2} \frac{2}{kj}$$

$$= \frac{(3800)(12)(2)}{(16)(5.9)^2(0.4)(0.85)}$$

$$= 482 \text{ psi}$$

Although f_m appears high, we have ignored the effect of the compressive reinforcing in the doubly reinforced member. The following is an approximate analysis based on the two moment theory with two No. 5 bars on each side of the column.

For the front wall it is reasonable to consider the use of a fully grouted wall because the pilaster and the end columns already constitute a considerable solid mass. For the fully grouted wall we may use $f_m' = 1500$ psi, and the allowable bending stress thus increases to

$$F_b = 1.33 \times 0.166 \times 1500$$

$$= 331 \text{ psi}$$

Assuming the axial load to be almost negligible compared to the moment, we analyze for the full moment effect only. With a maximum stress of 331 psi we first determine the moment capacity with tension reinforcing only as

$$M_1 = \frac{f_m(bd^2)(k)(j)}{2} \left(\frac{1}{12}\right)$$

$$= \frac{(331)(16)(5.9)^2(0.4)(0.85)}{2(12)}$$

$$= 2612 \text{ ft-lb}$$

This leaves a moment for the compressive reinforcing of

$$M_1 = 3800 - 2600$$

$$= 1200 \text{ ft-lb}$$

If the compressive reinforcing is two No. 5 bars, then

$$f_s' = \frac{M_2}{A_s'(d - d')}$$

$$= \frac{(1200)(12)}{(0.62)(4.125)}$$

$$= 5630 \text{ psi}$$

This is a reasonable stress even with the assumed low k value of 0.4. As shown in Fig 10.22, if k is 0.4 and f_m is 331, the compatible strain value for f_s' will be

$$f_s' = 2n(f_c) = (2)(40)(3.31)\left(\frac{0.61}{2.36}\right)$$

$$= 6844 \text{ psi}$$

As shown by the preceding calculation, the stress in the tension reinforcing will not be critical. This approximate analysis indicates that the column is reasonably adequate for the moment. The axial load capacity should also be checked, using the procedure shown later for the pilaster design.

Window Header. As shown in Fig. 10.20, the header consists of a 6-ft-deep section of wall. This section will have continuous reinforcing at the top of the wall

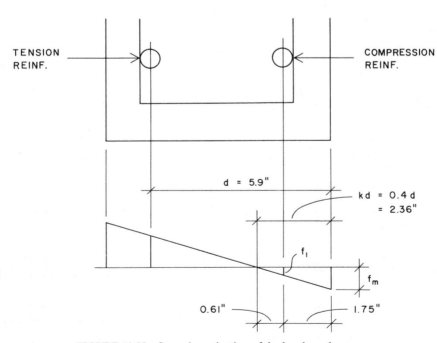

FIGURE 10.22 Stress investigation of the header column.

and at the bottom of the header. In addition there will be a continuous reinforced bond beam in the wall at the location of the steel ledger that supports the edge of the roof deck.

Using the loading previously determined, and an approximate design moment of $wL^2/10$, the steel area required for gravity alone will be

$$A_s = \frac{M}{f_s(jd)}$$

where

$$M = \frac{wL^2}{10} = \frac{(620)(15)^2}{10}$$

$$= 13,950 \text{ ft-lb}$$

$$d = \text{approximately 68 in.}$$

Then

$$A_s = \frac{(13.95)(12)}{(20)(0.85)(68)}$$

$$= 0.145 \text{ in.}^2$$

This indicates that the minimum reinforcing at the top of the wall may be two No. 3 bars or one No. 4 bar. This should be compared with the code requirement for minimum wall reinforcing, which is usually a minimum of 0.0007 times the gross cross-sectional area of the wall in either direction and a sum of 0.002 times the gross cross-sectional area of the wall in both directions. Thus

$$\text{minimum } A_2 = (0.007)(7.625)(12)$$

$$= 0.064 \text{ in.}^2/\text{ft of width or height}$$

With two No. 3 bars $A_s = 0.22 \text{ in.}^2$

$$\text{required spacing} = \frac{0.22}{0.064}$$

$$= 3.44 \text{ ft or 41.3 in.}$$

The minimum horizontal reinforcing would then be two No. 3 bars at 40 in., or every fifth block course.

At the bottom of the header there is also a horizontal force consisting of the previously calculated wind load plus some force from the cantilevered canopy. Estimating this total horizontal force to be 250 lb/ft, we add a horizontal moment as

$$M = \frac{wL^2}{10} = \frac{(0.25)(15)^2}{10}$$

$$= 5.625 \text{ k-ft}$$

for which we require that

$$A_2 = \frac{M}{f_s(jd)} = \frac{(5.625)(12)}{(26.7)(0.85)(5.9)}$$

$$= 0.504 \text{ in.}^2$$

This must be added to the previous area required for the vertical gravity loads:

$$\text{total } A_s = 0.504 + \frac{(\frac{1}{2})(0.145)}{1.33}$$

$$= 0.504 + 0.055$$

$$= 0.559 \text{ in.}^2$$

The requirement for vertical load is divided by two because it is shared by both bottom bars. It is divided by 1.33, since the previous calculation did not include the increase of allowable stresses for wind loading. If this total area is satisfied, the bottom bars in the header would have to be two No. 7s. An alternative would be to increase the width of the header at the bottom by using a 12-in.-wide block for the bottom course, as

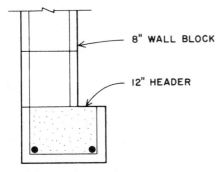

FIGURE 10.23 Alternate header detail.

shown in Fig. 10.23. This widened course would be made continuous in the wall.

Pilaster Column. To permit the wall construction to be continuous, the girder stops short of the inside of the wall and rests on the widened portion of the wall called a *pilaster*. As shown in Fig. 10.24, the pilaster and wall together form a 16-in. square column. The principal gravity loading on the column is due to the end reaction of the girder. Since this load is eccentrically placed, it produces both axial force and bending on the column. The parapet, canopy, and column weight add to the axial compression.

FIGURE 10.24 Pilaster column.

Because of its increased stiffness, the column tends to take a considerable portion of the wind pressure on the solid portion of the wall. We will assume it to take a 6-ft-wide strip of this load. As shown in Fig. 10.25, the direct wind pressure on the wall (pushing inward on the outer surface) causes a bending moment of opposite sign from that due to the eccentric girder load. The critical wind load is therefore due to the outward wind pressure (suction force) on the wall. For a conservative design we will take this to be equal to the inward pressure of 20 psf. The combined moments are thus

$$\text{wind moment} = \frac{wL^2}{8} = \frac{(20)(6)(13.33)^2}{8}$$

$$= 2665 \text{ ft-lb}$$

Assuming an eccentricity of 4 in. for the girder (see Fig. 10.18) and an end reaction due to dead load and one half live load of 23.5 kips,

$$\text{girder moment} = \frac{(23.5)(4)}{12}$$

$$= 7.833 \text{ kip-ft or } 7833 \text{ ft-lb}$$

For the combined wind plus gravity loading we have used only half the live load. With the allowable stress increase, it should be apparent that this loading condition is not critical, so we will design for the gravity loads only. For this we will redetermine the girder-induced moment with full live load assuming this to result in a reaction of 27.6 kips:

$$\text{girder } M = \frac{(27.6)(4)}{12}$$

$$= 9.2 \text{ kip-ft}$$

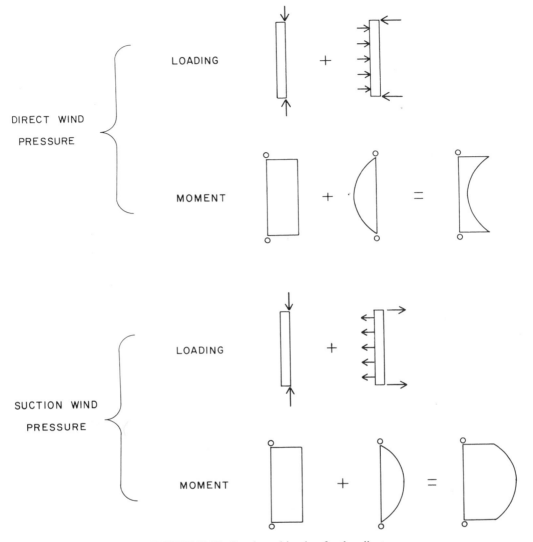

FIGURE 10.25 Load combination for the pilaster.

The gravity loads of the canopy, parapet, roof edge, and column must be added. We will therefore assume a total vertical design load of approximately 35 kips. With this total load the equivalent eccentricity for design will be

$$e = \frac{M}{N} = \frac{(9.2)(12)}{35} = 3.15 \text{ in.}$$

The usual requirement is for a minimum percentage of reinforcing of 0.005 of the gross column area. Thus

$$\text{minimum } A_s = (0.005)(16)^2$$
$$= 1.28 \text{ in.}^2$$

With four No. 7 bars $A_s = 2.40$ in.². Then the allowable axial load is determined as

$$P = (\tfrac{1}{2})(0.20 f_m' A_e + 0.65 A_s F_{sc})$$

$$\times \left[1 - \left(\frac{h'}{42t} \right)^3 \right]$$

where

A_e = total area of the fully grouted column

  = $(15.625)^2 = 244$ in.^{2}

F_{sc} = $0.40 F_y = 16$ ksi

h' = effective (unbraced) height of the column

  = 13.3 ft or 160 in.

P = $(\tfrac{1}{2})[(0.20 \times 1.5 \times 244)$

  $+ (0.65 \times 2.40 \times 16)]$

$$\times \left[1 - \left(\frac{160}{42(15.6)} \right)^3 \right]$$

  = 48.4 kips

Ignoring the compression steel, the approximate moment capacity is

$$M = A_s f_s (jd) = \frac{(1.20)(20)(0.85)(13.5)}{12}$$

  = 22.95 kip-ft

and for the combined effect

$$\frac{actual\ P}{allowable\ P} + \frac{actual\ M}{allowable\ M}$$

$$= \frac{27.6}{48.4} + \frac{9.2}{22.95}$$

$$= 0.55 + 0.40 = 0.95$$

Note that the one-half factor has been used with the axial load formula, assuming no special inspection during construction. Although a more exact analysis should be performed, this indicates generally that the

column is reasonably adequate for the axial load and the moment previously determined.

Shear Walls. The calculation of the loads applied to the roof diaphragm is shown in Table 10.5 the north–south direction the load is symmetrically placed, the shear walls are symmetrical in plan, and the long diaphragm is reasonably flexible, all of which results in very little potential torsion. Although the code requires that a minimum torsion be considered by placing the load off center by 5% of the building long dimension, the effect will be very little on the shear walls.

For the seismic forces we find that

$$V = \frac{ZIC}{R_W} W$$

where

Z = 0.4 (zone 4, *UBC* Table 16-I)

I = 1 (ordinary or standard occupancy, *UBC* Table 16-K)

C = 2.75 (maximum value, *UBC* Sec. 1628.2.1)

R_W = 6 (masonry shear wall, *UBC* Table 16-N)

W = building weight for direction considered

Thus, for the two axes of the building,

$$V = \frac{(0.4)(1)(2.75)}{6} W = 0.1833 W$$

Using the data from Table 10.5, we have

N–S V = (0.1833)(540) = 99 kips

E–W V = (0.1833)(417) 76.5 kips

At the ends of the building the max-

TABLE 10.5   LOADS TO THE ROOF DIAPHRAGM: EXAMPLE 10

Load Source and Calculations	North–South Load (kips)	East–West Load (kips)
Roof dead load		
$150 \times 60 \times 29$ psf	261	261
East and west exterior walls		
$50 \times 11 \times 70$ psf $\times 2$	0	77
$10.67 \times 7 \times 70$ psf $\times 2$	0	11
$10.67 \times 5 \times 10$ psf $\times 2$	0	1
North wall		
$150 \times 12 \times 70$ psf	126	0
South wall		
$65.3 \times 10 \times 70$ psf	46	0
$84 \times 6 \times 70$ psf	35	0
$84 \times 6 \times 10$ psf	5	0
Interior north–south partitions		
$60 \times 7 \times 10$ psf $\times 5$	21	21
Toilet walls		
Estimated $250 \times 7 \times 10$ psf	17	17
Canopy		
South: 150×100 plf	15	15
East and west: 40×100 plf	4	4
Rooftop HVAC units (estimate)	10	10
Total loads	540	417

imum shear stress in the roof diaphragm due to the north–south seismic force is

$$\text{maximum } v = \frac{49{,}500}{60}$$

$$= 825 \text{ plf}$$

This is a very high shear for the metal deck. It would require a heavy-gauge deck and considerable welding at the diaphragm edge. Although it would probably be wise to reconsider the general design and possibly use at least one permanent interior partition, we assume the deck to span the building length for the shear wall design.

In the other direction the shear in the roof deck will be considerably less:

$$\text{east–west total } V = 0.1833(410)$$

$$= 75.2 \text{ kips}$$

$$\text{maximum } v = \frac{37{,}600}{150}$$

$$= 251 \text{ plf}$$

This is very low for the deck, so if any interior shear walls are added, the deck gauge could probably be reduced to that required for the gravity loads only.

In the north–south direction, with no added shear walls, the end shear forces will be taken almost entirely by the long solid walls because of their relative stiffness. The shear force will be the sum of the end shear from the roof and the force due to the weight of the end wall. For the latter we compute the following:

Wall weight

$$= 18 \text{ ft} \times 50 \text{ ft} \times 70 \text{ psf}$$

$$\times 63{,}000 \text{ lb}$$

7 ft $\times$ 10.67 ft $\times$ 70 psf

$$= 5228 \text{ lb}$$

11 ft $\times$ 10.67 ft $\times$ 70 psf

$$= 1174 \text{ lb}$$

$$\text{Total} = 69,402 \text{ lb}$$

Lateral force $= 0.1833 W$

$$= 0.1833 \times 69.4 = 12.7 \text{ kips}$$

The total force on the wall is thus 12.7 + 49.5 = 62.2 kips and the unit shear force in the wall is

$$v = \frac{62,200}{44.67} = 1392 \text{ lb/ft}$$

Assuming a 60% solid wall with 8-in. blocks, the unit stress on the net area of the wall is thus

$$v = \frac{1392}{12 \times 7.625 \times 0.60}$$

$$= 25.4 \text{ psi}$$

With the reinforcing taking all shear and no special inspection, the allowable shear stress is dependent on the value of M/Vd for the wall. This is determined as

$$\frac{M}{Vd} = \frac{(49.5 \times 15) + (12.7 \times 9)}{62.2 \times 44.67}$$

$$= 0.308$$

Interpolating between the values for M/Vd of 0 and 1.0,

allowable $v = 35 + (0.69)(25) = 52$ psi

This may be increased by the usual one third for seismic load to (1.33)(52) = 69 psi. This indicates that the masonry stress is adequate, but we must check the wall rein-

forcing for its capacity as shear reinforcement. With the minimum horizontal reinforcing determined previously—No. 5 at 48 in.—the load on the bars is

$$V = (1392) \left(\frac{48}{12} \right)$$

$$= 5568 \text{ lb per bar}$$

and the required area for the bar is

$$A_s = \frac{V}{f_s} = \frac{5568}{26,667}$$

$$= 0.21 \text{ in.}^2$$

This indicates that the minimum reinforcing is adequate. Some additional stress will be placed on these walls by the effects of torsion, so that some increase in the horizontal reinforcing is probably advisable.

Overturn is not a problem for these walls because of their considerable dead weight and the natural tiedown provided by the doweling of the vertical wall reinforcing into the foundations. These dowels also provide the necessary resistance to horizontal sliding.

In the east–west direction the shear walls are not symmetrical in plan, which requires that a calculation be made to determine the location of the center of rigidity so that the torsional moment may be determined. The total loading is reasonably centered in this direction, so we will assume the center of gravity to be in the center of the plan.

The following analysis is based on the examples in the *Masonry Design Manual* (Ref. 16). The individual piers are assumed to be fixed at top and bottom and their stiffnesses are found from Table C.2 The stiffness of the piers and the total wall stiffnesses are determined in Fig. 10.26. For the location of the center of stiffness we use the

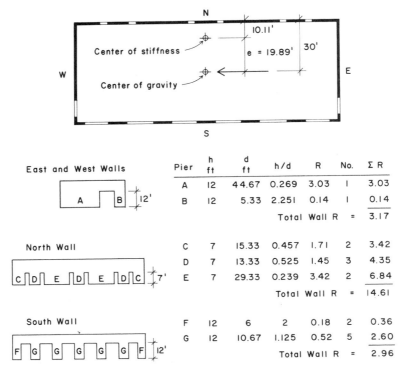

FIGURE 10.26 Stiffness analysis of the masonry piers. For stiffness factors (R) of the individual piers, see Appendix C.

values determined for the north and south walls:

$$\bar{y} = \frac{(R \text{ for S wall})(60 \text{ ft})}{(\text{sum of the } R \text{ values for the N and S walls})} = \frac{(2.96)(60)}{17.57}$$

$$= 10.11 \text{ ft}$$

The torsional resistance of the entire shear wall system is found as the sum of the products of the individual wall rigidities times the square of their distances from the center of stiffness. This summation is shown in Table 10.6. The torsional shear load for each wall is then found as

$$V_W = \frac{Tc}{J} R_i$$

where

V_W = total shear force on an individual wall

TABLE 10.6   TORSIONAL RESISTANCE OF THE MASONRY SHEAR WALLS

Wall	Total Wall R	Distance from Center of Stiffness (ft)	$R(d)^2$
South	2.96	49.89	7,367
North	14.61	10.11	1,495
East	3.17	75	17,831
West	3.17	75	17,831
Total torsional moment of inertia (J)			44,524

T = torsional moment = V times e

c = distance of the wall from the center of stiffness

R_i = resistance value for the wall

J = torsional moment of inertia, as determined by the sum of the Rd^2 values for all the walls

In the north–south direction *UBC* Sec. 1628.5 requires that the load be applied with a minimum eccentricity of 5% of the building length, or 7.5 ft. Although this produces less torsional moment than the east–west load, it is additive to the direct north–south shear and therefore critical for the end walls. The torsional load for the end walls is thus

$$V_W = \frac{(99)(7.5)(75)(3.17)}{44,524}$$

$$= 3.96 \text{ kips}$$

As mentioned previously, this should be added to the direct shear of 49,500 lb for the design of these wall:

For the north wall:

$$V_W = \frac{(76.5)(19.89)(10.11)(14.61)}{44,524}$$

$$= 5.05 \text{ kips}$$

This is actually opposite in direction to the direct shear, but the code does not allow the reduction and thus the direct shear only is used.

For the south wall:

$$V_W = \frac{(76.5)(19.89)(49.89)(2.96)}{44,524}$$

$$= 5.05 \text{ kips}$$

The total direct east–west shear will be distributed between the north and south walls in proportion to the wall stiffnesses:

For the north wall:

$$V_W = \frac{(76.5)(14.61)}{17.57} = 63.6 \text{ kips}$$

For the south wall:

$$V_W = \frac{(76.5)(2.96)}{17.57} = 12.9 \text{ kips}$$

The total shear loads on the walls are therefore

north: V = 63.6 kips

south: V = 5.05 + 12.9

$= 17.95$ kips

The loads on the individual piers are then distributed in proportion to the pier stiffnesses (R) as determined in Fig. 10.26. The calculation for this distribution and the determination of the unit shear stresses per foot of wall are shown in Table 10.7. A comparison with the previous calculations for the end walls will show that these stresses are not critical for the 8-in. block walls.

In most cases the stabilizing dead loads plus the doweling of the end reinforcing into the foundations will be sufficient to resist overturn effects. The heavy loading on the header columns and the pilasters will provide considerable resistance for most walls. The only wall not so loaded is wall *C*, for which the loading condition is shown in Fig. 10.27. The overturn analysis for this wall is as follows:

overturn M = (7440)(7.0)(1.5)

$= 78,120$ ft-lb

stabilizing M = $(23,000)\left(\dfrac{15.33}{2}\right)$

$= 176,295$ ft-lb

This indicates that the wall is stable with-

TABLE 10.7 SHEAR STRESSES IN THE MASONRY WALLS

Wall	Shear Force on Wall (kips)	Wall R	Pier	Pier R	Shear Force on Pier (kips)	Pier Length (ft)	Shear Stress in Pier (lb/ft)
North	63.6	14.61	C	1.71	7.44	15.33	485
			D	1.45	6.31	13.33	473
			E	3.42	14.89	29.33	508
South	17.95	2.96	F	0.18	1.09	6	182
			G	0.52	3.15	10.67	296

out any requirement for anchorage even though the wall weight in the plane of the shear wall was not included in computing the overturning moment.

Example 11, Building E

Building E is a three-story office building. Assuming the building is to be built for investment, with a speculative rental occupancy or a sale for undetermined purpose, a feature typically desired is the adaptability of the building to change. With regard to the structure, this usually means an emphasis on minimizing permanent elements of the construction—notably on the interior of the building. In this instance the permanent elements are limited to the exterior walls and the core elements (stairs, elevators, duct shafts, rest rooms) with a few interior freestanding columns (see Fig. 10.28).

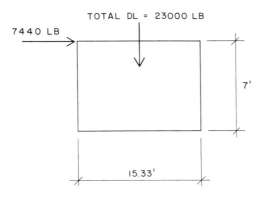

FIGURE 10.27 Stability of wall *C.*

The following will be used for consideration of lateral actions:

Wind: map speed = 80 mph; exposure B

Seismic: zone 3

Assumed construction loads:
    Floor finish = 5 psf
    Ceiling, lights, ducts = 15 psf
    Walls (average surface weight);
        Interior partitions = 25 psf
        Exterior curtain wall = 25 psf

Fire codes permitting, the most economical structure for the building will be one that makes the most use of light wood frame construction. It is unlikely that the building would use all wood construction of the type illustrated in Example 1, but a mixed system is quite possible. It is also possible to use steel, masonry, or concrete construction and eliminate wood, except for nonstructural uses. In addition to code requirements, consideration must be given to the building owners' preferences and to design criteria or standards for acoustic privacy, thermal control, and so on.

The plan as shown, with 30-ft square bays and a general open interior, is an ideal arrangement for a beam and column system in either steel or reinforced concrete. Other types of systems may be made more effective if some modifications of the basic plans are made. These changes may affect the planning of the building core, the plan

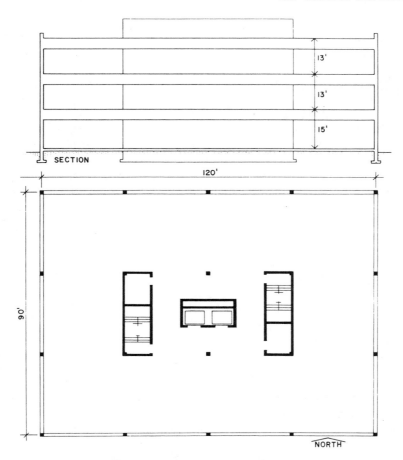

FIGURE 10.28 Example 11. Building E.

dimensions for the column locations, the articulation of the exterior wall, or the vertical distances between the levels of the building.

The general form and basic type of the structural system must relate to both the gravity and lateral force problems. Considerations for gravity require the development of the horizontal spanning systems for the roof and floors and the arrangement of the vertical elements (walls and columns) that provide support for the spanning structure. Vertical elements should be stacked, thus requiring coordinating the plans of the various levels.

The most common choices for the lateral bracing system would be the following (see Fig. 10.29):

1. *Core Shear Wall System* (Fig. 10.29*a*). This consists of using solid walls to produce a very rigid central core. The rest of the structure leans on this rigid interior portion, and the roof and floor construction outside the core—as well as the exterior walls—are somewhat more free of concerns for lateral forces as far as the structure as a whole is concerned.

2. *Truss-Braced Core.* This is similar in

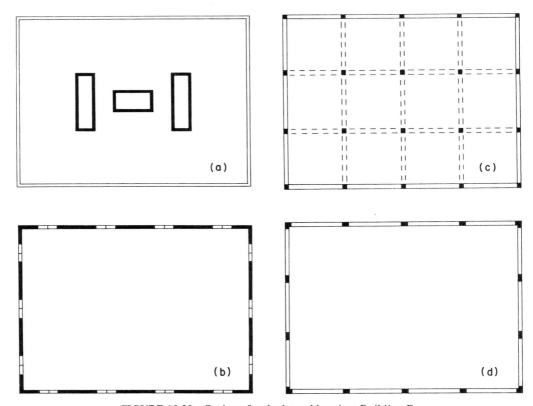

FIGURE 10.29   Options for the lateral bracing: Building E.

nature to the shear wall-brace core, and the planning considerations would be essentially similar. The solid walls would be replaced by bays of trussed framing (in vertical bents) using various possible patterns for the truss elements.

3. *Perimeter Shear Walls (Fig. 10.29b)*. This in essence makes the building into a tubelike structure. Because doors and windows must pierce the exterior, the perimeter shear walls usually consist of linked sets of individual walls (sometimes called piers).

4. *Mixed Exterior and Interior Shear Walls*. This is essentially a combination of the core and perimeter systems.

5. *Full Rigid Frame System (Fig. 10.29c)*. This is produced by using the vertical planes of columns and beams in each direction as a series of rigid bents. For this building there would thus be four bents for bracing in one direction and five for bracing in the other direction. This requires that the beam-to-column connections be moment resistive.

6. *Perimeter Rigid Frame System (Fig. 10.29d)*. This consists of using only the columns and beams in the exterior walls, resulting in only two bracing bents in each direction.

In the right circumstances any of these systems may be acceptable. Each has advantages and disadvantages from both structural design and architectural planning points of view. The braced core schemes were popular in the past, especially for buildings in which wind was the major concern. The core system allows for the greatest freedom in planning the exterior walls, which are obviously of major concern to the architect. The perimeter system, however, produces the most torsionally stiff building—an advantage for seismic resistance.

The rigid frame schemes permit the free planning of the interior and the greatest openness in the wall planes. The integrity of the bents must be maintained, however, which restricts column locations and planning of stairs, elevators, and duct shafts so as not to interrupt any of the column-line beams. If designed for lateral forces, columns are likely to be large, and thus offer more intrusion in the building plan.

Other solutions are also possible, limited only by the creative imagination of designers. In the sections that follow we will illustrate the design of two possible structures for the building. We do not propose that these are ideal solutions, but merely that they are feasible alternatives. They have been chosen primarily to permit illustrating the design of the elements of the construction.

A structural framing plan for one of the upper floors of Example 11 is shown in Fig. 10.30. The plan indicates major use of structural masonry walls for both bearing walls and shear walls. The lateral bracing scheme is a combination of the core and perimeter systems. The floor—and probably the roof—structures consist of deck–joist–beam systems that are supported mostly by the interior and exterior bearing walls. The material in this section consists of the design of the major elements of this system.

For wind it is necessary to establish the design wind pressure, defined by the code as

$$p = C_e C_q q_s I$$

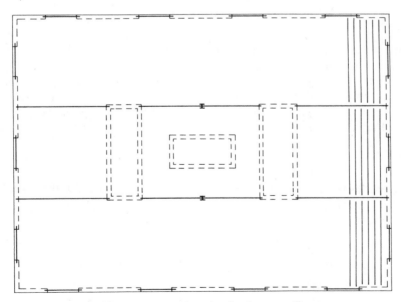

FIGURE 10.30   Framing plan for the upper floors.

where C_e is a combined factor including concerns for the height above grade, exposure conditions, and gusts. From *UBC* Table 16-G (see Appendix B) assuming exposure B:

$$C_2 = 0.7 \text{ from 0 to 20 ft above grade}$$
$$= 0.8 \text{ from 20 to 40 ft}$$
$$= 1.0 \text{ from 40 to 60 ft}$$

and C_q is the pressure coefficient. Using the projected area method (method 2), we find from *UBC* Table 16-H (see Appendix B) the following.

For vertical projected area:

$$C_q = 1.3 \text{ up to 40 ft above grade}$$
$$= 1.4 \text{ over 40 ft}$$

For horizontal projected area (roof surface):

$$C_q = 0.7 \text{ upward}$$

The symbol q_s is the wind stagnation pressure at the standard measuring height of 30 ft. From *UBC* Table 16-F the q_s value for a speed of 80 mph is 17 psf. For the importance factor I (*UBC* Table 16-K) we use a value of 1.0.

Table 10.8 summarizes the foregoing data for the determination of the wind pressures at the various height zones for Example 11. For the analysis of the horizontal wind effect on the building, the wind pressures are applied and translated into edge loadings for the horizontal diaphragms (roof and floors) as shown in Fig. 10.31. Note that we have rounded off the wind pressures from Table 10.8 for use in Figure 10.31.

Figure 10.32a shows a plan of the building with an indication of the masonry walls that offer potential as shear walls for resistance to north–south lateral force. The numbers on the plan are the approximate plan lengths of the walls. Note that although the core construction actually produces vertical tubular-shaped elements, we have considered only the walls parallel to the load direction. The walls shown in Fig. 10.32a will share the total wind load delivered by the diaphragms at the roof, third-floor, and second-floor levels (H_1, H_2, and H_3, respectively, as shown in Fig. 10.31). Assuming the building to be a total of 122-ft wide in the east–west direction, the forces at the three levels are:

$$H_1 = 186 \times 122 = 22,692 \text{ lb}$$
$$H_2 = 216 \times 122 = 26,352 \text{ lb}$$
$$H_3 = 194 \times 122 = 23,668 \text{ lb}$$

and the total wind force at the base of the shear walls is the sum of these loads, or 72,712 lb.

Although the distribution of shared load to masonry walls is usually done on the basis of a more sophisticated analysis for relative stiffness (as was done for Example

TABLE 10.8 DESIGN WIND PRESSURES FOR EXAMPLE 11 (EXPOSURE CONDITION B)[a]

Building Surface Zone	Height Above Ground (ft)	C_e	C_q	Pressure p (psf)
1	0–15	0.62	1.3	13.2
2	15–20	0.67	1.3	14.3
3	20–25	0.72	1.3	15.4
4	25–30	0.76	1.3	16.2
5	30–40	0.84	1.3	17.9
6	40–60	0.95	1.4	21.8

[a] Horizontally-directed pressure on vertical surface: $p = C_e \times C_q \times 16.4$ psf.

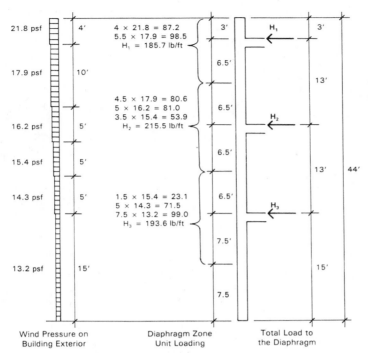

FIGURE 10.31 Generation of wind loads.

10), if we assume for the moment that the walls are stiff in proportion to their plan lengths (as is done with plywood walls), we may divide the maximum shear load at the base of the walls by the total of the wall plan lengths to obtain an approximate value for the maximum shear stress. Thus

$$\text{maximum shear: } v = \frac{72{,}712}{260}$$

$$= 280 \text{ lb/ft of wall length}$$

This is quite a low force for a reinforced masonry wall, which tells us that if wind alone is of concern we have considerable overkill in terms of total shear walls. However, because we will find that the seismic forces are considerably greater for our example, we will reserve final judgment on the structural scheme.

Table 10.9 presents the analysis for determining the building weight to be used for the computation of the seismic effects in the north–south direction. In the tabula-

tion we have included the weights of all the walls, which eliminates the necessity for adding the weights of the shear walls in any subsequent analysis of individual walls. Tabulations are done separately for the determination of loads to the three upper diaphragms (eventually producing three forces similar to H_1, H_2, and H_3, as determined for the wind loading). Except for the shear walls, the weight of the lower half of the first-story walls is assumed to be resisted by the first-floor-level construction (assumed to be a concrete structure poured directly on the ground) and is thus not part of the distribution to the shear wall system.

For the total seismic shear force to the shear walls, we note from the given data that

$Z = 0.3$ (zone 3, *UBC* Table 16-I)

$I = 1$ (standard occupancy, *UBC* Table 16-K)

$C = 2.75$ (maximum value, *UBC* Sec. 1628.5)

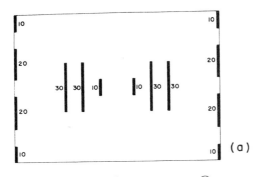

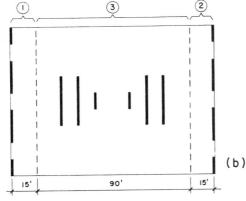

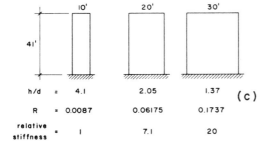

FIGURE 10.32 Masonry shear wall system: (a) north–south wall system; (b) load zone peripheries for the shear wall groups; (c) relative stiffness of the walls as vertical cantilevers.

$R_W = 6$ (masonry shear wall, *UBC* Table 16-N)

Thus

$$V = \frac{ZIC}{R_W}W = \frac{(0.3)(1.0)(2.75)}{6}W$$

$$= 0.1375\,W$$

And using the value from Table 10.9 for W, we have

$$V = (0.1375)(2817.9) = 388 \text{ kips}$$

This total force must be distributed to the roof and second floor in accordance with the requirements of Sec. 1628.4 of the *UBC*. The force at each level F_x is determined from Formula (28-8) as

$$F_x = (V)(w_x h_x)\Big/ \sum_{i=1}^{n} w_i h_i$$

where

F_x = force to be applied at each level x
w_x = total dead load at level x
h_x = height of level x above the base of the structure

(Notice that F_t has been omitted from the formula because T is less than 0.7 sec.)

The determination of the F_x values is shown in Table 10.10. Distribution of the total forces at each level to the individual shear walls requires two considerations. The primary concern is for the functioning of the horizontal diaphragms. First, if these are considered to be infinitely stiff, then the distribution to the individual walls will be strictly in terms of their relative stiffness or deflection. Second, if the horizontal diaphragms are considered to be quite flexible (in their diaphragm spanning actions), then the distribution to the shear walls will be on a peripheral basis.

Figure 10.32b shows the building plan with the north–south shear walls and a breakdown of peripheral distribution assuming the flexible horizontal diaphragm. On this basis, the end shear walls each carry one eighth of the total shear and the core walls carry three-fourths of the shear. In this approach the next step would be to consider the relative stiffness of the group

TABLE 10.9 DEAD LOAD FOR THE NORTH-SOUTH SEISMIC FORCE

Level	Source of Load	Unit Load (psf)	Load (kips)
Roof	Roof and ceiling	25	120 × 90 × 25 = 270
	Masonry walls	60	480 × 9.5 × 60 = 273.6
	Window walls	15	140 × 9.5 × 15 = 20.0
	Interior walls	10	300 × 5 × 12 = 18.0
	Penthouse + equipment (estimate total)		= 25.0
	Subtotal		606.6
Third floor	Floor	55	120 × 90 × 55 = 594.0
	Masonry walls	60	480 × 13 × 60 = 374.4
	Window wall	15	140 × 13 × 15 = 27.3
	Interior walls	10	300 × 9 × 12 = 32.4
	Subtotal		1028.1
Second floor	Floor	55	120 × 90 × 55 = 594.0
	Masonry walls	60	480 × 14 × 60 = 403.2
	Window walls	15	140 × 14 × 15 = 29.4
	Interior walls	10	300 × 10 × 12 = 39.6
	Subtotal		1066.2
First floor	Shear walls	60	260 × 7.5 × 60 = 117.0
	(Remainder of first floor direct to ground)		
	Total dead load for base shear		= 2817.9

of walls in each of the zones and to distribute forces to the individual walls.

In truth, the nature of the diaphragms is most likely somewhere between the two extremes described (just as most structural connections are neither pinned nor fully fixed but actually partially fixed). It is thus not uncommon in practice for designers to investigate both conditions and to incorporate data from both analyses into their designs.

For either approach it is necessary to consider the relative stiffness of the walls of various plan length. The most common means for doing this is the method illustrated in Example 10. Figure 10.32c shows an analysis for the relative stiffness of the walls with the three plan lengths of 10 ft, 20 ft, and 30 ft. The walls are assumed to be cantilevered from fixed bases and the distributions shown are for the roof load for which the wall height (h in the table and figure) is 41 ft. For a precise analysis separate distributions should be made for the distribution of the floor diaphragm loads using the shorter wall heights. If this is

TABLE 10.10 SEISMIC LOADS: EXAMPLE 11

Level	w_x (kips)	h_x (ft)	$w_x h_x$	F_x[a] (kips)
Roof	606.6	41	24,871	138.5
Third floor	1028.1	28	28,787	160.4
Second floor	1066.2	15	15,993	89.1
			69,651	

[a] $F_x = \dfrac{388}{69,651}(w_x h_x)$.

done, it will be found that the percent of load carried by the shorter walls will be considerably increased.

Referring to Fig. 10.32*b*, if we consider the group of core walls in peripheral zone 3, their total combined stiffness is

$$4 \times 0.1737 = 0.6948$$
$$2 \times 0.0087 = \underline{0.0174}$$
$$\text{Total} = 0.7122$$

The portion of load carried by a single 10-ft-long wall will thus be

$$\frac{0.0087}{0.7122} = 0.0122 \text{ or barely more than 1\%}$$

It is therefore reasonable to assume that the 30-ft walls carry the entire load to the core zone. For a single pair of walls constituting one stair plus a rest room tower, the portion of the full lateral load will thus be

$$\tfrac{1}{2} \times \tfrac{3}{4} \times F_x = \tfrac{3}{8} \times F_x$$

Referring to Table 10.10 and Fig. 10.33, the loads for a single tower are:

$$H_1 = \tfrac{3}{8} \times 138.5 = 52 \text{ kips}$$
$$H_2 = \tfrac{3}{8} \times 160.4 = 60 \text{ kips}$$
$$H_3 = \tfrac{3}{8} \times 89.1 = 33 \text{ kips}$$

and the total overturning moment about the base of the wall at the first-floor level is

$$H_1 \times 41 = 52 \times 41$$
$$= 2132 \text{ kip-ft}$$
$$H_2 \times 28 = 60 \times 28$$
$$= 1680$$

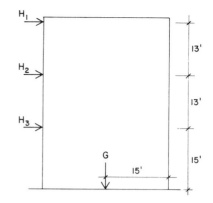

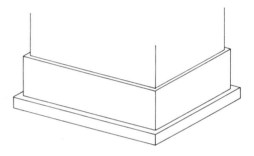

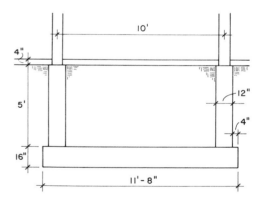

FIGURE 10.33 Loading and form of the stair tower and its foundation.

$$H_3 \times 15 = 33 \times 15$$
$$= \underline{495}$$
$$\text{Total} = 4307 \text{ kip-ft}$$

For the dead load moment that resists this overturn effect we make the following assumptions:

1. The walls are 8-in. concrete block weighing 60 psf of wall surface. For the entire tower the total weight is thus approximately

$$80 \times 41 \times 60 = 196.8 \text{ kips}$$

2. As bearing walls, the tower walls carry approximately 1800 ft^2 of roof or floor periphery, which results in a supported dead load of

$$55 \times 1800$$
$$= 99 \text{ kips/floor or 198 kips total}$$
$$25 \times 1800$$
$$= 45 \text{ kips of roof load}$$

This results in a total load (*G* in Fig. 10.33) of 439.8 kips and a restoring dead load moment of

$$439.8 \times 15 = 6597 \text{ kip-ft}$$

The safety factor against overturn is thus

$$\frac{6597}{4307} = 1.53$$

which indicates that there is no real need for a tiedown force.

There will, of course, be a considerable tiedown developed in the form of the doweling of the wall reinforcing into the foundation. Investigation of the foundation must include concerns for the overturning and sliding of the entire structure and the maximum soil pressure developed by the combination of gravity and lateral loads. These matters are discussed more fully in Sec. 8.6,

where example computations are also given for shear wall foundations similar to that shown in this example.

Because of their relative stiffness, the core walls carry the major portion of the lateral force in this building. If the building is designed only for the lower wind force, it would be possible to use the core alone and to eliminate the exterior shear walls. For seismic load, however, the exterior walls add significantly to the torsional resistance, thus making the core-braced scheme less desirable. Core bracing is common where wind is the critical lateral force, but is less used for seismic resistance.

Example 12, Building E

Figure 10.34 shows an elevation of the east and west walls for Building E with the structural masonry developed as a pierced wall. (See the discussion of Examples 1 and 2 in this section.) The wall is shown here with rows of windows aligned both vertically and horizontally.

Depending on details of the wall construction—most notably, the locations of control joints in the masonry—the wall could be made to behave as a continuous pierced wall or as a series of vertical piers consisting of the solid strips between the windows (see Fig. 10.14). In this case the solid strips between the windows are 44 ft high and only 9 ft wide, an aspect ratio of height to width not very feasible for isolated shear walls.

If the wall is treated as a single, solid piece 44 ft high and 92 ft long, it is extremely stiff as a vertical cantilever. Flexural action would be insignificant and the wall would be designed only for lateral shear plus vertical compression. This process has been illustrated in previous examples, so we will limit this example to some discussion of a few special problems.

For the one-piece, pierced wall, the criti-

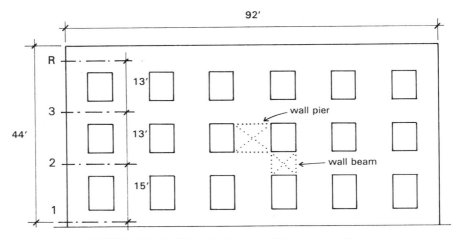

FIGURE 10.34 Building E: elevation of the pierced wall.

cal design elements are the small panels between the windows—both horizontally and vertically. As shown in Fig. 10.34, there are two basic elements. The solid portion between windows in the horizontal row is called a *wall pier* and is subjected to horizontal shearing force. As a flexural member, it is fixed at its top and bottom and functions as a doubly fixed beam. However, this action is not really significant due to the extreme stiffness of the element. It is therefore designed essentially only for the portion of the horizontal force that it carries in shearing action. Determining that portion, however, may be accomplished by considering the wall stiffness due to its dimensions and aspect ratio, as illustrated in the previous examples in this section.

The solid portion of wall between windows in the vertical row is described in architectural terms as the *spandrel*. In the pierced wall the spandrel panels function as *wall beams*, spanning across the window openings. For vertical gravity loads they may truly function as reinforced beams and be designed as such. Under lateral loads they will be subjected to a vertical shearing action, as described in Fig. 4.2.

The general effect of this shearing action is essentially similar to that for the wall piers.

In tension-weak materials, such as concrete, masonry, and plaster, a critical shear action typically results in diagonal tension cracking. Under lateral load from a single direction, the wall piers and spandrels will tend to develop these diagonal cracks. During an earthquake, as the load direction rapidly reverses, the typical form of cracking is an X-shape. This is illustrated in Fig. 10.34. Examples of such cracking are shown in Fig. 4.16.

With a nonstructural wall cladding of stucco or masonry veneer, the best means for reducing X-cracks is to use control joints in the cladding and cushion the cladding in some way from the movements of the structure. If the wall is truly to work as a single-piece, pierced shear wall, however, it will have to be adequately reinforced for the critical shear actions. Figure 10.35 shows some considerations for this reinforcement.

The individual piers and spandrels must actually be reinforced for the diagonal tension forces. However, since the reinforcement can usually only be placed vertically and horizontally in the masonry, the prac-

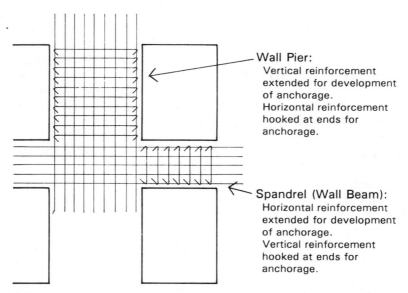

Wall Pier:
Vertical reinforcement
extended for development
of anchorage.
Horizontal reinforcement
hooked at ends for
anchorage.

Spandrel (Wall Beam):
Horizontal reinforcement
extended for development
of anchorage.
Vertical reinforcement
hooked at ends for
anchorage.

FIGURE 10.35 Layout of the reinforcement for development of maximum strength of the stiff piers and spandrels.

tical solution is to place identical vertical and horizontal amounts of reinforcement. This reinforcement must be fully developed within the boundaries of the piers and spandrels, which requires either extension into the adjacent wall or hooking the ends of the bars, as shown in Fig. 10.35. Some of the vertical bars in the piers and some of the horizontal bars in the spandrels will undoubtedly be continuous throughout the wall construction.

To reinforce the wall piers and spandrels most effectively, it would be desirable to place reinforcement in an X-shape pattern in direct resistance of the diagonal tension forces. Although not generally feasible in masonry walls, this is possible in cast concrete walls and is discussed in Sec. 10.3.

If this wall is developed as a shear wall, it will surely also be developed as a vertical support for the floor and roof structures. This will involve some attachment of the wall to those constructions for transfer of the loads to the wall. In developing the construction details for intersections of the wall and the horizontal construction, it is desirable to achieve three goals.

1. No bending in the wall due to the gravity loads; that is, the supported loads should be placed axially on the wall.

2. Attachment of the horizontal construction to achieve the lateral force transmission to the shear wall.

3. Attachment of the horizontal construction to achieve an anchorage of the wall against outward movement (generally described as *horizontal anchorage*).

These goals must be achieved by means that are compatible with the many other requirements for detailing of the wall construction, including interior and exterior finishes, water control, fire resistance, and ordinary requirements for the particular form of masonry.

10.3 CONCRETE SHEAR WALLS

Concrete shear walls represent the single strongest element for resistance to lateral shear force. When used for subgrade construction (basement walls) or for extensive walls in low-rise buildings, they indeed provide great stiffness and strength for the shear wall tasks. Their greatest strengths are generally developed with sitecast construction (concrete poured in forms at the site in the desired position). However, large precast walls are also capable of considerable bracing when properly developed with the total structure.

Of critical concern for concrete walls—and all reinforced concrete, for that matter—is the proper detailing of the steel reinforcement. Recommended details for this are specified in building codes and in the publications of various organizations in the concrete industry, including the American Institute of concrete (ACI), Portland Cement Association (PCA), and Concrete Reinforcing Steel Institute (CRSI).

Much of what has been said about masonry construction in the preceding section also generally applies to concrete construction. The structures produced are heavy and stiff and weak in tension. Required seismic shear forces are a maximum due to the combination of weight and overall stiffness of the structures. For earthquakes, concrete shear walls can work well, but proper detailing of the construction is very important. When used in combination with other structures (wood and steel framing, for example), adequate anchorage or effective separation must be used to provide for the differences in seismic movements.

For wind, the heavy, solid, stiff structure (concrete or masonry) is often an advantage, providing an anchor for lighter elements of the construction. Indeed, excessive weight is of opposite concern generally for wind and seismic effects. Concrete and masonry foundation walls are typically the direct anchors for structures of wood and steel.

An advantage of reinforced concrete over reinforced masonry (particularly concrete block construction) is the greater flexibility with regard to placement of the steel reinforcing bars. These can only be placed in the modular voids and mortar joints of masonry walls, but can be placed with more freedom in the concrete mass. Furthermore, the steel rods can only be vertical or horizontal in masonry, while the bars can take any direction in the concrete mass. In Fig. 10.35, for example, the bars can only be in the vertical/horizontal grid shown in the concrete block construction. However, additional diagonal bars could be used in a concrete wall, adding considerably greater strength in direct resistance to the diagonal tension stresses.

Most of the example buildings in the preceding section could as well be developed with concrete shear walls, either sitecast or precast. Modular dimensions would, of course, not strictly be required, and the full potential of the cast material could be explored for many sculptural purposes. Nevertheless, the heavy, stiff, tension-weak materials are essentially the same, and many of the concerns for structural behavior would be the same.

11

TRUSS SYSTEMS

General considerations for use of trussing (triangulated framing) for lateral bracing are treated in Sec. 5.2. In this chapter we present examples of various uses of trussing.

11.1 VERTICAL TRUSSES

The use of trussing for development of the vertical bracing elements in a lateral-force-resistive system is most commonly achieved by adding diagonal members to a beam and column framework. The following examples show some simple applications of this method.

Example 13, Building A

The general form of Building A is shown in Fig. 10.1, together with some details for a light wood frame structure. Various other constructions are possible for this build-

ing, including the masonry walls illustrated in Sec. 10.2. In this example we propose the use of a light steel frame, with lateral bracing achieved with X-bracing formed in some of the panels of framing in the exterior walls.

It is possible that much of the appearance of this building might be the same as the constructions illustrated previously, with the X-braces concealed within hollow wall construction. Thus planning for the X-bracing would be somewhat similar to planning for shear walls in the same locations. The X-braces might be left exposed for a general "high-tech" architectural style, but this typically occurs less often than the concealed structure.

Figure 11.1 shows a possible development for such a bracing system. For resistance in the north–south direction the east and west walls are braced with the diagonals in the same walls that were used as shear walls in Example 1. In the other

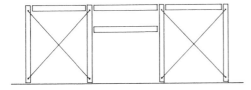

(a) east and west walls

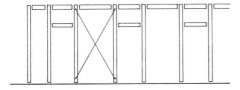

(b) north and south walls

FIGURE 11.1 Braced frame scheme: Example 13, Bulding A.

direction, however, only two walls on each side (north and south) are used so that there are a total of four walls with diagonals (see Fig. 11.1b). It is assumed that the diagonal members and their connections to the steel frame can be adequately developed with only four trussed bays in each direction.

Some considerations for the design of this system are as follows:

1. While still popular for wind bracing, this system is now held in disfavor for seismic resistance. A review of the R_W factors in *UBC* Table 16-N will show that of all the lateral bracing systems described, this system (1.3 in the table) has the lowest factor (4) and thus requires the largest base shear for design. The required design force is twice that for a plywood shear wall, two-and-one-half times that for an eccentric braced frame (EBF), and three times that for a ductile rigid frame (SMRF). For all of the code's penalties, however, the system may

still be feasible for particular situations, and is of course still viable when wind resistance is critical. It should also be noted that the 1994 *UBC* does not permit trussed concrete frames in seismic zones 3 and 4.

2. Connections stressed during lateral force actions should be tight and non-loosening (generally called *positive*), for the buffeting during wind storms and shaking during earthquakes will tend to shake them loose. Rigid frames will normally have such connections, but the simple pinned connections used routinely for trussing do not always have this character.

3. Although trussed structures are normally quite stiff, there are a number of things that may contribute to lateral deflection. These include the shortening and lengthening of the truss members, the deformation of connections, and the deformation of column anchorage. For the X-braced structure, a major contribution is that of the tension elongation of the X members, as these will be quite highly stressed and are usually the longest members of the truss system.

4. Planning of trussing is often a major architectural problem. Although trussing diagonals may be exposed if fire codes permit, they are frequently incorporated in wall constructions. Locating solid walls at points that are strategically useful to the lateral bracing system may be difficult.

For the purpose of this example, let us assume that the building weight that generates seismic force is the same as that in the wood framed structure in Example 1 of Sec. 10.1. Thus, from the preceding discussion, regarding the R_W factors, the seismic forces for this structure are twice those determined for Example 1. On this basis, we consider the situation for the bracing in the north and south walls as follows:

$$\frac{\text{total east–west}}{\text{seismic force}} = 27.6 \text{ kips}$$

$$\frac{\text{load to north}}{\text{and south walls}} = \frac{27.6}{2} = 13.8 \text{ kips}$$

With two bents in the wall, the load per bent is one half of the total load in the wall, or 6.9 kips. Assuming the bent layout as shown in Fig. 11.2, the tension force in the diagonal is thus

$$T = \frac{1}{0.53} \times 6.9 = 13 \text{ kips}$$

If a round rod of A36 steel is used (with allowable tension stress of 22 ksi), the area of the rod cross section must be

$$A = \frac{T}{1.33F_t} = \frac{13}{1.33 \times 22} = 0.44 \text{ in.}^2$$

which could be satisfied with a $\frac{3}{4}$-in. diameter rod.

However, *UBC* Sec. 2211.8.3.1 requires that the bracing connection be designed for a force that is $3(R_W/8)$ times that determined from the seismic effect. If an ordinary threaded end is used, the allowable stress at the threads is 0.33 times the ultimate stress limit for the steel. Assuming a low value of 58 ksi for the A36 rod, the required tensile stress area at the threads is thus

$$\text{For } T: 3 \left(\frac{R_W}{8}\right)T = (3)\left(\frac{4}{8}\right)(13) = 19.5 \text{ kips}$$

$$A = \frac{19.5}{(1.33)(0.33 \times 58)} = 0.77 \text{ in.}^2$$

This requires a rod with $1\frac{1}{4}$ in. diameter, with a net tensile stress area of 0.969 in.^{2} and a gross area of 1.227 in.^{2} (from the *AISC Manual*, Ref. 11).

If this size rod is used, the lateral deflection of the top of the bent due only to the stretching of the rod is (see Fig. 5.9).

$$d = \frac{TL}{AE \cos \theta} = \frac{13 \times (10 \times 12)/0.53}{1.227 \times 29,000 \times 1/0.53}$$

$$= 0.044 \text{ in.}$$

This indicates a very small deflection, although the actual deflection will be larger due to various other deformations in the frame and the connections. Still the deflection will be small—probably somewhere between that of the plywood shear wall and the masonry shear wall.

As discussed in Sec. 5.2, we do not recommend the use of the tension-only, X-braced frame for seismic zones 3 and 4, unless the braces are designed for a very low stress and are made quite stiff. This type of stiffened brace may also be produced by designing for compression, rather than tension, using the requirements for stress reduction, as given in *UBC* Sec. 2211.8.2.2. Designed for tension or compression, the diagonals may consist of various members, depending on their length, the required forces, and details of the construction.

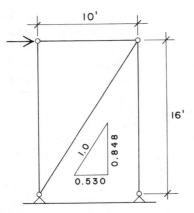

FIGURE 11.2 Layout of the east–west bent: Example 13.

The trussed bents must also be designed for overturning and sliding due to the lateral forces. Since the bent columns and horizontal members are likely to be parts of the gravity load resisting systems, the design for these effects must be done in combination with the load effects from gravity.

As has been discussed previously, the X-braced bent is still a valid bracing system for wind, even when constituted with light steel rods for diagonals. For major seismic forces, however, it is now used mostly only with quite stiff diagonals and is in many cases replaced by various forms of eccentric braced framing.

Example 14, Building D

This example presents a variation of Building D in which a steel column structure is used for the front (south) wall of the building. A typical reason for the use of such a structure is the desire for more open space in the wall than can be achieved with a shear wall system or a braced frame system.

Stability of the column system for lateral shear in the plane of the wall for this structure requires the development of moment in the columns. Options include the following:

> Fix the column bases, creating vertical cantilevers (as for shear walls).
> Develop moment-resisting connections at the tops of the columns, producing a rigid frame with the horizontal steel framing.
> Or—do both.

The proposed scheme, shown in Fig. 11.3, consists of a continuous trussed column bent, with parallel-chord trusses rigidly connected to the tops of the columns. We will assume that this system is used only

for the south wall, and that the rest of the walls are braced by shear walls or concentric braced frames.

For lateral load resistance, a problem that arises with such a mixed system is what to use for the R_W factor for the total shear on the building. This does not qualify as a *dual system*, since the single wall bracing does not interact with other systems in its own plane (see discussion in Sec. 5.5). Distribution of the total shear to the individual bracing elements may be done on the basis of relative stiffness or peripheries, but the problem remains as to what total shear should be distributed.

Inspection of *UBC* Table 16-N yields the following values for the R_W factor:

> Plywood shear walls: 8
> Masonry shear walls: 6
> Frame with masonry walls: 8
> Steel concentric braced frame: 8
> Steel eccentric braced frame: 10
> Ordinary steel rigid frame: 6

When in doubt it is usually best to use the lowest R_W value (yielding the highest value for design shear).

Let us assume that the construction is in general the same as for Example 10 in Sec. 10.2, except for the use of the trussed steel bents on the south wall. That involves a mixed system with masonry shear walls and what may arguably be classified as an "ordinary steel rigid frame," for both of which the R_W factor is 6. Using this factor for the building shear will produce the same east–west shear as in Example 10 in Sec. 10.2: a total of 76.5 kips. If a simple peripheral distribution is assumed, the total load to the six-span front bent is thus one-half of this, or 38.25 kips.

If the columns are pin based, and the intermediate columns are assumed to be 50% stiffer than the end columns, the bent

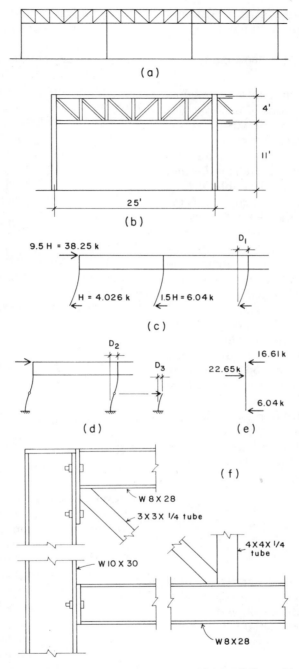

(a)

4'

11'

25'

(b)

9.5 H = 38.25 k

D₁

H = 4.026 k   1.5H = 6.04 k

(c)

D₂

D₃

(d)

16.61 k

22.65 k

6.04 k

(e)

(f)

W 8 X 28

3 X 3 X ¼ tube

W 10 X 30

4 X 4 X ¼ tube

W 8 X 28

FIGURE 11.3 Example 14, Building D: (*a*) layout of the south wall bents; (*b*) form of the single bent unit; (*c*) bent loading; (*d*) assumed function of the bents; (*e*) column loading; (*f*) bent details.

action is as shown in Fig. 11.3c. The moment on an intermediate column is thus

$$M = 6.04 \times 11 = 66.4 \text{ kip-ft}$$

Although the column must be designed for the combined actions of bending plus axial compression due to gravity forces, the bending due to the lateral load is by far the major action. If a section is chosen that is slightly stronger than required for this moment, it will serve as a reasonable first try. In fact, because the critical concern is the more likely for lateral deflection, stress concerns may be secondary. Let us try a column consisting of a $W 10 \times 30$ wide flange section, which has a moment capacity approximately one third larger than that required for the computed lateral moment alone. We then consider the lateral deflection (D_1, in Fig. 11.3c) based on the cantilever action of the intermediate column. Although there is also some lateral deflection due to the rotation at the bottom of the truss, the general stiffness of the bent will be primarily indicated by the cantilever action of the column. Using the $W 10 \times 30$ with an I_x of 170 in.[4], and the deflection formula for a simple fixed cantilever with a concentrated load at its end, we determine the deflection as

$$D_1 = \frac{PL^3}{3EI} = \frac{6.04 \times (11 \times 12)^3}{3 \times 29{,}000 \times 170}$$
$$= 0.939 \text{ in.}$$

Although there is no generally accepted limit for such deflection, this compares reasonably with a value of one-half of 1% of the story height ($0.005h$) that is frequently used for frame drift. In this case

$$0.005h = (0.005)(15 \times 12) \times 0.90 \text{ in.}$$

In this situation, however, it is possible that the drift of approximately 1 in. may

cause problems with the distortions resulting in the wall construction. In addition, with the mixed bent and shear wall system, it is usually desirable to maintain relatively rigid bents for a more favorable load sharing with the stiff shear walls. We may therefore consider stiffening the bent by one of two means. The first is to simply choose a stiffer section. Although the $W 10 \times 30$ is close to a square and generally of the usual form for a column, we may chose a section that is more of the proportions of a beam. This is not unreasonable as the major action is that of the lateral bending moment. Care should be used, however, to avoid having too low a stiffness on the minor axis (r_y) for the column action.

Let us consider the use of a $W 16 \times 31$ section, which has an I_x of 375 in.[4] and an r_y of 1.17. The increase in stiffness will reduce the deflection to

$$D_1 = \frac{170}{375} \times 0.939 = 0.426 \text{ in.}$$

The r_y value of the $W 16 \times 31$ is slightly lower than that of the $W 10 \times 30$, but is possibly still not a critical concern, as the axial load is probably quite low.

Another means for stiffening the bent is to develop a moment-resistive connection at the column base, as shown in Fig. 11.3d. If we assume the top and bottom of the column to be fully fixed, the column will inflect at midheight. The cantilever length of the column is thus one half the height and for comparison with the pin-based bent, we observe the following:

$$D_2 = \text{twice the deflection}$$
$$\text{of the half-height column}$$

$$D_3 = \frac{P(L/2)^3}{3EI} = \frac{PL^3}{24 \ EI} = \frac{1}{8}D_1$$

$$D_3 = 2D_3 = \tfrac{1}{4}D_1$$

This indicates a significant stiffening effect, although in reality the effect is lessened by some rotation at the top and at the base due to distortions in the connections. Nevertheless, the fixed base is a major means for stiffening the frame. It does, of course, require a foundation capable of developing the necessary resistance to the moments and shears caused by the fixed bases.

Investigation of the equilibrium of the intermediate column indicates the development of forces in the chords of the trusses as shown in Fig. 11.3e. To develop the lateral resistance of the bents, the trusses must be assembled as a continuous system. If we assume an approximate gravity load of 1500 lb/ft and a maximum moment of $wL^2/10$, the maximum chord force due to gravity load is

$$T = C = \frac{wL^2}{10d}$$

$$= \frac{(1.5)(25)^2}{10 \times 4} = 23.4 \text{ kips}$$

Adding this to the force determined in the bottom chord due to lateral load (Fig. 11.3e) produces a total of approximately 46 kips. Although the top chord of the truss will most likely be braced laterally by the roof system, the bottom chord may be unbraced for the entire distance between columns. With these assumptions, a possible choice for the bottom chord is a $W\,8 \times 28$.

Some possible details for the truss assembly are shown in Fig. 11.3f. The trusses are shop assembled with web members of tubular sections that are cut to fit and directly welded to the flanges of the wide flange chords. Column connections are achieved with plates welded to the ends of the chords of the trusses and field bolted to the column flanges. A cap plate at the top of the column and web stiffening plates at the bottom chord of the truss should be used to brace the thin web and flanges of the column.

Example 15, Building E

If Building E is built with a column and a beam frame, it is possible to use trussed bracing for lateral resistance. The planning of such bracing is in many ways similar to that required for the shear wall schemes, and planes of trussed bracing could be substituted for some of the shear walls indicated in the preceding example. Because the trussing is produced ordinarily by using the columns and beams of the gravity resistive system and simply adding diagonals, the planning of both systems must be tightly coordinated.

Figure 11.4 shows some possibilities for use of X-bracing for Building E. In Fig. 11.4a is shown the use of a large single brace in place of the 30-ft core shear walls. If overturning and anchorage can be solved, two of these trussed bents can easily brace the building in place of the shear wall towers in Example 11 in Sec. 10.2.

Figure 11.4b shows the use of a series of perimeter bents corresponding to the locations of the exterior shear walls shown in the plan in Fig. 10.30a. Again, if overturning can be dealt with, it would be possible to use fewer braced bents, with not every wall unit braced. As with the systems using perimeter shear walls, the location of perimeter X-bracing is significant to the development of torsion and an advantage in seismic force resistance.

A particularly strategic location for either shear walls or trussed bents on a building such as this is at the building's exterior corners. In Fig 11.4c, d, and e are three possibilities for the use of perimeter trussing at the building corners. Figure 11.4c shows a single width vertical bent—the simplest and most direct solution. Figure 11.4d shows

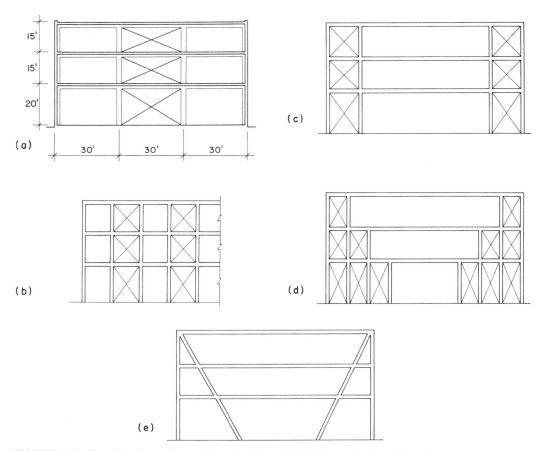

FIGURE 11.4 Considerations of trussed bracing: Example 15: (*a*) core bracing; (*b*) perimeter bents; (*c*) corner bents; (*d*) stepped corner bents; (*e*) multistory corner diagonal brace.

a scheme in which the number of bents is increased by one for each lower story. Although this adds some complexity, it produces a form which is reflective of the relative magnitude of the shear and the overturning effects at each level.

Figure 11.4*e* shows the use of a single large diagonal that extends from the corner at the roof level down through all three stories, producing a single triangular form. This is actually a very direct solution that is commonly used in structures other than those for buildings.

In most cases trussed bracing would not be used for both core and perimeter loca-

tions in the same building. However, there is nothing wrong with such an application if some reasonable means can be used to determine the distribution of forces to the individual bents. As was indicated with the examples of one-story buildings, it may also be possible to mix systems of shear walls and some trussed or rigid frame bents.

Trussed bracing in the form of X-bracing or other concentric systems is still widely used for wind bracing. However, for seismic design the favored systems are now eccentric bracing or heavy X-bracing (as opposed to tension only, light X-bracing). This issue is discussed in Sec. 5.2.

Example 16, Building E

The general form of building E is shown in Fig. 10.28. The plan as shown has a regular 30-ft grid of columns, which lends itself to a beam-and-column framing system. If such a system is developed with structural steel members, a possible solution is to use a core bracing system with vertical planar trussed bents placed at the locations of some of the core walls. A partial plan of a steel framing system for the core area is shown in Fig. 11.5. The openings on the plan are provided for the stairs and elevators.

Referring to Fig. 11.5, it may be noted that there are some extra columns in the framing plan at the location of the stair, rest room, and elevator walls. These columns will be used in conjunction with the regular columns and some of the horizontal framing to define vertical planes of framing for the development of the truss bracing system shown in Fig. 11.6. These frames will be braced for resistance to lateral loads by the addition of diagonal X-braces. For a simplified design, we will consider the diagonal members to function only in ten-

sion, making the vertical frames consist of statically determinate trusses. There are thus four vertically cantilevered trusses in each direction, placed symmetrically at the building core.

With the symmetrical building exterior form and the symmetrically placed core bracing, this is a reasonable system to use in conjunction with the horizontal roof and floor structures to develop resistance to horizontal forces due to wind or seismic actions. We will illustrate the design of the trussed bents for wind.

The wind load for this building was determined for Example 11 in Sec. 10.2. The basis for determination of the wind forces is illustrated in Fig. 10.31. The accumulated forces noted as H_1, H_2, and H_3, in Fig. 10.31 are shown applied to one of the vertical trussed bents in Fig. 11.7a. For the east–west bents, the loads will be as shown in Fig. 11.7. These loads are determined by multiplying the edge loadings for the diaphragms as shown in Fig. 10.31 by the 92-ft overall width of the building on the east and west sides. For a single bent, this total

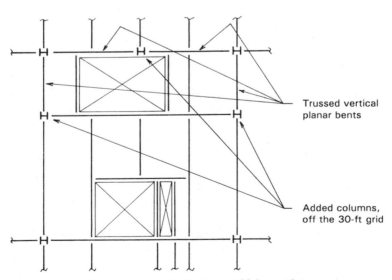

Trussed vertical
planar bents

Added columns,
off the 30-ft grid

FIGURE 11.5 Plan of revised core with bent columns.

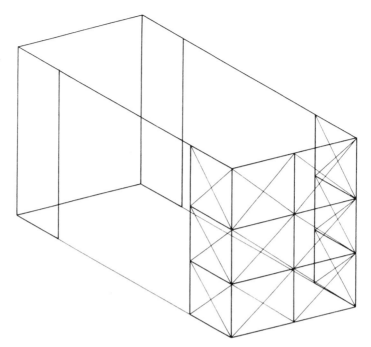

FIGURE 11.6 Development of the core-bracing system for lateral load resistance.

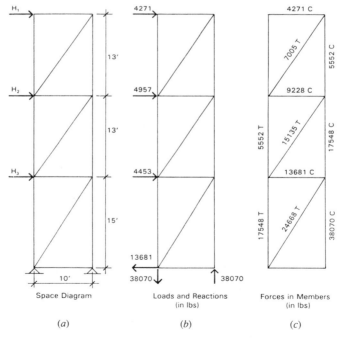

FIGURE 11.7 Investigation of one of the east–west trussed bents.

force is divided by four. Thus for H_1 the bent load is determined as

$$H_1 = 185.7 \times 92 \div 4 = 4271 \text{ lb}$$

Analyzed as a truss, ignoring the compression diagonals, the resulting internal forces in the bent are as shown in Fig. 11.7c. The forces in the diagonals may be used to design tension members, using the usual one-third increase in allowable stress. The forces in the vertical columns may be added to the gravity loads and checked for possible critical conditions for the columns previously designed for gravity load only. The anchorage force in tension (uplift) should be dealt with as for the shear walls in Chapter 10. This may reveal the need for tension-resistive anchor bolts and require some special considerations for the foundations.

The horizontal forces must be added to the beams in the core framing and an investigation done for the combined bending and axial compression. This can be critical, since beams are ordinarily quite weak on their minor axes (the y-axis), and it may be practical to add some additional horizontal framing members to reduce the lateral unbraced length of some of these beams.

Design of the diagonal members and of their connections to the frame must be developed with consideration of the form of the frame members and the general form of the wall construction that encloses the steel bents. Figure 11.8 shows some possible details for the diagonals for the bent analyzed in Fig. 11.7. A consideration to be made in the choice of the diagonal members is the necessity for the two diagonals to pass each other in the midheight of a bent level. If the most common truss members—double angles—are used, it will be necessary to use a joint at this crossing, and the added details for the bent are considerably increased.

An alternative to the double angles is to use either single angles or channel shapes.

These may cross each other with their flat sides back-to-back without any connection between the diagonals. However, this involves some degree of eccentricity in the member loadings and the connections, so they should be designed conservatively. It should also be carefully noted that the use of single members results in single shear loading on the bolted connections.

11.2 HORIZONTAL TRUSSING

The general use of trussing for horizontal bracing elements is discussed in Sec. 5.2. A common reason for using horizontal trussing is the absence of an adequate horizontal decking that can be used for diaphragm action.

Example 17, Building F

Figure 11.9 shows a large one-story building with a complex roof that consists of a regularly spaced series of sky windows. The roof structure consists of a series of clear-spanning trusses that define the planes for the windows. The remainder of the roof consists of sloped portions of decking that are supported by the top of one truss and the bottom of the next truss. This system has the common name *sawtooth roof*, due to its appearance in cross section, and has been widely used for large industrial facilities that are able to use the advantage of the potential daylighting and natural ventilation provided by the extensive windows.

Due both to its steep slope and its lack of continuity, the roof decking is not able to perform diaphragm actions for the building as a whole. It can, however, contribute to the lateral bracing of the spanning trusses. The usual diaphragm action in the general plane of the roof structure is provided by the X-bracing in the structural bays adjacent to the outside walls (see Fig. 11.9). This diagonal framing works with the bottom

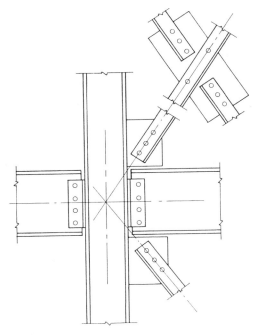

FIGURE 11.8 Joint details for the braced bent.

chords of the spanning trusses and the horizontal framing in the wall structures to constitute a series of trusses at the building edges.

Because of the length of the building, the plan indicates the use of a center row of columns that are used to develop a three-span trussed bent that serves to break the roof into two units for the horizontal span for lateral forces. Similar trussed bents are used at the building ends. It is possible that trussed bents might also be used along the long exterior walls (see Example 15 in Sec. 11.1), and the entire building lateral-force-resistive system would thus be developed with trussing. However, the exterior walls might also be developed with construction capable of the necessary shear wall actions.

We illustrate here only the development of the horizontal X-bracing along the long walls for resistance to wind load in the north–south direction. The loading condi-

tion for this is illustrated in Fig. 11.10, and the investigation of one of the X-braced trusses is shown in Fig. 11.11.

The member forces shown in Fig. 11.11 are relatively small for structural steel, and the X-members could be quite small with regard only to the required tension resistances. However, there are several considerations that must be made in selecting the X-bracing.

1. The general form of the construction of the spanning trusses, regarding member shapes and joint details.

2. A possible desire to reduce lateral movements by lowering the tension stress in the X-members. Thus slightly larger members may be used to reduce strain and lengthening and the deflection of the X-braced trusses.

3. The problem of sag, since the horizontal X-members are quite long (approximately 28 ft), and if essentially designed for tension will typically have low bending resistance on the horizontal span. This might be alleviated by using some suspension hangers from the roof construction above the centers of the X-braced bays.

Although we have shown the investigation of the horizontal trusses for their particular task, they are parts of the general structural system and their participation for other purposes—such as providing lateral bracing for the spanning trusses or support for lighting—will also affect considerations for selection of members and development of construction details.

In this example, with the use of the center trussed bents, the loading and span conditions for these trusses are actually the same as for the two-span trusses on the long sides. In other situations, the span

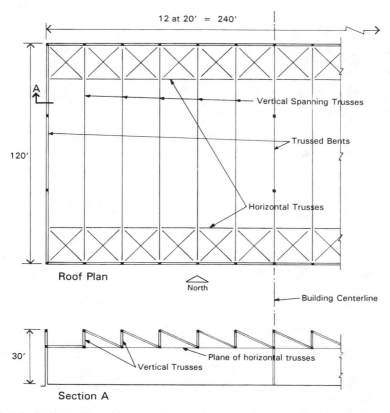

12 at 20' = 240'

A

120'

Vertical Spanning Trusses

Trussed Bents

Horizontal Trusses

Roof Plan

North

Building Centerline

30'

Plane of horizontal trusses

Vertical Trusses

Section A

FIGURE 11.9 Building F: roof framing plan layout and section through the sawtooth roof structure.

conditions may be different, but the general construction used for the trusses on the longer side would probably also be used here.

11.3 TRUSSED TOWERS

The tower-like structure is defined not so much by height or overall size as by its proportions of height to width. This height-to-width ratio is the surest measure of the development of critical structural responses. The most notable concern is typically for overturn, or simply the maximum moment at the base of the vertical cantilever.

On this basis, tower-like actions can be ascribed to slender poles with buried bottom ends (see Secs. 8.7 and 13.3), freestanding walls (see Sec. 6.2), and very tall shear walls (see Sec. 8.6). In fact, any structure with an aspect ratio of height to base width of more than 3 or so will begin to demonstrate tower-like responses to lateral forces. When the ratio exceeds 5 or so the critical problem of overturn may be expected to be predominant and a fixed base of some kind will be required.

As the actual height dimension increases, however, lateral forces are affected by the dimension itself. The actual height will determine the nature of wind pressures, as described in Chapter 1, and the general overall dimensions of a structure will relate to dynamic properties that will modify responses to earthquakes.

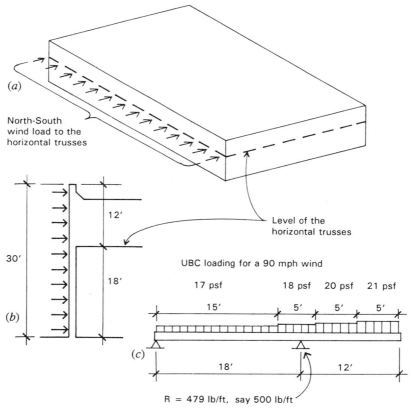

North-South
wind load to the
horizontal trusses

Level of the
horizontal trusses

UBC loading for a 90 mph wind

17 psf   18 psf   20 psf   21 psf

R = 479 lb/ft, say 500 lb/ft

FIGURE 11.10   Wind load distribution to the horizontal trusses.

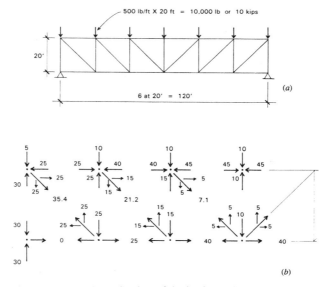

500 lb/ft X 20 ft = 10,000 lb or 10 kips

6 at 20' = 120'

FIGURE 11.11   Investigation of the horizontal bracing trusses.

238

While the actual size dimensions of a tower will establish various responses to lateral loads, the greater concern may be for the ratio of the height to the width of the base. In the case of a freestanding wall, this may apply to the width of the footing rather than simply to the dimensions at the bottom of the wall. If a wall is well anchored to its footing, the critical overturn may be that for the footing-to-soil interface.

A form commonly used to provide stability where overturn due to lateral effects is critical is one that achieves a spread base; examples being the tripod and the pyramid. Trussed towers often use this tapered form, as seen in the common form of support towers for electric power transmission lines (Fig. 11.12). A famous structure with this form is the Eiffel Tower in Paris.

Very tall high-rise buildings sometimes approach the aspect ratio of a tower-like structure. However, more common occurrences in buildings are the forms of individual shear walls, braced frame bents, or the general core-bracing structures. In Example 16 in Sec. 11.1, for instance, the building as a whole is quite squat in form, with a height/width ratio of less than 0.5 in the short dimension of the plan. However, the individual trussed towers in the core bracing have a height/width ratio of about 3, which is beginning to approach some tower-like action. The investigation process for that structure as illustrated in the example is a general description of the investigation for any trussed tower, so we will not present further examples here.

11.4 MIXED SYSTEMS

Trussed elements or some form of trussing may be used for part of a lateral bracing system, working with various other structural elements. The truss generally has a relative stiffness approaching that of a shear wall and may share loads with shear walls effectively in some cases. Thus, in the case of Example 17 in Sec. 11.2, the short end walls may be braced with shear walls

FIGURE 11.12 Highly familiar trussed tower structure: the cross-country electrical power transmission system. Forms derived on an almost purely functional basis for structural efficiency.

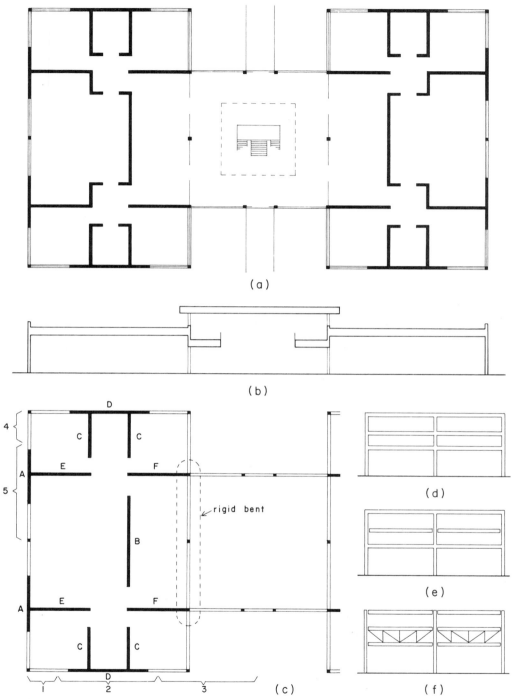

FIGURE 11.13 Example 1B, Building G: (*a*) building plan; (*b*) building section; (*c*) loading zones for the shear walls and bents; (*d*), (*e*) alternate forms for the rigid bents; (*f*) form of the trussed bent.

that share the north–south loading with the central trussed bent. Indeed, the trussed bent, as discussed in Sec. 5.2, is itself a mixed creature, having some of the nature of a truss and some of a rigid frame.

The following example illustrates a case where a separate vertical bracing element is utilized in combination with a shear wall system.

Example 18, Building G

The plan and section in Fig. 11.13*a* and *b* show a building with two symmetrical wings connected by a narrowed central portion. The presence of solid walls permits the consideration of the development of a shear wall system for lateral resistance, except for the center portion with respect to north–south loading. A possible solution in this situation is the use of a rigid frame bent at the east and west sides of the center section, as shown in the plan in Fig. 11.13*c*.

Unless the roof deck diaphragms are exceptionally rigid (as in the case of a poured-in-place concrete structural deck), the usual method of determining the distribution to the mixed vertical elements in this situation would be on the basis of peripheral distribution. For this distribution, the zones are as shown in Fig. 11.13*c*.

As the section shows, there are three upper levels of framing in the bent. The bents may thus take one of the two forms shown in Fig. 11.13*d* and *e*. In (*d*) the frame is developed as a three-story rigid bent, whereas in (*e*) the roof of the wings is developed with shear-type connections only and the bent is only two stories.

Figure 11.13*f* illustrates an alternative means of achieving the braced bent, that is, through the use of a trussed bent, which may be simpler for fabrication and more economical in general if the truss depth is adequate. Discussion of such a bent is presented in Example 2 in Sec. 11.1 and illustrated in Fig. 11.3. In this case, if the trussing is incorporated in a solid portion of the wall, it offers no intrusion in the architectural form or detail of the building.

As with other systems involving the mixing of shear walls and bents, it is necessary for the bents to be made quite stiff. Otherwise, the peripheral distribution will not be valid. The more rigid the shear walls, the more this condition is critical. If the walls are of reinforced masonry or concrete, and are quite long in plan with respect to their heights, the mixing for a peripheral distribution may be questionable.

We will not proceed with any computations or design of the elements for this example. Other examples show the methods for using the peripheral distribution and the design of shear walls and multistory rigid bents.

12

MOMENT-RESISTIVE FRAMES

Moment-resistive frames are assemblages of connected linear members in which both the members and the connections are capable of resistance to bending moments. The general considerations for such frames are discussed in Sec. 5.3. In this chapter we present some examples of moment-resistive frames (also called *rigid frames*) used for the vertical elements in lateral load resistive systems for buildings.

Two special cases of moment-resistive frames are the trussed bent, which is illustrated in Example 14 in Sec. 11.1, and the eccentrically braced frame, discussed in Sec. 5.4. Although present codes and references now commonly use the term *moment-resistive frame*, for brevity we will use the older term *rigid frame* in the following discussion.

12.1 VERTICAL PLANAR BENTS

One application of the rigid frame is for the creation of a planar bent, generally consist-

ing of the vertical supporting columns and horizontally spanning beams of a framing system. If connected with moment-resistive joints, such a frame can develop rigid frame actions. It can thus perform the same lateral-load-resisting functions as those of a shear wall or a planar trussed bent.

Example 19, Building A

This example uses the simple one-story box-shaped building, first described in Example 1 of Sec. 10.1 and shown in Fig. 10.1. Figure 12.1 shows a variation for Building A in which the roof structure consists of a series of rigid frame bents. There is a bent at each of the building ends and at 16-ft 8-in. intervals in the center portion of the building.

For this example the building section is slightly modified and the parapet is eliminated; thus the exterior wall framing is assumed to span the full wall height and the

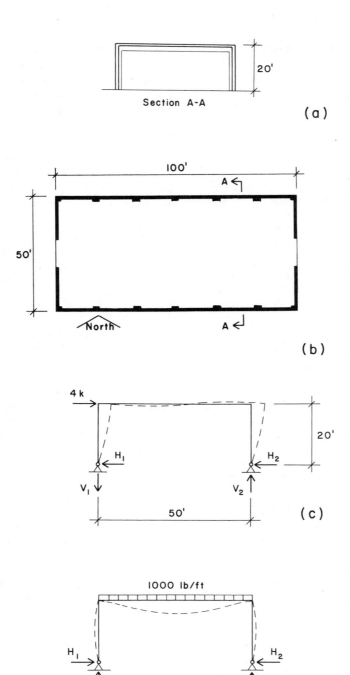

FIGURE 12.1 Example 19, Building A: (*a*) building section; (*b*) building plan; (*c*) lateral loading of the bents; (*d*) gravity loading of the bents.

load from wind to the edge of the roof diaphragm is modified to

$$\text{total } H = 27 \text{ psf} \times 100 \times \frac{20}{2}$$

$$= 27,000 \text{ lb or } 27 \text{ kips}$$

If the roof diaphragm is assumed to be reasonably stiff, it tends to distribute the loads to individual bents in proportion to their stiffness (see the discussion of stiffness of horizontal diaphragms in Sec. 6.1). In this case the end bents would be made quite stiff so as to have limited lateral deflection, which could cause problems for the end wall construction. Thus, although they carry only one half of the gravity load on the interior bents, they would most likely be of approximately equal size and stiffness. If this is the case, the load on an individual bent is

$$\frac{\text{total diaphragm load}}{\text{number of bents}} = \frac{27}{7} = 3.86 \text{ kips}$$

We assume this to be the case and will design the bents for a 4-kip horizontal force, applied at the top of the column.

The seismic forces in this case will be even lower than those for Example 1, due to the higher value of R_W for the rigid frame; thus the wind load will be critical for design of the bents.

Figure 12.1c shows the free-body diagram of the bent with the loads and reactions. The deflected shape under load is shown by the dashed line. Although this problem is essentially indeterminate, if we assume the bent to be symmetrical, we may reasonably assume the two horizontal reactions to be equal. In any event the vertical reactions are statically determinate and may be determined as follows:

$$V_1 = V_2 = \frac{4 \times 20}{50} = 1.6 \text{ kips}$$

On the basis of this analysis for the reactions, the distribution of internal forces is shown in the illustrations in Fig. 12.2. These forces must be combined with the forces caused by the gravity load to determine the critical design conditions for the bents.

Figure 12.1d shows the bent as loaded by a uniform load of 1000 lb/ft on top of the horizontal member. This effect is based on an assumption that the roof framing delivers an approximately uniform loading with a total dead plus live load of 40 psf as an average for the roof construction, including the weight of the horizontal bent member. As with the lateral load, the vertical reactions may be found on the basis of the bent symmetry to be

$$V_1 = V_2 = \frac{1 \times 50}{2} = 25 \text{ kips each}$$

Determination of the horizontal reactions in this case, however, is indeterminate and must consider the relative stiffness (I/L) of the bent members. For the pin-based columns it will be found that the horizontal reactions will each be

$$H = \frac{wL^3 I_c}{8 I_g h^2 + 12 I_c h L}$$

In this calculation we may use the relative, rather than actual, values of the column and girder stiffness (I_c and I_g in the formula). If we assume the girder to be approximately 1.5 times as stiff as the column, the horizontal reaction will be

$$H = \frac{(1)(50)^3(1)}{(8)(1.5)(20)^2 + 12(1)(20)(50)}$$

$$H = \frac{125,000}{4800 + 12,000} = 7.44 \text{ kips}$$

With these values for the reactions the free-body diagrams and distribution of in-

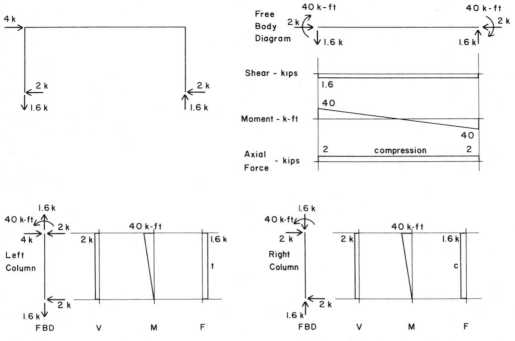

FIGURE 12.2 Investigation of the bents for lateral load.

ternal forces for the gravity loading are as shown in Fig. 12.3. These must next be combined with the previously determined lateral forces, as shown in Fig. 12.4. The design conditions for the individual bent members would be selected from the maximum values of Fig. 12.3 (gravity only) or three-quarters of the maximum values of the combined loading, as shown in Fig. 12.4. The adjustment of three-quarters for the comparison is based on the increased allowable stress for the combined forces that include the seismic load.

We may now proceed to make a preliminary design of the column and girder, based on these analyses. For the illustration we will design the bent in steel, using standard rolled sections, and then discuss other possibilities for its construction. For the column the critical condition is that of the leeward column, which must be de-

signed for the compression plus bending. We assume the following for the first trial design:

A36 steel, rolled WF sections.

Design for M = 148.8 kip-ft, axial compression = 25 kips.

(Combined loading is not critical; no increase in allowable stresses.)

Assume the wall to brace the column continuously on its y axis.

We may now try a section by picking one from the S_x tables in the *AISC Manual* (Ref. 11) with a moment capacity slightly higher than 148.8 kip-ft, based on the assumption that the axial load is not very important because of the relatively high stiffness on the unbraced x axis. We thus will try a W 16 × 57, with a listed moment capacity of 184

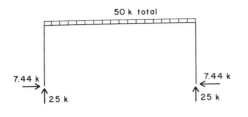

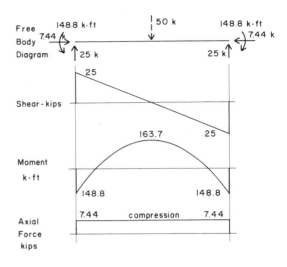

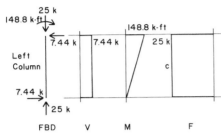

FIGURE 12.3 Investigation of the bent for gravity load.

kip-ft with $F_y = 36$ ksi. Checking this with formula 1.6–2 from the steel design specification of the *AISC Manual*, we get

$$\frac{f_a}{F_a} + \frac{f_b}{F_b} \leqslant 1$$

where

$$f_a = \frac{25}{6.8} = 1.49 \text{ ksi}$$

$$\frac{KL}{r_x} = \frac{(2)(240)}{6.72} = 71.4$$

$$F_a = 16.29 \text{ ksi (from } AISC \text{ } Manual, \text{ p. 5–74)}$$

$$f_b = \frac{M}{S} = \frac{148.8 \times 12}{92.2} = 19.36 \text{ ksi}$$

$$F_b = 24 \text{ ksi}$$

Then

$$\frac{f_a}{F_a} + \frac{f_b}{F_b} = \frac{1.49}{16.29} + \frac{19.36}{24}$$

$$= 0.09 + 0.81 = 0.90$$

which indicates that the section is adequate on a stress basis.

There are two additional design considerations for the column that have some importance. One has to do with the connection of the column to the girder. If the connection is to be a fully welded connection, the flange widths of the two members should be reasonably matched. The other consideration has to do with the lateral deflection, or drift, of the bent, which is discussed later.

For the girder design we assume the following:

A36 steel, rolled WF section.

Design for *M*: 163.7 kip-ft (positive moment at midspan).

Axial compression is negligible, based on the column design.

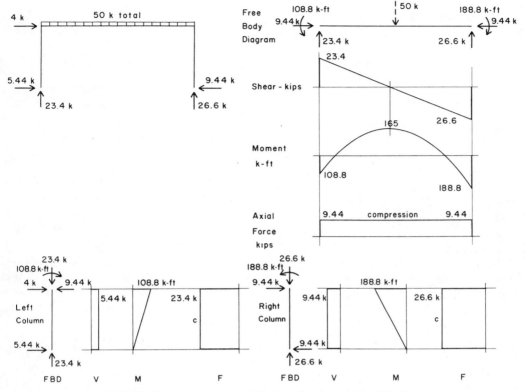

FIGURE 12.4 Investigation of the bent for the combined load.

Assume roof-framing braces the girder on 6-ft centers.

Once again using the S_x tables from the *AISC Manual*, we try a $W\,21 \times 50$ with a listed moment capacity of 189 kip-ft. This will make the f_b/F_b ratio in the combined stress formula

$$\frac{163.7}{189} = 0.866$$

Then

$$f_a = \frac{7.44}{14.7} = 0.506 \text{ ksi}$$

$$\frac{KL}{r_y} = \frac{(1)(72)}{1.30} = 55$$

$$F_a = 17.90 \text{ ksi}$$

which clearly indicates that the combined stress is not critical.

Two deflection problems should be considered. The first is that of the vertical deflection of the girder, which is most critical for the end bent because the details of the end wall construction must tolerate at least the deflection due to the live load. If we assume the live load to be approximately half of the total load, this deflection will be something less than

$$\frac{5WL^3}{384EI} = \frac{(5)(25)(50)^3(1728)}{(384)(29,000)(984)} = 2.4 \text{ in.}$$

which is the simple beam deflection based on no end moments.

The end moment will produce an upward deflection equal to $ML^2/8EI$. Using half of the calculated end moment for the live load deflection, this reduction amounts to

$$\frac{ML^2}{8EI} = \frac{(74.4)(50)^2(1728)}{(8)(29,000)(984)} = 1.4 \text{ in.}$$

resulting in a net midspan deflection of approximately 1 in.

The other deflection consideration is that of the lateral drift caused by wind, as mentioned previously. Again, this is most critical for the end wall construction. As shown in Fig. 12.5, the lateral deflection (Δ in the figure) may be calculated in two parts. The first part consists of the simple cantilever deflection of the column (t_1 the figure). The second part is caused by the rotation at the top of the column (t_2 in the figure). The determination of this combined deflection is

$$\Delta = t_1 + t_2$$

$$= \frac{Hh^3}{3EI_c} + \frac{Hh^2L}{8EI_g}$$

$$= \frac{(2)(20)^3(1728)}{(3)(29,000)(758)}$$

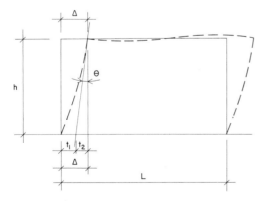

FIGURE 12.5   Lateral deflection of the bent.

$$+ \frac{(2)(20)^2(50)(1728)}{(8)(29,000)(984)}$$

$$= 0.41 + 0.30 = 0.71 \text{ in.}$$

Although these are live load deflections and consequently quite theoretical, the end wall construction, as well as any interior cross wall construction, should be designed to tolerate movements of this order of magnitude.

Although we do not illustrate their design, Figure 12.6 shows details for four other possible constructions of the bents. Figure 12.6a shows a bent built up from flat plates of steel, producing the same basic I-shaped cross section for the members. An advantage of this construction, as well as of the others in the figure, is that the members may be tapered in length. This offers the possibility of designing the girder for the required depths at the midpoint and ends and the possibility of providing for roof drainage while maintaining a flat bottom on the girder. And it offers the advantage of reducing the size of the column at the floor level where the moment capacity is not required.

Figure 12.6b and c illustrate construction of the bent in concrete with either conventional reinforcing or prestressing with steel cables. If the bents are cast flat on the site and tilted in place, they could be built in one piece.

Figure 12.6d shows the possibility of a plywood and timber box-type construction for the bent. Such bents would most likely be primarily shop-built, using field splices similar to those for the steel bents. The sketch in the figure shows a possibility for a joint placed near the girder inflection point.

A commonly used building form is one that uses a roof that is doubly pitched, as shown in Fig. 12.7. This roof is usually described as *gabled*, although a gable is actually the upper, triangular portion of the end wall. Correctly or not, this roof is described

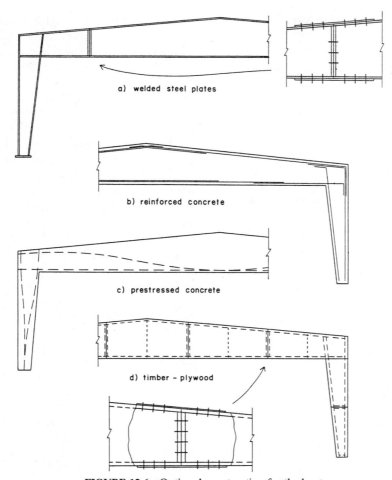

a) welded steel plates

b) reinforced concrete

c) prestressed concrete

d) timber - plywood

FIGURE 12.6 Optional construction for the bents.

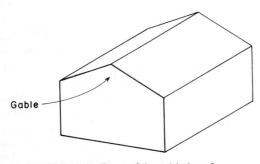

Gable

FIGURE 12.7 Form of the gabled roof.

as gabled, and the structure that forms it is described as having gabled rafters, a gabled truss, a gabled bent, and so on.

When spans for a gabled roof are short, the simplest structure is usually a pair of inclined rafters, as shown in Fig. 12.8a. If the two rafters are supported only by each other at the top, the structure requires the development of both vertical and horizontal support forces at the bottom ends of the rafters. Depending on the interior building

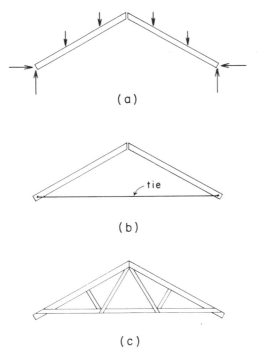

FIGURE 12.8 Structures for gabled roofs.

form and construction, there are various possibilities for the development of the horizontal reaction forces.

A simple solution for the inclined rafters is to provide a tie, as shown in Fig. 12.8*b*, which balances the two outward thrusts by opposing them against each other. The rafters and tie thus form a simple, single-triangle truss. If it is desired to have a flat ceiling in the space below, the ceiling joists (or attic floor joists) may be used to form the tie. If the span is long, or a clear span space beneath the roof is desired, a real truss may be formed, with sloping top chords and a flat bottom chord. A simple truss used commonly in wood-framed construction is that of the W-truss (named for the pattern of the interior members), as shown in Fig. 12.8*c*.

The sloped surfaces of the gabled roof are usually pitched at an angle sufficient to allow for the use of shingles or other forms of roofing that require fast draining. This usually requires a slope of at least 3 in 12 (1:4, 25%, or 14°) or steeper. This makes the use of the roof deck as a horizontal diaphragm somewhat questionable, although it is usually accepted for such function for slopes up to about 8 in 12 (about 34°).

If the roof deck functions as a horizontal diaphragm, the general development of lateral resistance for this building may be essentially similar to that for Example 1 for the flat-roof building. If the roof cannot be used for a horizontal diaphragm, there are two other options. The first is to use the horizontal structure that exists at the level of the flat ceiling (or the flat bottom chord of the truss). If an attic floor is used here, it may function as a floor diaphragm. If no floor or ceiling is used, the bottom chords of the trusses may be used to develop a horizontal braced frame, as described in Sec. 5.2.

In some buildings with a gable form roof it is desired to have a clear space beneath the roof, without the horizontal ties or the trusses. A possible solution in this case—for the development of resistance to both gravity and lateral forces—is a gabled bent. One form for such a bent is that shown in Fig. 12.9*a*, with a single, continuous frame, formed by four members that are rigidly connected (moment-resistive connections). Another option is to use a three-hinged frame, as shown in Fig. 12.9*b*.

When designing for wind, the *UBC* requires that the wind forces on the structure be determined by method 1 (see *UBC* Sec. 1619.2 and Table 16-H). By this method the forces are applied as either inward or outward pressures (direct or suction) to the individual building surfaces. Figure 12.10 illustrates the application of the required pressures. The forces on the walls and the leeward roof surface are of a constant value, but the force on the windward roof surface

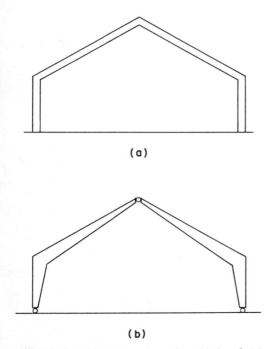

FIGURE 12.9   Rigid frame bents for gabled roofs: (*a*) continuous two-hinged; (*b*) two-part three-hinged.

varies for different slopes. The following example computation illustrates the application of these requirements.

Example 20, Building H

The building profile is as shown in Fig. 12.11*a*. Find the design wind forces for a basic wind speed of 100 mph. The building is 120 ft long in the direction perpendicular to the gabled profile section. Assume exposure condition C.

From *UBC* Table 16-F, $q_s = 26$ psf. From Table 16-G, $C_e = 1.2$ up to 20 ft and 1.3 from 20 to 40 ft. The design pressures will thus be

$$p = C_e C_q q_s I = C_e C_q (26)(1) = 26 C_e C_q$$

Thus, using the appropriate C_e values for height above ground and the appropriate

Wind direction for all cases

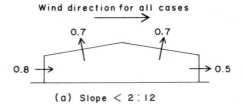

(a) Slope < 2 : 12

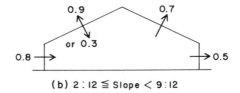

(b) 2 : 12 ≦ Slope < 9 : 12

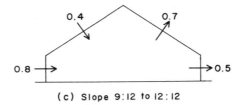

(c) Slope 9 : 12 to 12 : 12

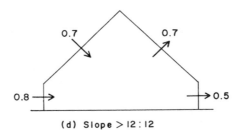

(d) Slope > 12 : 12

FIGURE 12.10   Wind pressure coefficients (C_q) for gabled bents from *UBC* Table 16-H (see Appendix B).

C_q values for the surfaces from Table 16-H, the design pressures for the surfaces are as summarized in Table 12.1.

Figure 12.11*b* shows the building profile with the pressures as determined in Table 12.1. The total effect of these pressures on the surfaces is illustrated by the resultant forces shown in Fig. 12.11*c*. These forces would be used for the design of the primary lateral bracing system and for investiga-

TABLE 12.1   DESIGN WIND PRESSURES: BUILDING H

Building Surface	C_e	C_q	$p = 26C_eC_q$ (psf)
Windward wall	1.2	0.8	25
Windward roof	1.3	0.4	13.5
Leeward roof	1.3	−0.7	23.7
Leeward wall	1.2	−0.5	15.6

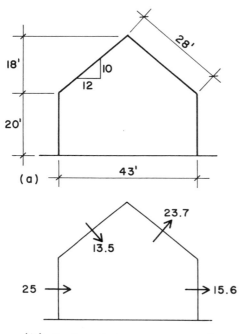

(a)

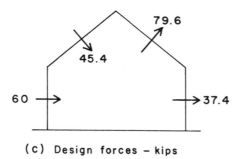

(b)   Design wind pressures - psf

(c)   Design forces - kips

FIGURE 12.11   Example 20, Building H: (*a*) layout of the bent; (*b*) design wind pressures in psf; (*c*) total design forces on the surfaces in kips.

tion of the total sliding or overturning effects on the building. Design of the structure for individual surfaces would be done with the individual C_q values given in *UBC* Table 16-H.

Seismic forces for the gabled building are determined in the usual manner. A variation here—as with wind force—is one relating to the roof slope. For seismic force, the distinction has to do with the structural function of the roof. If the slope is low and the roof is assumed to function as a horizontal diaphragm, the building functions as for the flat-roof structures in previous examples. Distribution of the seismic loads to the elements of the vertical bracing system would be as usual with a roof diaphragm.

If the roof is too steep to act as a horizontal diaphragm, the seismic loads due to the weight of the roof will be applied to the spanning roof structure in a manner similar to that for the gravity loads. That is, the roof will act like a wall does in transferring the horizontal seismic force due to its weight to a supporting structure.

Example 21, Building G

For the building shown in Fig. 11.13, one proposal for the interior bracing bents was shown in Fig. 11.13*d*, consisting of a planar rigid frame. For the example building it is most likely that this frame would be developed with steel shapes and welded connections.

A problem with any of the bent designs for this building is the obtaining of a relative stiffness for the bent that allows it to share loads equitably with the relatively stiff shear walls. This is most likely to involve some investigation of deflection of the loadsharing elements and an assignment of the shared load on the basis of their relative stiffnesses. This results in the establishment of a drift deflection limit for the bent to make it work in the system.

If column heights and beam spans are great, it will be necessary to have very large members for a rigid frame to have small lateral deflection. Thus the better solution is for many relatively short members rather than a few very long members, and where this can be done a lighter frame may be possible. However, the multiplicity of joints greatly increases the cost of assemblage of the frame, particularly with moment-resistive welded joints.

This is a case where the rigid frame is most likely to be used only if no other solution is capable of achieving the desired architectural design. It probably can be done but will probably be quite expensive.

Although planar elements can be developed as subsets within the structure for a building, the entire building framework is typically three-dimensional in nature. In some cases this full spatial nature of the frame may be developed, as discussed in Sec. 12.2.

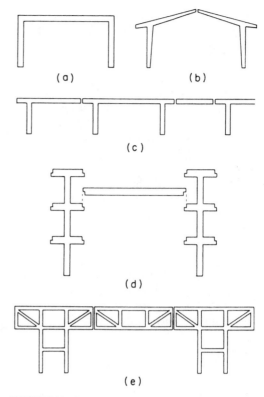

FIGURE 12.12 Use of precast concrete for structural elements.

Example 22, Precast Concrete Planar Bents

As discussed in Example 19, one option for the rigid bent for the gable roof consisted of precast concrete elements (see Fig. 12.6*b* and *c*). Figure 12.12 shows some of the possibilities for use of precast elements in planar arrangements that also develop some rigid frame actions.

Most precast elements for structures consist of single wall panels. columns, or spanning elements such as deck units or beams. However, the advantages of the casting operation may be utilized to produce single-piece units that comprise planar rigid frame bents or trusses. In this case some of the problems of effecting connections that are associated with wood and steel structures are eliminated. For the single-span bent shown in Fig. 12.12*a*, for example, the three elements of the bent are

monolithically formed with the cast material.

The single-span gabled bent shown in Fig. 12.12*b* can be produced as indicated with two halves cast in single pieces. This is a commonly used form of structure, particularly for small utilitarian buildings where the durability of the exposed concrete structure is an advantage.

For multispan or multistory buildings, it is sometimes possible to subdivide the entire structure into units that provide for a simpler form of assemblage. For the bents in Fig. 12.12*c*, the assemblage connections are reduced to simple pin-type connections. preserving both the continuity of the horizontal spans and the stability of the system for lateral loads. For the multistory bents in Fig. 12.12*d*, the need for moment connections and column splices is eliminated.

The concept of trussing seems somewhat inappropriate for concrete, since the idea of a concrete tension member is unreasonable. Nevertheless, trussing has been utilized in various ways for some structures. One possibility is the use of crisscrossed members, or X-bracing. In steel structures, this often means double tensioning, that is, providing two tension members, each responding to a different load direction. In concrete, the corresponding usage involves double compression. with each diagonal functioning in compression for a different load direction.

Tension is possible. of course, with either steel reinforceing or prestressing. Figure 12.12*e* shows a structure consisting of precast units in which the units have a composite trussed and rigid frame form. At a relatively large scale, where high-strength concrete and prestressing are feasible, this type of structure is quite reasonable.

Precast concrete elements are subject to many of the concerns of concrete structures in general. To their advantage, in some cases, is the ability to achieve some higher quality (including strength) of the concrete and some more accurate details of the form of members. One of the biggest problems is the achieving of good joints between separate cast elements. A particularly demanding requirement for joints comes from the need to resist major earthquake forces.

Almost every significant earthquake results in some spectacular failures of structures with precast concrete elements (Fig. 12.13). This commentary is not intended as a condemnation of the construction system, but it is an argument for the best available engineering design work, with serious review by other designers.

FIGURE 12.13 Structural collapse of the interior, precast concrete elements of this large parking structure occurred during the Northridge earthquake of January 1994. All three parking structures failed at this large shopping center, a short distance from the epicenter of the earthquake.

The heavy concrete structure adds to its own problems in an earthquake by the inertial force of its mass, and a correspondingly higher level of attention to the engineering design is necessary.

12.2   MOMENT-RESISTIVE SPACE FRAMES

Beam-and-column framing systems for multistory buildings are typically configured in a three-dimensional (or general spatial) system. The building codes have adopted the term *space frame* for this system, although architects have generally used the term to describe spatial trussed structures.

One means of stabilizing the beam and column frame is by using moment-resistive joints for some or all of the connections between the beams and columns. This yields the term *moment-resistive space frame*.

The two predominant forms for such a framework are:

Steel rolled sections with joints using welds or high-strength bolts for connections.

Sitecast concrete with the monolithic casting and the extension of reinforcement achieving the member continuity that creates the moment-resistive joints.

The following example illustrates some of the considerations in the development of a moment-resistive space frame structure. These systems require very thorough engineering design, due to their high indeterminate force resolution and the general complexities of the constructions with welded steel or reinforced concrete. The computational work here is very approximate, although reasonably useful for a preliminary design.

Example 23, Building E

Referring to the plan for Building E as shown in Fig. 10.28, observe that there is a series of columns arranged in rows in each direction defining a column bay system of 30-ft by 30-ft bays. Another way to visualize this system is in terms of the vertical planar bents that are defined by the columns in a single row associated with the beams that connect them at the three upper levels of the building: the second floor, the third floor, and the roof. These bents, as shown in Fig. 12.14, may be used to brace the building in both directions if the beam-to-column connections can be made moment resistive and the columns are vertically continuous. If the latter is achieved and the system is utilized for lateral bracing, it is in the words of the *UBC*, "a moment-resistive space frame," or, in the more commonly used reference, a rigid frame system.

For the structure as defined in Fig. 10.28, there are four bents in the east–west direction, each having five columns connected at each level by four beams. In the north–south direction there are five bents, each with four columns connected at each level by three beams. The form of the individual bents is as shown in Fig. 12.14*a*. The horizontal roof and the floor deck diaphragms deliver the lateral loads to the bents as shown for the north–south bent in Fig. 12.14*b*. The lateral loading develops shear, bending, and axial force in all the members of the frame; the investigation for these effects is a statically indeterminate problem.

There are many ways in which rigid frames can be utilized for lateral resistance in different situations. Individual frames may be employed selectively for part of the bracing system for the building, with the rest of the frame structure limited to simple post-and-beam functions. This can be achieved in steel framing by simply using moment-resistive connections only where moment resistance is desired. It is some-

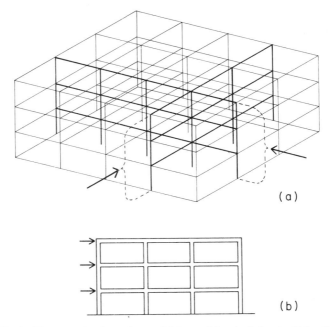

FIGURE 12.14 Vertical beam and column bents: (*a*) form of the single bents; (*b*) loading of a single bent.

what less able to be controlled in cast-in-place concrete frames, for moment continuity of the frame members is a natural attribute.

If the structure defined in Fig. 10.28 utilizes steel beams and columns, then any, all, or none of the frame bents can be made rigid" through the use of moment-resistive connections; the remainder of the frame functions as a post-and-beam system for gravity loads only. If the structure is of cast-in-place reinforced concrete, and the columns are all square and close to the same size, all of the defined bents will function as rigid frames for both lateral and gravity loading.

Plywood shear walls and X-bracing represent elements that function primarily for lateral load resistance, leaving other elements to develop gravity load resistance. Structural masonry walls, trussed bents, and rigid frames represent elements that are usually, by their nature and the means

of employing them, used for both vertical and lateral load resistance. Although a rigid frame may be visualized primarily as such for its lateral load resistance potential, it must also be designed for the combined effects that include the vertical load due to gravity.

For a building such as Building E, there are several possible schemes for using rigid frames for lateral load resistance. The basic techniques include the following:

1. *Full Rigid Frame.* This consists of using all of the bents defined by the alignment of columns and having continuous beams at all levels.

2. *Selected Frames.* This consists of choosing only some of the potential bents for use as lateral bracing. This may be controlled by use of the type of connections (in steel) or by the selection of the member sizes for stiffness. As in any load-sharing situation, the

stiffer bents will take the greater share of the total load.

3. *Braced Core.* As with shear wall or trussed frames, the permanent construction around stairs, elevators, duct shafts and pipe chases, or stacked rest rooms may become the location of lateral bracing.

4. *Perimeter Bracing.* This consists of using only the bents that occur in the planes of the exterior walls that are defined by the exterior columns and the spandrel beams.

Many structural and architectural issues are to be considered in choosing one of these schemes over the other. A currently popular system is one using the perimeter bents. Structurally, this has the principal advantage of providing the most torsional resistance—of particular concern for seismic load. Architecturally, it offers the freedom of the entire interior structure from concerns for lateral resistance. However, any of the other schemes—plus other more imaginative possibilities—may be logical for a given situation. The following example illustrates a design of a perimeter bent system for Building E, with no particular intention of prejudice.

Design of Perimeter Bent System. Figure 12.15*a* shows a structural-framing plan for a reinforced concrete beam and slab system for one of the upper floors of Building F. The construction detail section in Fig. 12.15*b* shows the condition at the building exterior, indicating the use of an exposed spandrel beam and column. The columns and column-line beams constitute a three-dimensional rigid frame, with the three typical column-beam joint conditions as shown in Fig. 12.15*c.*

The plan arrangements and beam layouts for this system permit considering using either a full rigid frame system or a perimeter bent system. For the cast-in-place concrete frame it is actually somewhat difficult to avoid the full rigid frame action. One technique that can be used is to manipulate the distribution of lateral forces to the individual bents by the choice of the member shapes and sizes. In this case, it has been decided to use the exterior wall bents (perimeter bents) as the main bracing system. To increase the stiffness of these bents the exterior columns are made oblong, with their large dimension in the direction of the wall plane and their interacting beams made exceptionally deep (see Fig. 12.15). With the square interior columns and the much shallower interior beams, the other bents will be much less stiff.

For an initial approximate design we will consider the perimeter bents to take the entire lateral load. This is not quite true, of course, and a final, more exact, investigation—preferably a true dynamic one—would indicate some minor forces in the other bents. However, as all the bents will be designed for gravity loading, it is reasonable to expect that the lateral forces on the interior bents may quite likely not be critical if the usual stress increase of one third is considered.

It is not reasonable, of course, to design the bents for lateral forces only. Although we do not present it, some values for internal forces generated by gravity loads must be added to the effects of the lateral loads, requiring a complete combined loading investigation. These values have been approximated and are included where required in the example.

The lateral force resisting systems for the building are shown in Fig. 12.16. For force in the east–west direction the resistive system consists of the horizontal roof and floor slabs and the exterior bents (columns and spandrel beams) on the north and south sides. For force in the north–south direction the system utilizes the bents on the east and west sides.

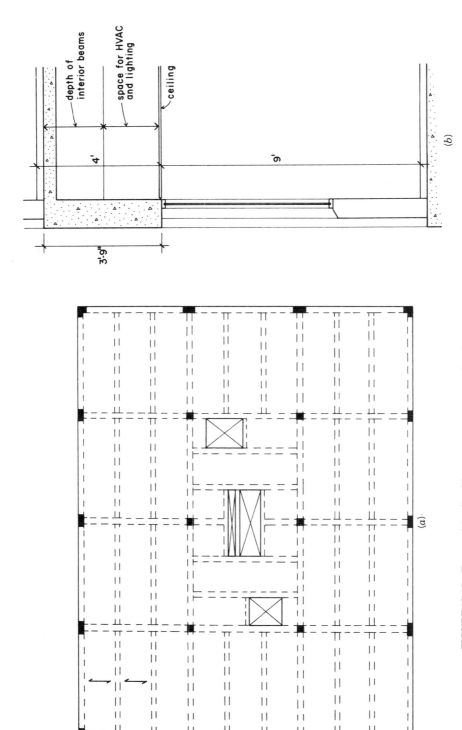

FIGURE 12.15 Form and details of the concrete structure: (*a*) framing plan for the upper floors; (*b*) section of the exterior wall and spandrel; (*c*) column-beam-typical cases.

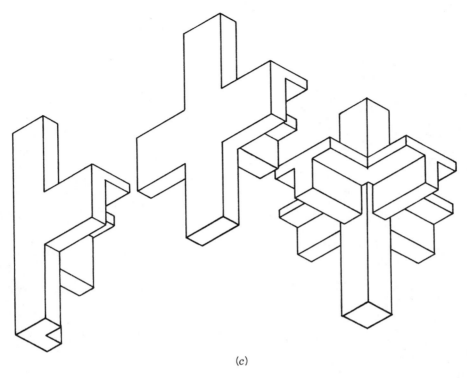

(c)

FIGURE 12.15 (*Continued*)

If the lateral load is the same in both directions, the stress in the slab (shear in the horizontal diaphragm) is critical for the north–south loading because the slab has less width for resistance to this loading. The loads are not equal, however. Wind force will be greater in the north and south directions because the building has a greater profile in this direction. This makes it even more obvious that this will be the loading critical for the slab in design for wind. However, for seismic load, a true dynamic analysis reveals that the load effect is greater in the east–west direction because the resistive bents are slightly stiffer in this direction. In any event, it is unlikely that the 5-in.-thick slab with properly anchored edge

reinforcing at the spandrels will be critically stressed for any loading.

Our considerations for lateral load will be limited to the seismic loading in the north–south direction and to investigations of the effects on the columns and spandrel beams on the east and west sides.

Determination of the Building Weight.
Figure 12.17 presents the analysis for determining the building weight to be used for computation of the seismic effects in the north–south direction. Tabulations are done separately for the determination of loads to the three upper diaphragms (eventually producing three forces similar to H_1, H_2, and H_3, as determined for the wind load-

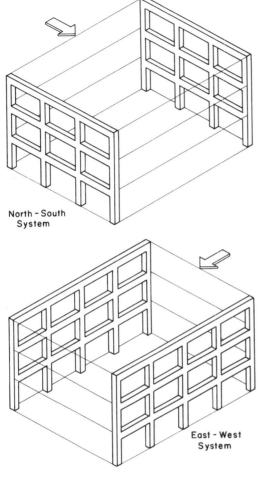

North – South
System

East – West
System

FIGURE 12.16   Perimeter bent system.

$C = 2.75$ (maximum value, *UBC* Sec. 1628.2.1)

$R_W = 12$ (special concrete frame, *UBC* Table 16-N)

Thus

$$V = \frac{ZIC}{R_W} W = \frac{(0.3)(1.0)(2.75)}{12} W$$

$$= 0.0.06875\, W$$

And using the value from Table 12.17 for W, we have

$$V = (0.06875)(5106) = 351 \text{ kips}$$

This total force must be distributed to the roof and upper floors in accordance with the requirements of Sec. 1628.4 of the *UBC*. The force at each level F_x is determined from Formula 28-8 as

$$F_x = (V)(w_x h_x) \Big/ \sum_{i=1}^{n} w_i h_i$$

where

F_x = force to be applied at each level x

w_x = total dead load at level x

h_x = height of level x above the base of the structure

(Notice that F_t has been omitted from the formula because T is less than 0.7 sec.) The determination of the F_x values is shown in Fig. 12.18.

For an approximate analysis we consider the individual stories of the bent to behave as shown in Fig. 12.19*b*, with the columns developing an inflection point at their mid-height points. Because the columns all move the same distance, the shear load in a single column may be assumed to be equal to the cantilever deflecting load and the individual shears to be propor-

ing). The weight of the lower half of the first-story walls is assumed to be resisted by the first-floor-level construction (assumed to be a concrete structure poured directly on the ground) and is thus not part of the distribution to the rigid bent system.

For the total seismic shear force to the bents, we note from the given data that

$Z = 0.3$ (zone 3, *UBC* Table 16-I)

$I =$ (standard occupancy, *UBC* Table 16-K)

Level	Source of Load	Unit Load (psf)		Load (kips)
Roof	Roof and ceiling	140	120 X 90 X 140 =	1512
	Columns at 0.6 k/ft		0.6 X 6 X 20 =	72
	Window walls	15	400 X 4.5 X 15 =	27
	Interior walls	10	200 X 5 X 10 =	10
	Penthouse + equipment (estimate total load)			25
	Subtotal			1646
Third floor	Floor	140	120 X 90 X 140 =	1512
	Columns		0.6 X 11 X 20 =	132
	Window walls	15	400 X 9 X 15 =	54
	Interior walls	10	200 X 9 X 10 =	18
	Subtotal			1716
Second floor	Floor	140	120 X 90 X 140 =	1512
	Columns		0.6 X 12 X 20 =	144
	Window walls	15	400 X 11 X 15 =	66
	Interior walls	10	200 X 11 X 10 =	22
	Subtotal			1744
Total dead load for base shear				5106

FIGURE 12.17 Determination of the building weight for the seismic shears.

Building Level	Gravity Load at Level - w_x (kips)	Height of Level Above Base - h_x (ft)	$(w_x)(h_x)$	F_x (kips)
Roof	1646	41	67,486	167
Third floor	1716	28	48,048	119
Second floor	1744	15	26,160	65
		Total:	141,694	

See text: $F_x = (351/141{,}694)(w_x h_x) = 0.024772(w_x h_x)$

From this calculation, the bent loads used for design are as shown below.

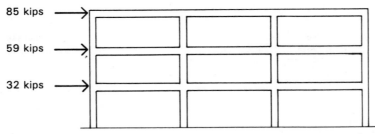

85 kips

59 kips

32 kips

FIGURE 12.18 Determination of the lateral forces at the upper levels of the structure by redistribution of the total base shear.

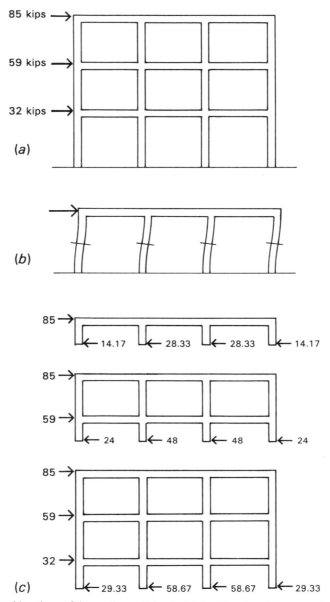

FIGURE 12.19 Considerations of the north–south bents: (a) bent loading; (b) assumed column deformation with midheight column inflection; (c) loading for the story shears. (*Note:* Load shown is the total load on the building, to be distributed to the individual bents.)

tionate to the stiffnesses of the columns. If the columns are all of equal stiffness in this case, the total load would be simply divided by four. However, the end columns are slightly less restrained as there is a beam on only one side. We will assume the net stiffness of the end columns to be one-half that of the interior columns. Thus the shear force in the end columns will be one-sixth of the load and that in the interior columns one-third of the load. The column shears for each of the three stories is thus as shown in Fig. 12.19c.

The column shear forces produce moments in the columns. With the column inflection points assumed at midheight, the moment produced by a single shear force is simply the product of the force and half the column height. These moments must be resisted by the end moments in the rigidly attached beams, and the actions are as shown in Fig. 12.20. These effects due to the lateral loads may now be combined with the previously determined effects of gravity loads for an approximate design of the columns and beams.

For the columns, we combine the axial compression forces with any gravity-induced moments and first determine that the load condition without lateral effects is not critical. We may then add the effects of the moments caused by lateral loading and investigate the combined loading condition, for which we may use the one-third

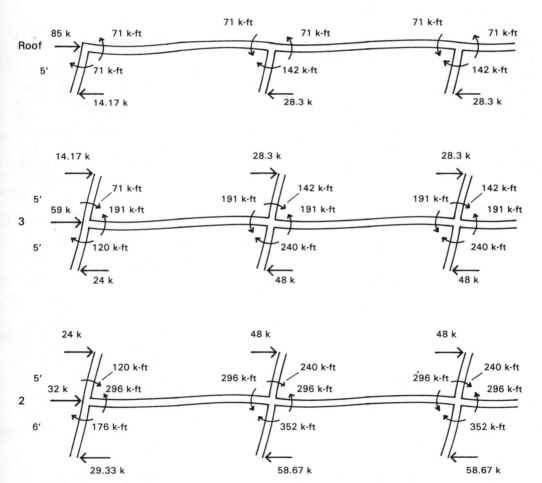

FIGURE 12.20 Investigation for the column shears and the column and girder moments for lateral load. (*Note:* Values shown are for the total load on the building, to be distributed to the individual bents.)

increase in allowable stress. Gravity-induced beam moments are taken from a separate analysis not shown here and are assumed to induce column moments as shown in Fig. 12.21. The summary of design conditions for the corner and interior column is shown in Fig. 12.22. The design values for axial load and moment and approximate sizes and reinforcing are shown in Fig. 12.23. Column sizes and reinforcing were obtained from the tables in the *CRSI Handbook* (Ref. 15) using concrete with f_c' = 4 ksi and grade 60 reinforcing.

The spandrel beams (or girders) must be designed for the combined shears and moments due to gravity and lateral effects. Using the values for gravity-induced moments and the values for lateral load moments from Fig. 12.20, the combined moment conditions are shown in Fig. 12.24a. For design we must consider both the gravity only moment and the combined effect. For the combined effect we use three-fourths of the total combined values to reflect the allowable stress increase of one-third.

Figure 12.24b presents a summary of the design of the reinforcing for the spandrel beam at the third floor. If the construction that was shown in Fig. 12.15b is retained with the exposed spandrel beams, the beam is quite deep. Its width should be approximately the same as that of the column, without producing too massive a section. The section shown is probably adequate, but several additional considerations must be made as will be discussed later.

For computation of the required steel areas we assume an effective depth of approximately 40 in. and use

$$A_s = \frac{M}{f_s jd} = \frac{M(12)}{(24)(0.9)(40)} = 0.0139M$$

A special problem that must be considered with concrete and masonry structures is the careful proportioning of the sizes of members that share load. For this building there are two such relations that need some attention. The first deals with the relative stiffness of the columns. The lateral stiffness of the bents will largely be

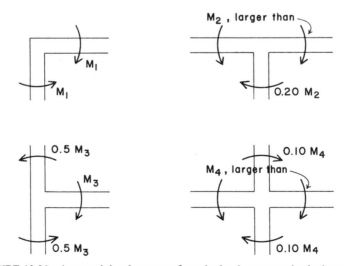

FIGURE 12.21   Assumed development of gravity load moments in the bent columns.

	Column	
	Intermediate	Corner

	Intermediate	Corner
Axial gravity design load (kips)*		
Third Story	90	55
Second Story	179	117
First Story	277	176
Assumed gravity moment (kip-ft)*		
Third Story	60	120
Second Story	40	100
First Story	40	100
Moment from lateral force (kip-ft) from Figure 12.20:		
Third Story	142	71
Second Story	240	120
First Story	352	176

* From computations not shown here.

FIGURE 12.22 Design values for the bent columns.

	Intermediate Column					Corner Column				
	Axial Load (kips)	Moment (k-ft)	e (in.)	Column Size (in.)	Reinforcing No. - Size	Axial Load (kips)	Moment (k-ft)	e (in.)	Column Size (in.)	Reinforcing No. - Size
Roof										
	90 X 3/4 = 68	202 X 3/4 = 152	27	20 X 28	6 - 9	55 X 3/4 = 41	191 X 3/4 = 143	42	20 X 24	6 - 10
3										
	179 X 3/4 = 134	280 X 3/4 = 210	19	20 X 28	6 - 9	117 X 3/4 = 88	220 X 3/4 = 165	23	20 X 24	6 - 10
2										
	277 X 3/4 = 208	392 X 3/4 = 294	17	20 X 28	6 - 11	176 X 3/4 = 132	276 X 3/4 = 207	19	20 X 24	6 - 11
1										

FIGURE 12.23 Design of the columns for the combined loading.

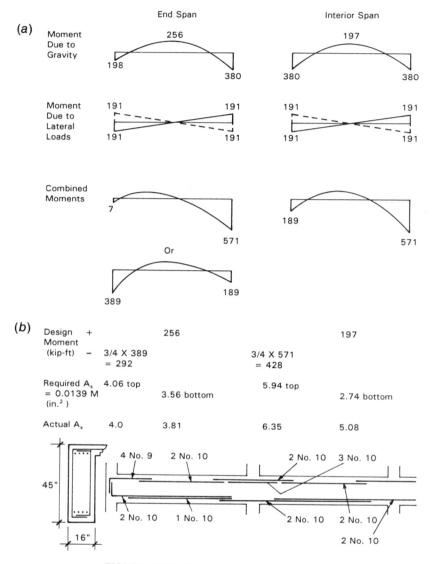

FIGURE 12.24 Design of the third-floor spandrel girder.

determined by the column stiffnesses. To assure that the exterior (perimeter) bents actually perform the major lateral bracing work, these bents must be stiffer than those in the building interior. Stiffness in this case is mostly determined by the dimensions of the concrete column cross sections.

For the ordinary cast-in-place concrete frame structure, any column-beam bent will have some reserve capacity for lateral load resistance. Extension, continuity, and anchorage of the reinforcement will provide some lateral bent action. Even if not designed for, some lateral load moments can be resisted due to the allowable stress increase of one-third. Thus if lateral load is minor, any bent can be assigned some portion of the total lateral force on the building. In this case, the interior bents could probably be designed for their minor share of the lateral forces, with little or no modification of the gravity load design.

The perimeter bents, however, would be designed for the entire lateral force. Thus when the interior bents yield under lateral

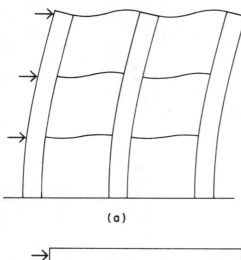

(a)

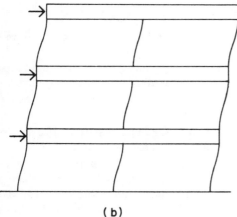

(b)

FIGURE 12.25 Lateral deformation of frames with members of disproportionate stiffnesses: (*a*) stiff columns and flexible beams; (*b*) stiff beams and flexible columns.

force, the perimeter bents have the full reserve capacity for the lateral loads.

A second consideration regarding stiffnesses has to do with the relative stiffnesses of the members in a single bent. It is generally desirable that all of the members of a bent that are connected at a single joint have close to the same relative stiffness. This is to assure that the bent deformations in Figs. 5.10*d* and 12.20 will actually occur. If columns are excessively stiff, the deformation will tend to be more of the character shown in Fig. 12.25*a*. If horizontal members (beams or trusses) are excessively stiff, deformation will tend to be more of the character shown in Fig. 12.25*b*. In this situation relative stiffness refers to the ratio of I/L for individual members. In this example the columns are approximately half as long as the girders; thus equal stiffness will occur if the girder sections have twice the moment of inertia of the column sections. Exact equality is not to be expected, but when connected members have I/L ratios of more than about 3 or 4 to 1 in difference, full flexing of the individual members may be doubtful.

Almost every major earthquake in recent times has brought some new lessons for concrete structures. For the sitecast concrete frame, such as that illustrated in Example 23, some major lessons pointed out from the Northridge earthquake were the following:

1. Concrete frames designed prior to the mid-1970s may have high vulnerability to major earthquake motions. The special requirements for ductile frames that have now been incorporated into the codes were mostly not in place then, and some spectacular failures have occurred (see Fig. 4.25). This has mostly to do with the steel reinforcement, including its design and detailing.

FIGURE 12.26   The effective column length for lateral deformation was dramatically changed in the exterior columns of this highrise apartment building by the addition of very stiff balcony railing/panels. The very short and stiff columns thus produced failed during an earthquake in a pure shear action, with X-cracking as frequently seen in pierced wall construction. Columns so altered by enclosing construction are now referred to as *captured columns*.

2. The assumption that certain parts of the frame structure do not function significantly to brace the building may be valid on a relative stiffness analysis basis, as was illustrated for Example 23, with the perimeter bents assumed to brace the structure. However, the rest of the building is along for the ride, and the effects of the lateral deflections on the whole construction must be accounted for. If other building parts fail at a deformation less than that projected for the ductile resistance of the bracing, their premature collapse can pull down the whole structure. This is one

theory put forward for the collapse sequence of the structure in Fig. 13.2 and may have been significant for some of the other parking garages that failed in the Northridge earthquake.

3. Failure mechanisms of concrete frames are predicted on a deformation analysis basis. This includes the general assumptions for the deflected shape of the frame; such as was illustrated in Figures 12.19 and 12.20 for Example 23. This is a critical consideration for the behavior of indeterminate structures in general. If the frame members are unintentionally

FIGURE 12.26   (*Continued*)

restrained in their deformation—which can occur when relatively strong construction is used for infill walls or spandrels, for example—the internal forces in members may be entirely different from what they were designed for. The *captured column*, as illustrated in Fig. 12.26, is a classic case. Simple remedies include using separation joints and movement-tolerating connections.

All of this is a matter of accumulating experience. And hopefully learning from it so that history does not repeat itself extensively.

PART IV

STRENGTHENING OF BUILDING STRUCTURES

As time passes, our understanding of the right way to design and construct buildings improves steadily. Although sometimes slow in happening, this eventually affects our design and construction practices. Evidence of this can be seen in the publication of successive editions of the standards and codes used in the building industry. In theory, this means that new buildings are simply built better in terms of their resistance to wind and earthquakes. Unfortunately, we know that older buildings have specific, demonstrated vulnerabilities. In this part we treat the topic of strengthening of older buildings for known weaknesses as demonstrated by the damage observed in disastrous windstorms and earthquakes.

13

NEED FOR STRENGTHENING

Need for strengthening may be generated in many ways. In some cases it is legally mandated by ordinances that require owners of older buildings to provide strengthening or to have their buildings demolished. Such has been the case with various cities with regard to old, unreinforced masonry buildings, because of their undisputed vulnerability to seismic shocks. However, building owners may act on their own to provide reassurance or to enhance the property value of their buildings by having them strengthened. In this chapter we deal with some of the reasons for strengthening and some experience with various means for providing existing buildings with improved resistance to both windstorms and earthquakes.

13.1 LEARNING FROM EXPERIENCE

Whenever a major windstorm or earthquake occurs and affects a significant number of buildings, many observers will inspect the damage. One direct benefit of this forensic study is the steady improvement of our judgment about correct design and construction practices; and a second benefit of note is our conclusion with regard to weaknesses of existing buildings.

Thus each major windstorm and earthquake typically brings some changes in regulations and practices. Some of the lessons are simply reaffirmations of old lessons. How many times do we need to be shown, for example, that unreinforced masonry is highly vulnerable to earthquakes?

(See Fig. 13.1.) Or that ordinarily con-nected, light wood frames can easily be pulled apart by high winds? Yet, for various reasons, we still have large existing stocks of these highly vulnerable buildings in high-risk areas.

For real advancement of knowledge, however, the more significant lessons from each disaster have to do with the perfor-mance of recently constructed buildings. When structures that have been designed and built in accordance with the latest standards and codes fail, they get consider-able study. These occurrences have the most effect on the development of new practices, while observations of more of the older types of failures simply strengthen our con-victions and resolve.

Failures, of course, are not always due to flaws in our theories or our judgments re-garding structural behaviors. A major fac-tor—seemingly on the increase—has to do with the gaps that can occur routinely be-tween design and construction. What de-signers determine to be required must go through several stages of communication and development before construction oc-curs. A typical scenario for this is the following:

FIGURE 13.1   Failure of an unreinforced masonry building in the January 17, 1994, Northridge earth-quake. A classic, predictable form of failure caused by the strong wall on one side pushing the wall around the corner in a direction perpendicular to the wall, and vice versa, until the entire corner breaks off. Similar photos can be found in the reports of prac-tically every major earthquake where such buildings have been affected.

1. A structural designer does the neces-sary computations for the members of a reinforced concrete frame struc-ture: determining member dimen-sion, required amounts of reinforce-ment, and other essential properties for the structure.

2. A graphic specialist (drafter or CADD operator?)—probably not the struc-tural designer who did the compu-tations—translates the designer's notes into construction plans and de-tails.

3. A specifications writer—probably neither the structural designer nor the drafter—prepares the specifica-tions that largely treat the specific materials and other factors that con-trol the quality of the construction.

4. Various persons on behalf of or in the employ of a builder interpret the con-tract documents (drawings and spe-cifications) and prepare derivative drawings and instructions for the fab-ricators, suppliers, and workers who will do the actual construction. The construction workers may thus never really see the original design docu-ments.

5. Construction is performed with various forms of inspection (or with none) and is in varying degrees highly vulnerable to the skill and/or integrity of the workers and contractors. The original designer is frequently not involved in inspection activity.

The opportunities for inadequate construction in this situation are obviously many. If the original designer does not thoroughly review the drawings and specifications, the contractor's drawings and instructions (typically called *shop drawings*), and has no part in inspection of the construction, the chances of the final construction being what was in the designer's mind are slim.

Major lessons from disasters often have to do with flaws in the process and not merely the theoretically regulated practices of design and construction (Fig. 13.2). Changes in standards and codes may do little to correct these situations.

Nevertheless, it is to be expected that we can learn better ways of designing and building from the damaged structures in any major windstorm or earthquake, and have a basis for determining what needs strengthening in existing buildings. As time goes on, we also derive some knowledge from the behavior of supposedly strengthened buildings and may find out that we were wrong in these cases also (Fig. 13.3).

13.2 EXTRA SAFETY

Although it is an oversimplification, the basic means of adding strength to a build-

FIGURE 13.2 Multistory concrete frame parking garage on the campus of California State University at Northridge. This large structure, a short distance from the epicenter of the January 17, 1994, earthquake, collapsed when interior girders pulled loose from supporting columns. A late design review recommended strengthening of the connections, but was passed over to permit construction to proceed without delay since conformance with present building codes was claimed. Real causes of the failure are still under study, but the entire design and construction process and its complex layers of review and approval are surely partly to blame.

FIGURE 13.3   Failure of masonry structure in the January 17, 1994, Northridge earthquake. This was a supposedly strengthened building, as can be observed from the exterior bolt details at the roof and upper floor levels. Indeed, the strengthening may have prevented total collapse and thus accomplished a basic life safety goal. However, the shear strength of the upper pierced wall was still insufficient, and repair seems unlikely.

ing structure is simply to increase the safety factor. A simple definition of the factor of safety is as follows:

$$FS = \frac{\text{actual capacity of the structure}}{\text{required capacity of the structure}}$$

In practice, this may be achieved by actually using higher safety factors (directly in the strength method or by reducing design allowable stresses in the stress method) or by increasing the design loads. For an example of the latter, the base shear as determined by the UBC formulas might be increased by 50%.

A common procedure in the design process for a building is to offer alternatives to the design client. These alternatives come with a price tag, and the basic question posed by the designer is: How much do you want and are willing to pay for? More expensive carpet? More space in the building entry? A more lush landscape?

Suppose that the opportunity could be offered to the client in a similar way to choose the safety factor or performance level for various physical responses and behaviors: for example, the response to windstorms and earthquakes. Who says that it should be left to the architect to decide what the safety factor for design should be? Or to the structural engineer? Or to the writers and publishers of building codes?

Structural computations are done with increasing accuracy in this age of the computer. Never mind that the data may be incomplete, speculative, or even flawed. The design of the structure will be done with considerable precision and a component of that mathematics is the setting of a specific value for a level of safety that represents a margin of reserve strength. We make the best investigation of structural behavior that we can and then make the structure x times stronger than it needs to be just to barely carry the loads.

But who sets the value of x? Who says that the structure should be twice, three times, 1.47 times as strong as the failure model? Why not let the design clients set the value? Give them the same kind of choices they get for carpet, interior space, and landscaping. Give them alternatives with a lateral-force-resisting system designed for safety factors of 2 (the all-time average from the good old days), 3, 4, or 5. Let them buy the safety they want—then enjoy the carpet, space, and nice site with a level of peace of mind they feel good about.

Satisfying the building code requirements—which is what is routinely done by designers—means acceptance of a *minimal* design. The minimum safety factor. If that level of safety is assumed to be represented by an average safety factor of 2, then using a factor of 3 means giving 50% greater safety. How much more would it cost for a structure with 50% more resistance? Probably not much for the average building.

For most buildings the structure is usually not more than about 15% of the total building cost. And not all of that is for the lateral resistive structural elements. With most of the gravity-resistive system remaining unchanged, increasing only the lateral resistive system by 50% should not raise the cost of the average building structure by more than a few percent.

If these cost increases are used with respect to the entire building project cost, the cost of increased safety becomes even less significant. In areas with high property values (most urban areas, in other words) the building cost may be half or less of the entire cost for the land and its development.

Strengthening might thus be viewed as an increase in safety achieved simply by providing something more than the minimum resistance required by legally enforced codes. However, more intelligent strengthening would relate to the demonstrated major weaknesses of particular materials and systems. For the latter form of strengthening, some particular behaviors of a specific structure would be given special attention and some higher degree of strengthening. This focused or concentrated strengthening is likely to be indicated even more strongly for existing buildings, for which demonstrated failures of similar structures have been observed.

13.3 TYPE OF STRENGTHENING NEEDED

Reduction of the potential for damage in existing buildings may take several directions. Many of these have to do with changes in the lateral resistive structure, generally visualized as strengthening it for added force resistance. The type of strengthening required will relate to various aspects of the materials and form of the structure, and possibly to the general form and construction of the building.

Reduction of Original Construction

In some cases the lack of strength may be due to deterioration or misuse of the building. Wear of exposed structural elements or the usual ravages of time, moisture, or other effects can reduce the physical capacity of many structures. Strengthening, in these cases, essentially means repair, although other reasons for strengthening may also be present. Steel bolts or ties may be rusted, masonry may be heavily cracked by various effects, essential wood framing may be rotted, and some major structural members may have been altered or even removed in remodeling. Possibilities for these problems should be investigated first in any effort to improve the structure for an older building.

Quality of Original Construction

Strength may also be reduced by omissions or poor construction. Recent studies of buildings under construction, as well as forensic examinations of damaged buildings, have indicated that this problem is quite widespread. The potential for this is inherent in typical design and construction processes, as discussed in Sec. 13.1. Before proceeding with plans to *improve on* the existing structure as designed, it should be determined that the existing construction is truly in conformance with the original design. This is relatively easy in some cases, especially where structural elements are exposed to view. However, it is not so easy in other situations; for example, the determination of whether reinforcement is properly placed inside reinforced concrete or masonry construction.

Reducing the Forces

It may also be possible to make alterations of the building that may partially relieve the forces on the structure: for example, removing elements that protrude from the building surface or that otherwise result in added wind forces, or replacing heavy nonstructural elements with lighter ones to reduce the building mass for seismic forces. Thus the structure is essentially "strengthened" with regard to its task, with no change in the structure itself.

After all of these considerations, if real strengthening is still required, it will relate to what is presently perceived as the potential shortcomings of a particular structure. These shortcomings may be discovered by theoretical studies, by laboratory experiments, or by forensic examinations of damaged buildings. Damage may be actual structural failure of some form or may be related to extensive nonstructural loss, such as glass breakage, dropped ceilings, burst piping, and so on.

In fact, the term *strengthening* is not quite the best general description for improving the behavior of a lateral resistive structure. Extensive nonstructural damage often occurs where no significant damage to the structure is involved. So-called *cosmetic damage* is often due to excessive movement of the building, which may be due to a lack of stiffness on the part of the structure. This relates, of course, to how success or failure of the structure is defined, which is an issue discussed in Chapter 2. However, where potential nonstructural damage is of great concern, the stiffness of the structure may be as important for design as its strength. It may even be a life safety problem if the nonstructural damage presents some hazard, such as with falling glass from upper-story windows, broken gas piping, dropped ceilings, and so on.

Adding Strength to the Lateral Resistive System

When real additional strength (basically, more force resistance) is needed, the process for developing it involves several considerations, as follows:

1. What form of strengthening is significant to the particular materials and form of construction of the structure?

2. How much additional strength is required, and what is the basis for its quantification?

3. Should the existing structure be improved or reinforced, or should it be replaced?

4. What available technology or means in general can be used for strengthening a particular type of structure?

A partial answer to question 2 may simply be to bring an existing building up to the level of current building code requirements. This issue is discussed in the next section. Of more significance, however, may be the observed success of previous strengthening efforts on buildings that were subsequently subjected to major windstorms or earthquakes.

Modifications of Stiffness

When a number of structural elements share the task of resisting a single force, the initial share taken by each element will be determined by the relative stiffness of the elements. This issue is discussed at length in Chapters 5 and 6 with regard to the planning and development of the parts of a general lateral resisting system.

A problem arises when the stiffer elements are not necessarily the stronger ones. Such may be the case when the bracing system is a relatively light rigid frame that is connected by construction to an extensive solid wall system. Even though the walls are not supposed to take the lateral forces, their relative stiffness will result in the load being taken by the walls—possibly up to the point where they fail before any significant part of the load is taken by the frame.

The situation just described is a common cause of nonstructural damage, which occurs when the walls are rigid but not strong, as in the case of stucco or brick veneer applied to a frame or even to a plywood shear wall. A possible solution in this situation is to reduce either the stiffness of the walls or the ability of the load transfers through the connections between the walls and the support structure. Where details of the construction can achieve this, the intended bracing system may be made to do its job in the first place, thus avoiding the nonstructural damage.

Some situations involving soft-story failures occur because of this relationship between a frame and the building wall construction. In Fig. 4.21 the building may actually have a rigid frame for its full height, but a rigidly attached upper wall construc-

tion may achieve the actual bracing of the upper levels, particularly in dynamic response to seismic forces. The result may be either extensive damage to the nonstructural walls or a soft-story failure at the open story, or possibly both.

Another type of element that frequently attracts unintended lateral force is window glass. In its own plane, a large sheet of glass has the same stiffness character as a wall surface of plaster, stucco, or masonry veneer. In addition, most types of glass have the same low capacity for tension, shock, and deformation as concrete, masonry, and plaster. Glass breakage in windstorms usually occurs due to out-of-plane wind pressures, but in earthquakes it is usually due to in-plane lateral forces.

Improvement of the lateral resisting structure—whether viewed as strengthening or not—can be achieved partly by improving the relationships between the bracing structure and the rest of the building construction. The building thus becomes more resistive in general to the load conditions, with possibly no actual increase in the strength of the bracing system.

For remedial purposes, this improved relationship between the bracing and the rest of the building may be achieved in one of two basic ways:

1. Reduce the proportionate stiffness and potential for attraction of loads by the nonstructural elements.
2. Increase the stiffness of the bracing system.

13.4 BUILDING CODE RETROFIT

Building codes tend to reflect a consensus of necessary standards. To be enforced, they must be enacted as ordinances by a governmental body (usually at the city or county level). As such, they must accept a role of interest in the public welfare, but must also respond to all the usual pressures of any ordinance. Nobody who understands the process expects a building code to be on the leading edge of developments in any area.

Still, advances are made, often in response to public pressures brought on by major disasters, particularly disastrous fires, floods, windstorms, and earthquakes. These events are usually necessary to overcome a general lethargy and inertia that resists changes, as well as the pressure of various groups that have vested interests (building trade unions, developers, building product manufacturers).

When significant changes are made in building codes, they usually apply only to future construction. Making owners of existing buildings do something to improve their properties is usually very difficult. Thus older buildings are generally affected only by changes in the codes when some significant remodeling or change of occupancy (building use) is proposed, in which case the construction must generally be brought up to the standards of the current codes where necessary.

In this situation the status of any existing building is such that it stands in some particular condition with regard to potential shortcomings. For any changes in the building code that consist of higher standards, the existing stock of buildings is date-stamped with some form of shortcoming. As time goes on, the list of shortcomings grows with each change in the code.

An example of the general situation for code requirements for wind and earthquakes an example may be made using the *Uniform Building Code.* This model code has been issued in a new edition with regular frequency, lately every three years. At the time of publication of this book the current edition is the 1994 edition, with preceding editions in 1991, 1988, 1985, 1982, 1979, 1976, 1973, 1971, and so on. This means that a particular building—constructed in 1972, for example—was probably built in conformance with the requirements of the 1971 *UBC*, if that was the code of jurisdiction. As it stands today, therefore, the building cannot be expected to have any of the features that were required by the eight upgrades of the code since 1971.

Positioning a building in time with regard to code changes may be used to anticipate what might be required for a code retrofit. In the end, of course, the building must comply to the present code if the retrofit is actually required—as for a remodeling, for example. However, if upgrading is desired simply for its benefits, the needs may be more specifically identified by anticipating the construction and its design at the time that work was done.

Since codes are clearly documented, the significant changes for a particular building can be determined relatively easily. However, code changes are not the entire picture; other factors may be of equal or greater significance. All the points raised in Sec. 13.3 should be considered if real improvement—not just getting a permit for remodeling—is the goal.

Of course, not all code changes have necessarily resulted in improved resistance to wind or earthquakes. Adjustments in structural requirements also reflect evidence from research and changes in the availability and use of building materials and products over time. Some examples are the following.

1. The commonly used grade of steel prior to World War II was A7 steel, with a basic allowable bending stress of 20,000 psi. Shortly after the war, the common grade became A36 steel, with a basic allowable bending stress of 24,000 psi. This resulted in a gen-

eral increase in tolerable bending moments some 20% higher, the resulting use of smaller members, and some accompanying increased flexibility (reduced stiffness) of structures in general. As a result, steel structures after about 1945 can be expected to have slightly bouncier floors, skinnier columns, and possibly more sideway (drift) from lateral loads, and potentially, some increased possibilities for buckling failures, which are part of the load limit condition for many steel frames.

2. Starting with the 1963 edition of the ACI Code, reinforced concrete structures have increasingly been designed by what is now called the *strength method* versus the older *working stress method.* In various ways, the strength method permits smaller concrete dimensions and use of more steel as well as higher grades of steel (with higher usable stress). Thus, as with steel structures, increased flexibility is a possibility, as is some increased amount of cracking in structural actions since concrete remains essentially a brittle (tension-crack-developing) material.

3. In the 1960s, increasing pressures from the building industry, unions, and others resulted in building code acknowledgments of the usable diaphragm capacities of wood frame surfacing materials other than plywood. Thus use for shear walls was permitted for gypsum drywall, plaster, stucco, and wood fiber panels. Eventually, wood fiber panels were permitted for roof and floor diaphragms by some codes. This is now viewed as highly questionable by many engineers on the basis of the performance of buildings in windstorms and earthquakes in the past two decades, notably buildings built since the early 1960s.

A conclusion that can be drawn from this is that positioning of a building with respect to its date of construction must be comprehensively developed if real improvement is desired in any structural modifications. Changes in the way that we build should truly be done cautiously so as not to lose really significant advantages in favor of perceived improvements.

13.5 MEANS OF STRENGTHENING

The idea of gaining strength begins with a structure already existing or designed. Strengthening is therefore a form of quantified improvement of the structure's resistance to the forces being considered. The general means for achieving this are the following:

1. Increasing the strength of individual elements of the structure by making them larger or otherwise better for the necessary structural performance (bending, compression, shear, etc.). If this is done for all the elements, some collective gain for the entire structure is inevitable.

2. Redevelopment of the general structural system for better overall performance. Examples: aligning columns to produce bents; adding columns to stiffen a bent; rearranging members of a trussed system for better force distribution; using separation joints to control internal distributions in the whole system; and planning for less rotation effect in a complex, unsymmetrical plan.

3. Improving details for the load-transferring connections between elements

in the system; use of positive connections, and so on.

4. Adding secondary elements for better performance of a system. Examples: adding shear walls or eccentric trussed bracing to a rigid frame.

As discussed previously, structural modifications may improve on stiffness or basic stability of the system, as well as adding pure, raw, force-resistive strength.

The actual means of achieving improvements depends first on whether the building already exists or is still being designed. In design stages it may still be possible to reconsider factors relating to the general building size and form, the exact placement of individual elements of the lateral resistive system, and the basic construction materials. Gains in these areas may conceivably result in a better structure for the general task by improving the actual loading conditions or form of structural responses.

A second major consideration is for the materials and form of the lateral resistive system. Is the basic structural material wood, steel, concrete, or masonry? Is the system a box type, trussed, rigid frame, or some other basic form? Changing materials, basic structural type, and various details of the system may bring improvements before consideration is given to the strength of individual members or connections.

Finally is the consideration for achieving more strength from individual members of the system. Based on their individual tasks, what constitutes the most significant type of strength gain? Adding more mass of material may not be the issue, but rather, improvement in buckling resistance, frame member relative stiffness, or the arrangement of internal elements (such as reinforcement in concrete or masonry elements).

14

STRENGTHENING OF OLDER BUILDINGS

When any major disaster causes significant damage to buildings, the general vulnerability of similar buildings is effectively demonstrated. With time and the accumulation of evidence, this eventually leads to the conclusion that a particular building—of a certain age and type of construction—has identifiable vulnerabilities for various risks, including windstorms and earthquakes. The general idea of adding strength to lateral resistive structures was discussed in Chapter 13. In this chapter we treat the case of existing buildings, for which the work is usually one of remedial alteration of the construction.

14.1 PREDICTABLE VULNERABILITY

The nature and degree of predictable, potential damage to a building depends on the following:

1. The materials of construction and the date of construction. Based on the general technology at any time, the building code in force, and typical construction practices, some reasonable expectations can be forecast.

2. The exact location of the building. This relates to general risk for the region, but also very specifically to the site conditions for an individual building. of special concern are general problems of instability for the site and situations such as a potential for soil liquefaction in an earthquake. There may also be a building/site interaction potential that was not considered in the original structural design.

3. The general condition of the structure. Rusted steel, rotted timbers, cracked masonry or concrete, or other forms of deterioration may exist, so

that the structure is not as good as when it was built.

4. The general effects of winds or earthquakes on similar structures of approximately the same age in the immediate area. This is probably the most reliable indicator of potential problems.

With all of this information, plus a review of the original structural design and construction drawings (if they are available), a reasonable projection of potential damage can be made. The work to be done to reduce the hazard will depend on circumstances discussed in the remaining sections of this chapter.

14.2 PRESERVATION VERSUS REMODELLING

Saving old buildings may be of interest strictly because of their value as real estate or because of continuing need for the facility. In this case, any remedial structural modifications come under general remodeling and would probably be tied to other adjustments in the general construction. The investment in structural strengthening will probably not be made unless the building is also otherwise upgraded, and possibly altered for a different use.

Since remodeling may be done with less concern for cosmetic changes (different exposed materials, different visible details, etc.), a somewhat freer hand may be had with structural alterations. Indeed, an entirely different structure may in some cases be installed, providing support for the old, weaker structure as well as for the general construction.

However, when the building is a historic landmark, or otherwise has value to the community as part of its cultural heritage, it will probably be important not to alter its appearance in any significant way. In this case the most desirable approach is to reinforce, restore, repair, or otherwise fix up the existing structure. If this is not really possible, replacement with newer—but basically the same—materials will be considered. If all that is not feasible, it may be necessary to install an entirely new backup structure, but it will usually be quite difficult to accomplish with no significant changes in the building appearance, both inside and out.

In general, timber and steel structures are the easiest to strengthen, since the structural elements can more easily be replaced or reinforced. This may be made especially easier when structural members are exposed to view, reducing the need to tear up a lot of finished construction.

On the other hand, as discussed in Sec. 14.1, structures of masonry and concrete generally offer more challenge for strengthening efforts. This is further complicated when the structure is exposed and an objective is to preserve its appearance. Restoration may have to give way to replacement, with some form of imitation in some cases.

Despite our most valiant efforts, major windstorms and earthquakes usually result in the loss of some of our cherished heritage of old buildings. Short of completely removing them to some location of less risk, there is a limit to what can be done to preserve highly vulnerable old buildings. Old buildings of unreinforced masonry and vintage concrete construction steadily dwindle away in high-risk windstorm and earthquake zones.

14.3 METHODS OF STRENGTHENING

The following are some methods used to add lateral resistive strength to existing buildings.

Improving the Existing Structure

This may consist of repairing damaged or deteriorated parts. It might also consist of replacing some weak elements with stronger ones: of better material, greater size, or better cross-sectional properties. If there is nothing fundamentally wrong with the existing structure, this approach may be reasonable. It will be easier, of course, for some structures—such as for steel frames versus ones of sitecast concrete.

It may be possible to alter the existing structure to improve its lateral resistance. An example would be to add triangulated elements to a rectilinear framework, giving it a potential for developing trussed resistance for lateral forces. Another example would be to replace a surfacing material on a wood frame with a different surfacing material that has a higher shear wall potential. The idea here is to stick with the basic structure and work to improve it—not replace it.

Adding a Separate Lateral Resistive Structure

An existing structure may be adequate for gravity loads but is not basically constituted to develop resistance to lateral forces. It may be possible to add some form of special bracing that does the job with some degree of independence from the gravity-resisting structure: for example, the construction of a core-bracing system on the interior of a building, with no change in the peripheral structure. This is sometimes the best choice for historic preservation where the building exterior is the most significant feature to be preserved.

Supplement the Existing Structure

The basic lateral resistive system may not be wrong; there is just not enough of it. There may simply not be enough shear walls in the building plan layout, for example, or possibly there are some critical locations where no shear walls exist. Keeping the existing shear walls and adding some more may be a solution.

This may be the case for various forms of vertical bracing, including trussed bracing and rigid frames. Besides strengthening the existing elements or installing an entirely new system, there is the possibility of simply helping out the existing system with some new elements to share the loads.

This method may be a special possibility where major remodeling occurs as part of the upgrading of the old building. The general development of the new construction can include the installation of new bracing elements without touching the old ones.

Better Tying or Anchoring of Existing Elements

Many failures of old buildings consist of major parts of the construction becoming detached. A perfectly good shear wall may come apart from the roof or foundation due to lack of capacity of the connecting construction to transfer the necessary forces. Whole roofs depart the premises and whole buildings slide or lift from their foundations in every major windstorm and earthquake.

Good engineering design for lateral force resistance includes following through very thoroughly with the transfer of forces throughout the structure, all the way into the ground. Major strengthening of many older buildings can often be significantly achieved simply by tying perfectly good existing parts together for better anchorage and load transfers.

These methods must relate to the particular form of vulnerability of a given structure and to the feasibility of its upgrading.

If a structure cannot feasibly be improved, it must be replaced or have other structural elements installed to do its job.

14.4 PERFORMANCE OF STRENGTHENING

Interest in strengthening of older buildings has become increasingly important in recent years. This applies to both remodeling for investment savings versus demolition and replacement, and the preservation of cherished landmarks. Often, structural upgrading is a minor factor compared to cosmetic improvements, new plumbing, adding air conditioning, or restoring highly crafted finishes. Nevertheless, a lot of intensive engineering design has been expended on improving the structural performance of existing buildings, most of it for better resistance to windstorms and earthquakes.

As the stock of strengthened buildings increases steadily, so does the potential for its testing in major disasters. As with the design of new structures, this feeds back to improve judgments about the true effectiveness of various attempts at strengthening.

Anyone approaching the design for the strengthening of an older building is well advised to find out how various strengthening methods performed in recent events. Although the buildings may be old, their strengthening is still a young art, and the latest lessons should receive a lot of attention.

GLOSSARY

The material presented in this glossary constitutes a brief dictionary of words and terms frequently encountered in discussions of the design of building structures. Many of the words and terms have reasonably well-established meanings; in those cases we have tried to be consistent with the accepted usage. In some cases, however, words and terms are given different meanings by different authors or by groups that work in different fields. In these situations we have given the definition as used for this book so that the reader may be clear as to our meaning.

In some cases words or terms are commonly misused with regard to their precise meaning. We have generally used such words and terms as they are broadly understood, but we have given both the correct and popular definitions in some cases. Words and terms are often used differently in different topic areas. Where such is the case, we have given all the uses, identified by the topic areas.

To be clear in its requirements, a legal document such as a building code often defines some words and terms. Reference should be made to such definitions in interpreting such a document. For a fuller explanation of some of the words and terms given here, as well as definitions not given here, the reader should use the index to find the related discussion in the text.

Abutment. Originally, the end support of an arch or vault. Now, any support that receives both vertical and lateral loading.

Acceleration. Rate of change of the velocity, expressed as the first derivative of the velocity (dv/dt) or as the second derivative of the displacement (d^2s/dt^2). Acceleration of the ground surface is more significant than its displacement during an earthquake, as it relates more directly to the force effect. $F = ma$ as a dynamic force.

Accidental Torsion. Minimum torsional effect on buildings required by the *UBC* in some instances, even when there is no actual computed torsional effect. *See also* Torsion.

Active Lateral Pressure. *See* Lateral Pressure.

Adequate. Just enough; sufficient. Indicates a quality of bracketed acceptability—on the one hand, not insufficient on the other hand; not superlative or excessive.

Adobe. Masonry construction that uses unburned (not fired) clay units.

Aerodynamic. Fluid-flow effects of air, similar to current effects in running water.

Allowable Stress. *See* Stress.

Amplitude. *See* Vibration.

Analysis. Separation into constituent parts. In engineering, the investigative determination of the detail aspects of a particular phenomenon. May be qualitative, meaning a general evaluation of the nature of the phenomenon, or quantitative, meaning the numerical determination of the magnitude of the phenomenon.

Anchorage. Refers to attachment for resistance to movement; usually a result of uplift, overturn, sliding, or horizontal separation. Tiedown, or hold-down, refers to anchorage against uplift or overturn. Positive anchorage generally refers to direct fastening that does not easily loosen.

Aseismic. Correct word for description of resistance to seismic effects. Building design is actually *aseismic design*, although the term *seismic design* is more commonly used.

Assemblage. Something put together from parts. A random, unordered assemblage is called a *gathering*. An order assemblage is called a *system*.

Backfill. *See* Fill.

Base Shear. See the *UBC* definition of *base* in Sec. 1625.

Battering. Describes the effect that occurs when two elements in separate motion bump into each other repeatedly, such as two adjacent parts of a structure during an earthquake. Also called *hammering* or *pounding*.

Bent. Planar framework, or some portion of one, that is designed for resistance to both vertical and horizontal forces in the plane of the frame. May be achieved as a *rigid frame* with moment-resisting joints or as a *braced frame* with trussing.

Box System. Structural system in which lateral loads are not resisted by a vertical load-bearing space frame but rather by shear walls or a braced frame.

Braced Frame. Literally, any framework braced against lateral forces. Codes use the term for a frame braced by trussing (triangulation).

Bracing. In structural design usually refers to provision for the resistance to movements caused by lateral forces or by the effects of buckling, torsional rotation, sliding, and so on.

Brittle Fracture. Sudden ultimate failure in tension or shear. The basic structural behavior of so-called brittle materials.

Buckling. Collapse, in the form of sudden sideways deflection, of a slender element subjected to compression.

Buffeting. Wind effect caused by turbulent airflow or by changes in the wind direction that result in whipping, rocking, and so on.

Building Code. *See* Code.

Cavity Wall. Wall built of two or more wythes of masonry units so arranged as to provide a continuous airspace within the wall. The facing and backing, outer wythes, are tied together with non-corrosive ties (e.g., brick or wire).

Centroid. Geometric center of an object, usually analogous to the center of gravity. The point at which the entire mass of the object may be considered to be concentrated when considering moment of the mass.

CMU. Concrete masonry unit; or, good old concrete blocks.

Code. Typically used to refer to the document that provides requirements for design, construction, or manufacture of materials and systems for buildings; that is, the *Building Code.*

Collector. Force transfer element that functions to collect loads from a horizontal diaphragm and distribute them to the vertical elements of the lateral resistive system.

Compaction. Action that tends to lower the void ratio and increase the density of a soil mass. When produced by artificial means, the degree of compaction obtained is measured in percent with reference to the theoretical minimum volume of the soil.

Confined Concrete. Concrete mass that is sufficiently wrapped by constraining reinforcement (ties, spirals, stirrups, etc.) so as to develop three-dimensional stress conditions.

Connection. Union or joining of two or more distinct elements. In a structure, the connection itself often becomes an entity. Thus actions of the parts on each other may be visualized in terms of their actions on the connection.

Consolidation. Volume reduction in a soil mass produced by a lowering of the void ratio. The effect resulting from compaction, shrinkage, and so on.

Continuity. Most often used to describe structures or parts of structures that have behavior characteristics influenced by the monolithic, continuous nature of adjacent elements, such as continuous, vertical multistory columns; continuous, multispan beams; and rigid frames.

Core Bracing. Vertical elements of a lateral-bracing system developed at the location of permanent interior walls for stairs, elevators, duct shafts, or rest rooms.

Crawl Space. Space between the underside of the floor construction and the ground surface that occurs when a framed floor is held above the ground but there is no basement.

Critical Damping. Amount of damping that will result in a return from initial deformation to the neutral position without reversal.

Curtain Wall. Exterior building wall that is supported entirely by the frame of the building rather that being self-supporting or load bearing.

Cut. Removal of existing soil deposits during the recontouring (or grading) of the ground surface. *See also* Grade.

Damping. *See* Vibration.

Dead Load. *See* Load.

Deep Foundation. Foundation used to achieve a considerable extension of the bearing effect of a supported structure below the ground surface. Elements most commonly used are *piles* or *piers.*

Deflection. Generally refers to the lateral movement of a structure caused by loads, such as the vertical sag of a beam, the bowing of a surface under wind pressure, or the lateral sway of a tower. *See also* Drift.

Degree of Freedom. *See* Freedom.

Density. *See* Mass.

Design. Conception, contrivance, or planning of a work (verb). Descriptive image (picture, model, etc.) of a proposed work (noun).

Determinate. Having defined limits; definite. In a structure, the condition of having the exact sufficiency of stability externally and internally, therefore being determinable by the resolution of forces alone. An excess of stability conditions produces a structure characterized as indeterminate.

Diaphragm. Surface element (deck, wall, etc.) used to resist forces in its own plane by spanning or cantilevering. *See also* Horizontal Diaphragm *and* Shear Wall.

Displacement. Movement away from some fixed reference point. Motion is described mathematically as a displacement–time function. *See also* Acceleration *and* Velocity.

Drag. Generally refers to wind effects on surfaces parallel to the wind direction. Ground drag refers to the effect of the ground surface in slowing the wind velocity near ground level.

Drag Strut. Structural member used to transfer lateral load across the building and into some part of the vertical system. *See also* Collector.

Drift. Generally refers to lateral deflection. *Story drift* refers to lateral movement of one level of a structure with respect to another.

Dual Bracing System. Combination of a moment-resisting space frame and shear walls or braced frames, with the combined systems designed to share the lateral loads.

Ductile. Describes the load–strain behavior that results from the plastic yielding of materials or connections. To be significant, the plastic strain prior to failure should be considerably more than the elastic strain up to the point of plastic yield.

Ductile Moment-Resisting Space Frame. Rigid frame structure that complies with code requirements intended to assure a ductile yielding form of response to seismic forces.

Dynamic. Generally used to characterize load effects or structural behaviors that are nonstatic in nature. That is, they involve time-related considerations such as vibrations, energy effects versus simple force, and so on.

Earthquake. Common term used to describe sensible ground movements, usually caused by subterranean faults or explosions. The point on the ground surface immediately above the subterranean shock is called the *epicenter*. The magnitude of the energy released at the location of the shock is the basis for the rating of the shock on the *Richter scale*.

Eccentric Braced Frame. Braced frame in which the bracing members do not connect to the joints of the beam-and-column frame, thus resulting in axial force in the braces, but bending and shear in the frame members. Forms include: kneebrace, K-brace, chevron-brace (two forms: V-brace and inverted V-brace).

Elastic. Used to describe two aspects of stress–strain behavior. The first is a constant stress–strain proportionality, or constant modulus of elasticity, as represented by a straight-line form of the stress–strain graph. The second is the limit within which all the strain is recoverable; that is, there is no permanent deformation. The latter phenomenon may occur even though the stress–strain relationship is nonlinear.

Element. Component or constituent part of a whole. Usually a distinct, separate entity.

Energy. Capacity for doing work; what is used up when work is done. Occurs in various forms; mechanical, heat, chemical, electrical, and so on.

Epicenter. *See* Earthquake.

Equalized Settlement. Design method in

which a related set of foundations is designed for equal settlement under dead load rather than for a uniform bearing pressure under total load.

Equilibrium. Balanced state or condition, usually used to describe a situation in which opposed effects neutralize each other to produce a net effect of zero.

Equivalent Static Force Analysis. Technique by which a dynamic effect is translated into a hypothetical (equivalent) static effect that produces a similar result.

Erosion. Progressive removal of a soil mass due to water, wind, or other effects.

Essential Facilities. Building code term for a building that should remain functional after a disaster such as a windstorm or major earthquake; affects establishment of the *I* factor for base shear or design wind pressure.

Excavation. Removal of soil mass to permit construction.

Factored Load. Percentage of the actual service load (usually an increase) used for design by strength methods. *See also* Load.

Failure. The condition of becoming incapable of a particular function. May have partial as well as total connotations in structures. For example, a single connection may fail, but the structure might not collapse because of its ability to redistribute the load.

Fatigue. A structural failure that occurs as the result of a load applied and removed (or reversed) repeatedly through a large number of cycles.

Fault. Subterranean effect that produces an earthquake. Usually a slippage, cracking, sudden strain release, and so on.

Feasible. Capable of being or likely to be accomplished.

Fill. Usually refers to a soil deposit produced by other than natural effects. *Backfill* is soil deposited in the excessive part of an excavation after completion of the construction.

Flutter. Flapping, vibration type of movement of an object in a high wind. Essentially a resonant behavior. *See also* Vibration.

Footing. Shallow, bearing-type foundation element consisting typically of concrete that is poured directly into an excavation.

Force. Effort that tends to change the shape or the state of motion of an object.

Fracture. Break, usually resulting in actual separation of the material. A characteristic result of tension failure.

Freedom. In structures, usually refers to the lack of some type of resistance or constraint. In static analysis, the connections between members and the supports of the structure are qualified as to type, or degree, of freedom. Thus the terms *fixed support*, *pinned support*, and *sliding support* are used to qualify the types of movement resisted. In dynamic analysis the degree of freedom is an important factor in determining the dynamic response of a structure.

Freestanding Wall. *See* Wall.

Frequency. In harmonic motion (bouncing springs, vibrating strings, and swinging pendulums, etc.), the number of complete cycles of motion per unit of time. *See also* Vibration.

Friction. Resistance to sliding developed at the contact face between two surfaces.

Fundamental Period. *See* Period.

Gable Roof. Double-sloping roof formed by joined rafters or rigid frames with a ridge or peak at the top. A *gable* is the upper triangular portion of a wall at the end of the roof.

Geophysical. Refers to the physical behavior characteristics of the ground surface and of subterranean masses.

Geotechnical. General term now used to refer to the engineering field dealing with soils and the behavior of structures in and on the ground.

Grade. 1. Level of the ground surface. 2. Rated quality of material used for wood and steel.

Grade Beam. Horizontal element in a foundation system that serves some spanning or load-distributing function.

Grain. 1. Discrete particle of the material that constitutes a loose material, such as soil. 2. Fibrous orientation of wood.

Grout. Lean concrete (predominantly water, cement, and sand) used as a filler in the voids of hollow masonry units, under column base plates, and so on.

Grouted Masonry. Masonry of hollow units in which the voids are filled with grout.

Gust. Increase, or surge, of short duration in the wind velocity.

Hammering. *See* Battering.

Header. Horizontal element over an opening in a wall or at the edge of an opening in a roof or floor.

Hertz. Cycles per second.

Hold-Down. *See* Anchorage.

Horizontal Bracing System. Truss system in a horizontal plane that functions as a horizontal diaphragm.

Horizontal Diaphragm. *See* Diaphragm. Usually a roof or floor deck used as part of the lateral bracing system.

Impact. Action of striking or hitting.

Impulse. Impelling force action, characterized by rapid acceleration or deceleration. *Examples*: gust of wind; violent thrust from an earthquake.

Indeterminate. *See* Determinate.

Inelastic. *See* Stress–Strain Behavior.

Inertia. *See* Mass.

Interrupted Shear Wall. Shear wall that is not continuous to its foundation.

Irregular Structure. *See* Regular Structure.

Kern Limit. Limiting dimension for the eccentricity of a compression force if tension is to be avoided.

Key. Slot or protrusion developed to resist shear, as in the manner of a tongue-and-groove joint.

Lateral. Literally means to the side or from the side. Often used to refer to something that is perpendicular to a major axis or direction. With reference to the vertical direction of the gravity forces, wind, earthquakes, and horizontally directed soil pressures are called *lateral effects.*

Lateral Pressure. Horizontal soil pressure of two kinds: 1. *Active* lateral pressure is that exerted by a retained soil upon the retaining structure. 2. *Passive* lateral pressure is that exerted by soil against an object that is attempting to move in a horizontal direction.

Let-In Bracing. Diagonal boards nailed to studs to provide trussed bracing in the wall plane. In order not to interfere with the surfacing materials of the wall, they are usually notched in, or let in, to the stud faces.

Lintel. Beam placed over an opening in a wall.

Live Load. *See* Load.

Load. Active force (or combination of forces) exerted on a structure. *Dead load* is permanent load due to gravity, which includes the weight of the structure itself. *Live load* is any load component that is not permanent, including those due to wind, seismic effects, temperature changes, or shrinkage; but the term is most often used for gravity loads that are not permanent. *Service load* is the total load combination that the structure is expected to experience in use. *Factored* *load* is the service load multiplied by some increase factor for use in strength design.

Macro. Implies upper limits of scale; large, excessive. *See also* Micro.

Masonry Unit. Brick, stone, concrete block, glass block, or hollow clay tile intended to be laid in mortar.

Mass. Dynamic property of an object that causes it to resist changes in its state of motion. This resistance is called *inertia.* The magnitude of the mass per unit volume of the object is called its *density.* Dynamic force is defined by $F = ma$ or force equals mass times acceleration. Weight is defined as the force produced by the acceleration of gravity; thus $W = mg.$

Mat Foundation. Very large bearing-type foundation. When the entire bottom of a building is constituted as a single mat, it is also called a *raft foundation.*

Maximum Density. Theoretical density of a soil mass achieved when the void is reduced to the minimum possible.

Member. One of the distinct elements of an assemblage.

Micro. Implies lower limit of scale. Precise meaning: "very small." *See also* Macro.

Mercali Scale. Measurement system used to determine the location of the epicenter of an earthquake on the basis of defining of zones of relative intensity of observed damage and experiences of persons during the quake. This is essentially the basis for the determination of the zones of intensity of risk in the *UBC.*

Modified Mercali Scale. *See* Mercali Scale. Present version of the system in use.

Moment-Resisting Space Frame. A vertical load-bearing framework in which members are capable of resisting forces primarily by flexure (*UBC* definition). *See also* Rigid Frame.

Natural Period. *See* Period.

Normal. 1. Ordinary, usual, unmodified state of something. 2. Perpendicular, such as pressure normal to a surface, stress normal to a cross section, and so on.

Occupancy. In building code language, refers to the use of a building as a residence, school, office, and so on.

Occupancy Importance Factor (*I*). *UBC* term used in the basic equation for lateral force. Accounts for possible increased concern for certain occupancies.

Optimal. Best; most satisfying. The best solution to a set of criteria is the optimal one. When the criteria have opposed values, there may be no single optimal solution, except by the superiority of a single criterion (e.g., the lightest, the strongest, the cheapest, and so on).

Organic. Refers to material of biological (plant or animal) origin.

Overturn. Toppling, or tipping over, effect of lateral loads.

Parapet. Extension of a wall plane or the roof edge facing above the roof level.

Particle. Minute part. In structures, usually a very small piece of material, slightly bigger than molecular size. *See also* Grain.

Passive Soil Pressure. *See* Lateral pressure.

P–Delta Effect. Secondary effect on members of a frame, induced by the vertical loads acting on the laterally displaced frame.

Pedestal. Short pier or upright compression member. Is actually a short column with a ratio of unsupported height to least lateral dimension of 3 or less.

Perimeter Bracing. Vertical elements of a lateral bracing system located at the building perimeter (outer walls).

Period (of Vibration). Total elapsed time for one full cycle of vibration. For an elastic structure in simple, single-mode vibration, the period is a constant (called the *natural* or *fundamental period*) and is independent of the magnitude of the amplitude, of the number of cycles, and of most damping or resonance effects. *See also* Vibration.

Pier. 1. Short, stocky column with height not greater than three times its least lateral dimension. The *UBC* defines a masonry wall as a pier if its plan length is less than three times the wall thickness. 2. Deep foundation element that is placed in an excavation rather than being driven as a pile. Although it actually refers to a particular method of excavation, the term *caisson* is also used to describe a pier foundation.

Pilaster. Integral portion of the wall that projects on one or both sides and acts as a vertical beam, a column, an architectural feature, or any combination thereof.

Pile. Deep foundation element, consisting of a linear, shaftlike member that is placed by being driven dynamically into the ground. *Friction piles* develop resistance to both downward load and upward (pullout) load through friction between the soil and the surface of the pile shaft. *End-bearing piles* are driven so that their ends are seated in low-lying strata of rock or very hard soil.

Plastic. 1. Usually, a synthetic material of organic origin, including many types of resins, polymers, cellulose derivatives, and casein materials. 2. In structural investigation, the type of stress response that occurs in ductile behavior, usually resulting in considerable, permanent deformation.

Plastic Hinge. Region where the ultimate moment strength of a ductile member may be developed and maintained with corresponding significant rotation as a result of the local yielding of the material.

Polar Moment of Inertia. Second moment of an area about a line that is perpendicular to the plane of the area. Of significance in investigation of response to torsion.

Positive Anchorage. *See* Anchorage.

Pounding. *See* Battering.

Preconsolidation. Condition of a highly compressed soil, usually referring to a condition produced by the weight of soil above on some lower soil strata. May also refer to a condition produced by other than natural causes, such as piling up of soil on the ground surface, vibration, or saturating to dissolve soil bonding.

Pressure. Force distributed over, and normal to, a surface.

Presumptive Soil Pressure. Value for the allowable vertical bearing pressure that is permitted to be used in the absence of extensive investigation and testing. Requires a minimum of soil identification and is usually quite conservative.

Quick Soil. Soil deposit that is reasonably stable if undisturbed, but suddenly becomes quite loose and fluidlike when disturbed.

Raft Foundation. *See* Mat Foundation.

Reaction. Response. In structures, the response of the structure to the loads; the response of the supports to the actions of the structure. The term *reactions* usually refers to the components of force developed at the supports.

Reentrant Corner. Exterior corner in a building plan having a form such as that at the junction of the web and flange of a T.

Regular Structure. Structure (building) having no significant discontinuities in plan or in vertical configuration or in its lateral-force-resisting systems such as those described for irregular structures. *UBC* definition: Sec. 1625.

Reinforce. To strengthen, usually by adding something.

Relative Stiffness. *See* Stiffness.

Reserve Energy. Energy that a ductile system is capable of absorbing by plastic strains.

Resilience. Measurement of the absorption of energy by a structure without permanent deformation or fracture. *See* Toughness.

Resonance. *See* Vibration.

Restoring Moment. The resistance to overturn due to the weight of the affected object.

Retaining Wall. Structure used to brace a vertical cut, or a change in elevation of the ground surface. The term is usually used to refer to a *cantilever retaining wall*, which is a freestanding structure consisting only of a wall and its footing, although basement walls also serve a retaining function.

Retrofit. Usually refers to the task of bringing an existing object (building, etc.) into conformance with recent, typically more stringent, standards.

Richter Scale. Log-based measuring system for evaluation of the relative energy level of an earthquake at its center of origin.

Rigid Bent. *See* Rigid Frame.

Rigid Frame. Framed structure in which the joints between the members are made to transmit moments between the ends of the members. Called a *bent* when the frame is planar.

Rigidity. Quality of resistance to deformation. Structures that are not rigid are called *flexible*.

Risk. Degree of probability of loss due to some potential hazard. The risk of an earthquake in a particular location is the basis for the Z factor in the *UBC* equation for seismic base shear.

Rotation. Motion in a circular path. Also used to describe twisting, or torsional, effect. *See also* Torsion.

Safety. Relative unlikelihood of failure, absence of danger. The *safety factor* is the ratio of the resisting capacity of a structure to the actual demand on the structure.

Saturation. The condition that exists when the void in the soil is completely filled with water. A condition of *partial saturation* exists when the void is only partly filled with water. *Oversaturation*, or *supersaturation*, occurs when the soil contains water in excess of the normal volume of the void, which usually results in some flotation or suspension of the solid particles in the soil.

Section. The two-dimensional profile or area obtained by passing a plane through a form. *Cross section* implies a section at right angles to another section or to the linear axis of an object.

Seismic. Pertaining to ground shock. *See also* Aseismic.

Separation. Often used in structural investigation to denote situations in which parts of a structure are made to act independently. Partial separation refers to a selective, controlled situation that allows for some types of interaction while permitting independence of motion in other ways, for example—a connection that provides for vertical support but permits rotation and horizontal movements.

Service Load. *See* Load.

Setback. Abrupt discontinuity in the vertical configuration of a building or its lateral force resisting structure, consisting of a reduction in the total width of the profile.

Settlement. General downward movement of a foundation caused by the loads and the reactions of the supporting soil materials.

Shear. Force effect that is lateral (perpendicular) to the major axis of a structure, or one that involves a slipping effect, as opposed to a push-pull effect. Wind and earthquake forces are sometimes visualized as shear effects on a building because they are perpendicular to the major vertical (gravity) axis of the building.

Shear Wall. Vertical diaphragm.

Shoring. Bracing; usually refers to the bracing of a cut, a shoreline, or some structure that is in danger from erosion, undercutting, unequal settlements, and so on.

Site Coefficient (S). *UBC* term used in the basic equations for base shear to account for the effect of the period of the ground mass under the building.

Slenderness. Relative thinness. In structures, the quality of flexibility or lack of buckling resistance is inferred by extreme slenderness.

Soft Story. Story level in a multistory structure in which the lateral stiffness of the story is significantly less than that of stories above. See *UBC* definition in Table 16-L.

Space Frame. Ambiguous term used variously to describe three-dimensional structures. The *UBC* has a particular definition used in classifying structures for response to seismic effects.

Specific Gravity. Relative indication of density, using the weight of water as a reference. A specific gravity of 2 indicates a density (weight) of twice that of water.

Spectrum. In seismic analysis generally refers to the curve that describes the ac-

tual dynamic force effect on a structure as a function of variation in its fundamental period. Response spectra are the family of curves produced by various degrees of damping. This represents the basis for determining the building's general response to seismic force.

Stability. Inherent capability of a structure to develop force resistance as a property of its form, orientation, articulation of its parts, type of connections, methods of support, and so on. Stability is not directly related to quantified strength or stiffness, except when the actions involve the buckling of elements of the structure.

Static. State that occurs when the velocity is zero; thus no motion is occurring; generally used to refer to situations in which no change is occurring.

Stiffness. In structures, refers to resistance to deformation, as opposed to strength, which refers to resistance to force. A lack of stiffness indicates a flexible structure. Relative stiffness usually refers to the comparative deformation of two or more structural elements that share a load.

Story Drift. Total lateral displacement occurring in a single story of a multistory structure.

Strain. Deformation resulting from stress; usually measured as a percentage of deformation, called *unit strain* or *unit deformation*, and is dimensionless.

Strength. Capacity to resist force.

Strength Design. One of the two fundamental design techniques for assuring a margin of safety for a structure. *Stress design*, also called *working stress design*, is performed by analyzing stresses produced by the estimated actual usage loads, and assigning limits for the stresses that are below the ultimate capacity of the materials by some margin. *Strength design*, also called *ultimate strength de-*

sign, is performed by multiplying the actual loads by the desired factor of safety (the universal average factor being 2) and proceeding to design a structure that will have that load as its ultimate failure load.

Strengthening. Adding strength to a structure in one of two ways: (1) by reinforcing of an existing construction, or (2) by adding elements as required for an additional load consideration for a structure designed for other loads. Thus an existing building may be strengthened with additional bracing for earthquakes, or a designed structure may be modified before construction for additional strength development.

Stress. The mechanism of force within the material of a structure; visualized as a pressure effect (tension or compression) or a shear effect on the surface of a unit of the material and quantified in units of force per unit area. *Allowable*, *permissible*, or *working* stress refers to a stress limit that is used in stress design methods. *Ultimate* stress refers to the maximum stress that is developed just prior to failure of the material.

Stress Design. *See* Strength Design.

Stress–Strain Behavior. The relation of stress to strain in a material or a structure. Is usually visually represented by a stress–strain graph covering the range from no load to failure. Various aspects of the form of the graph define particular behavioral properties. A straight line indicates an elastic relationship; a bend or curve indicates inelastic behavior. A sudden bend in the graph usually indicates a plastic strain or yield that results in some permanent deformation. The slope of the graph (or of a tangent to the curve) is defined as the modulus of elasticity of the material.

Stud. One of a set of small, closely spaced columns used to produce a framed wall.

Subsidence. Settlement of a soil mass, usually manifested by the sinking of the ground surface.

Surcharge. Vertical load applied at the ground surface or simply above the level of the bottom of a footing. The weight of soil above the bottom of a footing is surcharge for that footing.

Tiedown. *See* Anchorage.

Torsion. Moment effect involving twisting or rotation that is in a plane perpendicular to the major axis of an element. Lateral loads produce torsion on a building when they tend to twist it about its vertical axis. This occurs when the centroid of the load does not coincide with the center of stiffness of the vertical elements of the lateral load-resisting structural system.

Toughness. The measurement of the total dynamic energy capacity of a structure, up to the point of complete failure. *See also* Resilience.

Ultimate Strength. Usually used to refer to the maximum static force resistance of a structure at the time of failure. This limit is the basis for the so-called strength design methods, as compared to the stress design methods that use some established stress limit, called the design stress, working stress, permissible stress, and so on.

Underpinning. Propping up of a structure that is in danger of, or has already experienced, some support failure; notably that due to excessive settlement, undermining, erosion, and so on.

Upheaval. Pushing upward of a soil mass.

Uplift. Net upward force effect; may be due to wind, to overturning moment, or to upward seismic acceleration.

Velocity. Time rate of a motion, also commonly called *speed.*

Veneer. Masonry facing that is attached to the backup but not so bonded as to act with it under load.

Vertical Diaphragm. *See* Diaphragm. Also called a *shear wall.*

Vibration. Cyclic, rhythmic motion of a body such as a spring. Occurs when the body is displaced from a neutral position and seeks to restore itself to a state of equilibrium when released. In its pure form it occurs as a harmonic motion with a characteristic behavior described by the cosine form of the displacement–time graph of the motion. The magnitude of linear displacement from the neutral position is called the *amplitude.* The time elapsed for one full cycle of motion is called the *period.* The number of cycles occurring in 1 second is called the *frequency.* Effects that tend to reduce the amplitude of succeeding cycles are called *damping.* The increase of amplitude in successive cycles is called a *resonant effect.*

Viscosity. General measurement of the mobility, or free-flowing character, of a fluid or semifluid mass. Heavy oil is viscous (has high viscosity); water has low viscosity.

Void. Open space within an object. In soils, refers to the portion of the volume not occupied by the solid materials, but filled partly with gas (air, etc.) and partly with liquid (water, oil, etc.).

Void ratio. Term most commonly used to indicate the amount of void in a soil mass, expressed as the ratio of volume of the void to the volume of the solids.

Wall. Vertical, planar building element. *Foundation* walls are those that are partly or totally below ground. *Bearing* walls are used to carry vertical loads in direct compression. *Grade* walls are those that are used to achieve the transition between the building that is above the ground and the foundations that are below it. (*See also* Grade Beam.) *Shear* walls are those used to brace the building against horizontal forces due to wind or seismic shock. *Freestanding* walls are walls whose tops are not laterally braced. *Retaining* walls are walls that resist horizontal soil pressure.

Weak Story. Story level in a multistory structure in which the lateral strength is significantly less than that of stories above. See *UBC* definition in Table 16-L.

Wind Stagnation Pressure. Reference pressure established by the basic wind speed for the region; used in determining design wind pressures.

Working Stress. *See* Stress.

Working Stress Design. *See* Strength Design.

Yield. *See* Stress–strain Behavior.

Zone. Usually refers to a bounded area on a surface, such as the ground surface or the plan of a level of a building.

APPENDIX A

DYNAMIC EFFECTS

A good lab course in physics should provide a reasonable understanding of the basic ideas and relationships involved in dynamic behavior. A better preparation is a course in engineering dynamics that focuses on the topics in an applied fashion, dealing directly with their applications in various engineering problems. The material in this section consists of a brief summary of basic concepts in dynamics that will be useful to those with a limited background and that will serve as a refresher for those who have studied the topics before.

A.1 KINEMATICS

The general field of dynamics may be divided into the areas of *kinetics* and *kinematics*. *Kinematics* deals exclusively with motion, that is, with time–displacement relationships and the geometry of move-

ments. *Kinetics* adds the consideration of the forces that produce or resist motion.

Motion can be visualized in terms of a moving point, or in terms of the motion of a related set of points that constitute a body. The motion can be qualified geometrically and quantified dimensionally. In Fig. A.1a the point is seen to move along a path (its geometric character) a particular distance. The distance traveled by the point between any two separate locations on its path is called *displacement*. The idea of motion is that this displacement occurs over time, and the general mathematical expression for the time–displacement function is

$$s = f(t)$$

Velocity is defined as the rate of change of the displacement with respect to time. As an instantaneous value, the velocity is expressed as the ratio of an increment of dis-

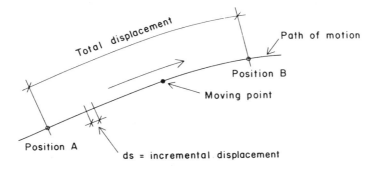

(a) Motion of a Point

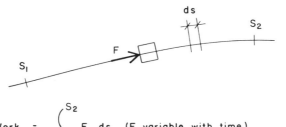

$$\text{Work} = \int_{S_1}^{S_2} F_t \, ds \quad \text{(F variable with time)}$$

$$= F(S_2 - S_1) \quad \text{(F constant with time)}$$

(b) Kinetics of a Moving Object

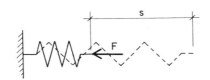

Potential (stored) energy:

$$E = F \cdot k \cdot s$$

$$k = \text{spring constant}$$

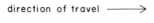

direction of travel ⟶

Kinetic energy:

$$E = \tfrac{1}{2} m (v_1^2 - v_2^2)$$

or, if $v_2 = 0$:

$$E = \tfrac{1}{2} m v_1^2$$

(c) Forms of Mechanical Energy

FIGURE A.1 Aspects of dynamic actions.

placement (*ds*) divided by the increment of time (*dt*) elapsed during the displacement. Using the calculus, the velocity is thus defined as

$$v = \frac{ds}{dt}$$

That is, the velocity is the first derivative of the displacement.

If the displacement occurs at a constant rate with respect to time, it is said to have *constant velocity*. In this case the velocity may be expressed more simply without the calculus as

$$v = \frac{\text{total displacement}}{\text{total elapsed time}}$$

When the velocity changes over time, its rate of change is called the *acceleration* (*a*). Thus, as an instantaneous change,

$$a = \frac{dv}{dt} = \frac{d^2s}{dt^2}$$

That is the acceleration is the first derivative of the velocity or the second derivative of the displacement with respect to time.

Except for the simplest cases, the derivation of the equations of motion for an object generally requires the use of the calculus in the operation of these basic relationships. Once derived, however, motion equations are generally in algebraic form and can be used without the calculus for application to problems. An example is the set of equations that describe the motion of a free-falling object acted on by the earth's gravity field. Under idealized conditions (ignoring air friction, etc.) the distance of fall from a rest position will be

$$s = f(t) = 16.1t^2 \ (s \text{ in ft}, \ t \text{ in sec})$$

This equation indicates that the rate of fall (the velocity) is not a constant but increases with the elapsed time, so that the velocity at any instant of time may be expressed as

$$v = \frac{ds}{dt} = \frac{d(16.1t^2)}{dt} = 32.2t \ (v \text{ in ft/sec})$$

and the acceleration as

$$a = \frac{dv}{dt} = \frac{d(32.2t)}{dt} = 32.2 \text{ ft/sec}^2$$

which is the acceleration of gravity.

Kinematics also includes the study of the various forms of motion: translation, rotation, plane motion, motion of deformable bodies, and so on. A study of the mechanics of motion is very useful in the visualization of the deformation of a structure by static as well as dynamic forces.

A.2   KINETICS

As stated previously, kinetics includes the additional consideration of the forces that cause motion. This means that in addition to the variables of displacement and time, we must consider the mass of moving objects. From Newtonian physics the simple definition of mechanical force is

$$F = ma = \text{mass} \times \text{acceleration}$$

Mass is the measure of the property of inertia, which is what causes an object to resist change in its state of motion. The more common term for dealing with mass is *weight*, which is the force defined as

$$W = mg$$

where *g* is the constant acceleration of gravity (32.2 ft/sec^2).

Weight is literally a dynamic force, although it is the standard means of measurement of force in statics when the velocity is assumed to be zero. Thus, in static analysis we express forces simply as

$$F = W$$

and in dynamic analysis, when using weight as the measure of mass, we express force as

$$F = ma = \frac{W}{g} a$$

If a force moves an object, work is done. *Work* is defined as the product of the force multiplied by the displacement (distance traveled). If the force is constant during the displacement, work may be simply expressed as

$$w = Fs = \text{force} \times \text{total distance traveled}$$

If the force varies with time, the relationship is more generally expressed with the calculus as

$$w = \int_{s_2}^{s_1} F_t \, ds$$

indicating that the displacement is from position s_1 to position s_2, and the force varies in some manner with respect to time.

Figure A.1*b* illustrates these basic relationships. In dynamic analysis of structures the dynamic "load" is often translated into work units in which the distance traveled is actually the deformation of the structure.

Energy may be defined as the capacity to do work. Energy exists in various forms such as heat, mechanical, and chemical. For structural analysis the concern is with mechanical energy, which occurs in one of two forms. *Potential energy* is stored energy, such as that in a compressed spring or an elevated weight. Work is done when the spring is released or the weight is dropped. *Kinetic energy* is possessed by bodies in motion; work is required to change their state of motion, that is, to slow them down or speed them up (Fig. A.1*c*).

In structural analysis energy is considered to be indestructible, that is, it cannot be destroyed, although it can be transferred or transformed. The potential energy in the compressed spring can be transferred into kinetic energy if the spring is used to propel an object. In a steam engine the chemical energy in the fuel is transformed into heat and then into pressure of the steam and finally into mechanical energy delivered as the engine's output.

An essential idea is that of the conservation of energy, which is a statement of its indestructibility in terms of input and output. This idea can be stated in terms of work by saying that the work done on an object is totally used and that it should therefore be equal to the work accomplished plus any losses due to heat, air friction, and so on. In structural analysis we make use of this concept by using a *work equilibrium* relationship similar to the static force equilibrium relationship. Just as all the forces must be in balance for static equilibrium, so the work input must equal the work output (plus losses) for work equilibrium.

A.3   HARMONIC MOTION

A special kinematic problem of major concern in structural analysis for dynamic effects is that of *harmonic motion*. The two elements generally used to illustrate this type of motion are the swinging pendulum and the bouncing spring. Both the pendulum and the spring have a neutral position where they will remain at rest in static

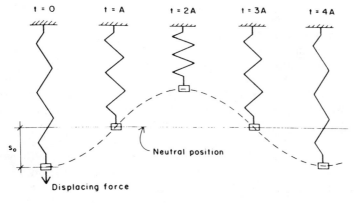

$t = 0$ $t = A$ $t = 2A$ $t = 3A$ $t = 4A$

S_o

Neutral position

Displacing force

(a) The Moving Spring

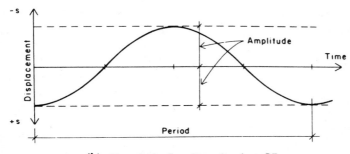

−s

Displacement

Amplitude

Time

+s

Period

(b) Plot of the Equation: $S = A \cos BT$

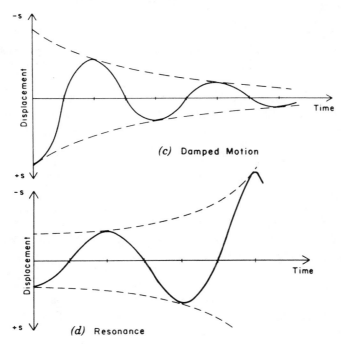

−s

Displacement

Time

(c) Damped Motion

+s

−s

Displacement

Time

(d) Resonance

+s

FIGURE A.2 Aspects of harmonic behavior.

equilibrium. If one displaces either of them from this neutral position by pulling the pendulum sideways or compressing or stretching the spring, they will tend to move back to the neutral position. Instead of stopping at the neutral position, however, they will be carried past it by their momentum to a position of displacement in the opposite direction. This sets up a cyclic form of motion (swinging of the pendulum; bouncing of the spring) that has some basic characteristics.

Figure A.2a illustrates the typical motion of a bouncing spring. Using the calculus and the basic motion and force equations, the displacement–time relationship may be derived as

$$s = A \cos Bt$$

The cosine function produces the basic form of the graph, as shown in Fig. A.2b. The maximum displacement from the neutral position is called the *amplitude.* The time elapsed for one full cycle is called the *period.* The number of full cycles in a given unit of time is called the *frequency* (usually expressed in cycles per second) and is equal to the inverse of the period. Every object subject to harmonic motion has a fundamental period (also called natural period), which is determined by its weight, stiffness, size, and so on.

Any influence that tends to reduce the amplitude in successive cycles is called a *damping effect.* Heat loss in friction, air resistance, and so on, are damping effects. Shock absorbers, counterbalances, cushioning materials, and other devices can also be used to damp the amplitude. Figure A.2c shows the form of a damped harmonic motion, which is the normal form of most such motions, because perpetual motion is not possible without a continuous reapplication of the original displacing force.

Resonance is the effect produced when the displacing effort is itself harmonic with a cyclic nature that corresponds with the period of the impelled object. An example is someone bouncing on a diving board in rhythm with the board's fundamental period, thus causing a reinforcement, or amplification, of the board's free motion. This form of motion is illustrated in Fig. A.2d. Unrestrained resonant effects can result in intolerable amplitudes, producing destruction or damage of the moving object or its supports. A balance of damping and resonant effects can sometimes produce a constant motion with a flat profile of the amplitude peaks.

Loaded structures tend to act like springs. Within the elastic stress range of the materials, they can be displaced from a neutral (unloaded) position and, when released, will go into a form of harmonic motion. The fundamental period of the structure as a whole, as well as the periods of its parts, are major properties that affect responses to dynamic loads.

A.4   DYNAMIC EFFECTS IN BUILDINGS

Load sources that involve motion, such as wind, earthquakes, walking people, moving vehicles, and vibrating heavy machinery, have the potential to cause dynamic effects on structures. Analyzing for their effects requires consideration of essential dynamic properties of the structure. These properties are determined by the size, weight, relative stiffness, fundamental period, type of support, and degree of elasticity of the materials of the structure and by various damping influences that may be present.

Dynamic load sources deliver an energy load to the structure that may be in the form of an impact, such as that caused by the moving air bumping into the stationary building. In this case the energy load is derived from the kinetic energy of the moving air, which is a product of its mass and velocity. In the case of an earthquake, or the vibration of heavy machinery, the load source is not a force as such but rather something that induces motion of the structure, in which case the mass of the building is actually the load source.

An important point to note is that the effects of a dynamic load on a structure are determined by the structure's response as well as by the nature of the load. Thus the same dynamic load can produce different effects in different structures. Two buildings standing side by side can have significantly different responses to the same earthquake shock if they have major differences in their dynamic properties.

Dynamic effects on structures may be of several types. Some of the principal effects are the following:

1. *Total energy load* is the balance between the peak magnitude of the load and the maximum work required by the structure and is known as the *work equilibrium concept*. Work done to the structure by the load equals the work done by the structure in resisting the load.

2. *Unstabilizing effects* occur if the dynamic load produces a stability failure of the structure. Thus a free-standing wall may topple over, an unbraced post and beam system may collapse sideways, and so forth, because of the combined effects of gravity and the dynamic load.

3. *Harmonic effects* of various types may be set up in the structure, especially if the load source is cyclic in nature, such as the footsteps of marching troops. Earthquake motions are basically cyclic, in the form of vibration or shaking of the surface of the ground. Relations between these motions and the harmonic properties of the structure can result in various effects, such as the

flutter of objects at a particular wind velocity, the resonant bouncing of floors, and the resonant reinforcing of the swaying of buildings during an earthquake.

4. *Using up of the structure's energy capacity* can occur if the energy of the load exceeds the limit for the structure. Actually, there are several degrees, or stages, of energy capacity, instead of a single limit. Four significant stages are the following:

a. The *resilience limit*, or the limit beyond which some form of permanent damage—however slight—may occur.

b. The *minor damage limit*, the damage being either relatively insignificant or easily repairable.

c. The *major damage limit*, short of total destruction but with loss of some minor elements. The structure as a whole remains intact, but some major repairs may be required to restore it to its original level of capacity.

d. The *toughness limit*, or the maximum, ultimate capacity represented by the destruction of the structure.

5. *Failure under repeated loadings* can result in some cases when structures progressively use up their dynamic resistance. The structures may successfully resist a single peak load of some dynamic effort, only to fail later under a similar, or even smaller, loading. This failure is usually due to the fact that the first loading used up some degree of structural failure, such as ductile yielding or brittle cracking, which absorbed enough energy to prevent total failure but was only a one time usable strength.

A major consideration in design for dynamic loads is what the response of the structure means to the building as a whole. Thus, although the structure may remain intact, that may be only a minor accomplishment if there is significant damage to the building as a whole. A high-rise building may swing and sway in an earthquake without there being any significant damage to the structure, but if the occupants are tossed about, the ceilings fall, the windows shatter, the partitions and curtain walls collapse, the plumbing bursts, and the elevators derail, it can hardly be said that the building was adequately designed.

In many cases analysis and design for dynamic effects are not done by working directly with the dynamic relationships but simply by using recommendations and rules of thumb that have been established by experience. Some testing or theoretical analysis may have helped in deriving ideas or data, but much of what is used is based on the observations and records from previous disasters. Even when actual calculations are performed, they are mostly done with data and relationships that have been translated into simpler static terms—so-called equivalent static analysis and design. The reasons for this practice have to do frankly with the degree of complexity of dynamic analysis. Even with the use of programmable calculators or computers, the work is quite laborious in all but the simplest of situations.

A.5 EQUIVALENT STATIC EFFECTS

Use of equivalent static effects essentially permits simpler analysis and design by eliminating the complex procedures of dynamic analysis. To make this possible the load effects and the structure's responses must be translated into static terms.

For wind load the primary translation consists of converting the kinetic energy of the wind into an equivalent static pressure, which is then treated in a manner similar to that for a distributed gravity load. Additional considerations are made for various aerodynamic effects, such as ground sur-

face drag, building shape, and suction, but these do not change the basic static nature of the work.

For earthquake effects the primary translation consists of establishing a hypothetical horizontal static force that is applied to the structure to simulate the effects of sideward motions during ground movements. This force is calculated as some percentage of the dead weight of the building, which is the actual source of the kinetic energy loading once the building is in motion—just as the weight of the pendulum and the spring keeps them moving after the initial displacement and release. The specific percentage used is determined by a number of factors, including some of the dynamic response characteristics of the structure.

An apparently lower safety factor is used when designing for the effects of wind and earthquake because an increase of one third is permitted in allowable stresses. This is actually not a matter of a less-safe design but is merely a way of compensating for the fact that one is actually adding static (gravity) effects and *equivalent* static effects. The total stresses thus calculated are really quite hypothetical because in reality one is adding static strength effects to dynamic strength effects, in which case 2 + 2 does not necessarily make 4.

Regardless of the number of modifying factors and translations, there are some limits to the ability of an equivalent static analysis to account for dynamic behavior. Many effects of damping and resonance cannot be accounted for. The true energy capacity of the structure cannot be accurately measured in terms of the magnitudes of stresses and strains. There are some situations, therefore, in which a true dynamic analysis is desirable, whether it is performed by mathematics or by physical testing. These situations are actually quite rare, however. The vast majority of building designs present situations for which a great deal of experience exists. This experience permits generalizations on most occasions that the potential dynamic effects are really insignificant or that they will be adequately accounted for by design for gravity alone or with use of the equivalent static techniques.

Use of the lateral force design criteria of the *UBC* (Ref. 1)—most of which is keyed to an analysis for equivalent static effects—is explained in detail in Chapters 1 and 2.

EXCERPTS FROM THE 1994 *UBC*

The materials in this section consist of reprints from the 1994 edition of the *Uniform Building Code* (*UBC*, Ref. 1). These excerpts are reproduced with permission of the publishers, International Conference of Building Officials, Whittier, California. Like all codes, the *UBC* is subject to continual revision. In recent times, it has been reissued in a new edition every three years.

TABLE 16-F—WIND STAGNATION PRESSURE (q_s) AT STANDARD HEIGHT OF 33 FEET

Basic wind speed (mph)[1] ($\times$ 1.61 for km/h)	70	80	90	100	110	120	130
Pressure q_s (psf) ($\times$ 0.0479 for kN/m^2)	12.6	16.4	20.8	25.6	31.0	36.9	43.3

[1]Wind speed from Section 1615.

TABLE 16-G—COMBINED HEIGHT, EXPOSURE AND GUST FACTOR COEFFICIENT (C_e)[1]

HEIGHT ABOVE AVERAGE LEVEL OF ADJOINING GROUND (feet) $\times$ 304.8 for mm	EXPOSURE D	EXPOSURE C	EXPOSURE B
0-15	1.39	1.06	0.62
20	1.45	1.13	0.67
25	1.50	1.19	0.72
30	1.54	1.23	0.76
40	1.62	1.31	0.84
60	1.73	1.43	0.95
80	1.81	1.53	1.04
100	1.88	1.61	1.13
120	1.93	1.67	1.20
160	2.02	1.79	1.31
200	2.10	1.87	1.42
300	2.23	2.05	1.63
400	2.34	2.19	1.80

[1]Values for intermediate heights above 15 feet (4572 mm) may be interpolated.

FOOTNOTES TO TABLE 16-H

[1]For one story or the top story of multistory partially enclosed structures, an additional value of 0.5 shall be added to the outward C_q. The most critical combination shall be used for design. For definition of open structures, see Section 1613.

[2]C_q values listed are for 10-square-foot (0.93 m^2) tributary areas. For tributary areas of 100 square feet (9.29 m^2), the value of 0.3 may be subtracted from C_q, except for areas at discontinuities with slopes less than 7 units vertical in 12 units horizontal (58.3% slope) where the value of 0.8 may be subtracted from C_q. Interpolation may be used for tributary areas between 10 and 100 square feet (0.93 m^2 and 9.29 m^2). For tributary areas greater than 1,000 square feet (92.9 m^2), use primary frame values.

[3]For slopes greater than 12 units vertical in 12 units horizontal (100% slope), use wall element values.

[4]Local pressures shall apply over a distance from the discontinuity of 10 feet (3048 mm) or 0.1 times the least width of the structure, whichever is smaller.

[5]Discontinuities at wall corners or roof ridges are defined as discontinuous breaks in the surface where the included interior angle measures 170 degrees or less.

[6]Load is to be applied on either side of discontinuity but not simultaneously on both sides.

[7]Wind pressures shall be applied to the total normal projected area of all elements on one face. The forces shall be assumed to act parallel to the wind direction.

[8]Factors for cylindrical elements are two thirds of those for flat or angular elements.

1994 UNIFORM BUILDING CODE

TABLE 16-H—PRESSURE COEFFICIENTS (C_q)

STRUCTURE OR PART THEREOF	DESCRIPTION	C_q FACTOR
1. Primary frames and systems	**Method 1** (Normal force method) Walls:	
	Windward wall	0.8 inward
	Leeward wall	0.5 outward
	Roofs[1]:	
	Wind perpendicular to ridge	
	Leeward roof or flat roof	0.7 outward
	Windward roof	
	less than 2:12 (16.7%)	0.7 outward
	Slope 2:12 (16.7%) to less than 9:12 (75%)	0.9 outward or 0.3 inward
	Slope 9:12 (75%) to 12:12 (100%)	0.4 inward
	Slope > 12:12 (100%)	0.7 inward
	Wind parallel to ridge and flat roofs	0.7 outward
	Method 2 (Projected area method) On vertical projected area	1.3 horizontal any direction
	Structures 40 feet (12 192 mm) or less in height	1.4 horizontal any direction
	Structures over 40 feet (12 192 mm) in height On horizontal projected area[1]	0.7 upward
2. Elements and components not in areas of discontinuity[2]	**Wall elements**	
	All structures	1.2 inward
	Enclosed and unenclosed structures	1.2 outward
	Partially enclosed structures	1.6 outward
	Parapets walls	1.3 inward or outward
	Roof elements[3]	
	Enclosed and unenclosed structures	
	Slope < 7:12 (58.3%)	1.3 outward
	Slope 7:12 (58.3%) to 12:12 (100%)	1.3 outward or inward
	Partially enclosed structures	
	Slope < 2:12 (16.7%)	1.7 outward
	Slope 2:12 (16.7%) to 7:12 (58.3%)	1.6 outward or 0.8 inward
	Slope > 7:12 (58.3%) to 12:12 (100%)	1.7 outward or inward
3. Elements and components in areas of discontinuities[2,4,5]	Wall corners[6]	1.5 outward or 1.2 inward
	Roof eaves, rakes or ridges without overhangs[6]	
	Slope < 2:12 (16.7%)	2.3 upward
	Slope 2:12 (16.7%) to 7:12 (58.3%)	2.6 outward
	Slope > 7:12 (58.3%) to 12:12 (100%)	1.6 outward
	For slopes less than 2:12 (16.7%)	
	Overhangs at roof eaves, rakes or ridges, and canopies	0.5 added to values above
4. Chimneys, tanks and solid towers	Square or rectangular	1.4 any direction
	Hexagonal or octagonal	1.1 any direction
	Round or elliptical	0.8 any direction
5. Open-frame towers[7,8]	Square and rectangular	
	Diagonal	4.0
	Normal	3.6
	Triangular	3.2
6. Tower accessories (such as ladders, conduit, lights and elevators)	Cylindrical members	
	2 inches (51 mm) or less in diameter	1.0
	Over 2 inches (51 mm) in diameter	0.8
	Flat or angular members	1.3
7. Signs, flagpoles, lightpoles, minor structures[8]		1.4 any direction

(Continued)

TABLE 16-I—SEISMIC ZONE FACTOR Z

ZONE	1	2A	2B	3	4
Z	0.075	0.15	0.20	0.30	0.40

The zone shall be determined from the seismic zone map in Figure 16-2.

TABLE 16-J—SITE COEFFICIENTS[1]

TYPE	DESCRIPTION	S FACTOR
S_1	A soil profile with either: (a) A rock-like material characterized by a shear-wave velocity greater than 2,500 feet per second (762 m/s) or by other suitable means of classification, or (b) Medium-dense to dense or medium-stiff to stiff soil conditions, where soil depth is less than 200 feet (60 960 mm).	1.0
S_2	A soil profile with predominantly medium-dense to dense or medium-stiff to stiff soil conditions, where the soil depth exceeds 200 feet (60 960 mm).	1.2
S_3	A soil profile containing more than 20 feet (6096 mm) of soft to medium-stiff clay but not more than 40 feet (12 192 mm) of soft clay.	1.5
S_4	A soil profile containing more than 40 feet (12 192 mm) of soft clay characterized by a shear wave velocity less than 500 feet per second (152.4 m/s).	2.0

[1]The site factor shall be established from properly substantiated geotechnical data. In locations where the soil properties are not known in sufficient detail to determine the soil profile type, soil profile S_3 shall be used. Soil profile S_4 need not be assumed unless the building official determines that soil profile S_4 may be present at the site, or in the event that soil profile S_4 is established by geotechnical data.

1994 UNIFORM BUILDING CODE

TABLE 16-K—OCCUPANCY CATEGORY

OCCUPANCY CATEGORY	OCCUPANCY OR FUNCTIONS OF STRUCTURE	SEISMIC IMPOR-TANCE FACTOR, I	SEISMIC IMPOR-TANCE[1] FACTOR, I_p	WIND IMPOR-TANCE FACTOR, I_w
1. Essential facilities[2]	Group I, Division 1 Occupancies having surgery and emergency treatment areas Fire and police stations Garages and shelters for emergency vehicles and emergency aircraft Structures and shelters in emergency-preparedness centers Aviation control towers Structures and equipment in government communication centers and other facilities required for emergency response Standby power-generating equipment for Category I facilities Tanks or other structures containing housing or supporting water or other fire-suppression material or equipment required for the protection of Category I, II or III structures	1.25	1.50	1.15
2. Hazardous facilities	Group H, Divisions 1, 2, 6 and 7 Occupancies and structures therein housing or supporting toxic or explosive chemicals or substances Nonbuilding structures housing, supporting or containing quantities of toxic or explosive substances which, if contained within a building, would cause that building to be classified as a Group H, Division 1, 2 or 7 Occupancy	1.25	1.50	1.15
3. Special occupancy structures[3]	Group A, Divisions 1, 2 and 2.1 Occupancies Buildings housing Group E, Divisions 1 and 3 Occupancies with a capacity greater than 300 students Buildings housing Group B Occupancies used for college or adult education with a capacity greater than 500 students Group I, Divisions 1 and 2 Occupancies with 50 or more resident incapacitated patients, but not included in Category I Group I, Division 3 Occupancies All structures with an occupancy greater than 5,000 persons Structures and equipment in power-generating stations; and other public utility facilities not included in Category I or Category II above, and required for continued operation	1.00	1.00	1.00
4. Standard occupancy structures[4]	All structures housing occupancies or having functions not listed in Category I, II or III and Group U Occupancy towers	1.00	1.00	1.00
5. Miscellaneous structures	Group U Occupancies except for towers	1.00	1.00	1.00

[1]The limitation of I_p for panel connections in Section 1631.2.4 shall be 1.0 for the entire connector.
[2]Structural observation requirements are given in Sections 108, 1701 and 1702.
[3]For anchorage of machinery and equipment required for life-safety systems the value of I_p shall be taken as 1.5.

TABLE 16-L—VERTICAL STRUCTURAL IRREGULARITIES

IRREGULARITY TYPE AND DEFINITION	REFERENCE SECTION
1. **Stiffness irregularity—soft story** A soft story is one in which the lateral stiffness is less than 70 percent of that in the story above or less than 80 percent of the average stiffness of the three stories above.	1627.8.3, Item 2
2. **Weight (mass) irregularity** Mass irregularity shall be considered to exist where the effective mass of any story is more than 150 percent of the effective mass of an adjacent story. A roof which is lighter than the floor below need not be considered.	1627.8.3, Item 2
3. **Vertical geometric irregularity** Vertical irregularity shall be considered to exist where the horizontal dimension of the lateral force-resisting system in any story is more than 130 percent of that in an adjacent story. One-story penthouses need not be considered.	1627.8.3, Item 2
4. **In-plane discontinuity in vertical lateral-force-resisting element** An in-plane offset of the lateral load-resisting elements greater than the length of those elements.	1628.7
5. **Discontinuity in capacity—weak story** A weak story is one in which the story strength is less than 80 percent of that in the story above. The story strength is the total strength of all seismic-resisting elements sharing the story shear for the direction under consideration.	1627.9.1

TABLE 16-M—PLAN STRUCTURAL IRREGULARITIES

IRREGULARITY TYPE AND DEFINITION	REFERENCE SECTION
1. **Torsional irregularity—to be considered when diaphragms are not flexible** Torsional irregularity shall be considered to exist when the maximum story drift, computed including accidental torsion, at one end of the structure transverse to an axis is more than 1.2 times the average of the story drifts of the two ends of the structure.	1631.2.9, Item 6
2. **Reentrant corners** Plan configurations of a structure and its lateral force-resisting system contain reentrant corners, where both projections of the structure beyond a reentrant corner are greater than 15 percent of the plan dimension of the structure in the given direction.	1631.2.9, Items 6 and 7
3. **Diaphragm discontinuity** Diaphragms with abrupt discontinuities or variations in stiffness, including those having cutout or open areas greater than 50 percent of the gross enclosed area of the diaphragm, or changes in effective diaphragm stiffness of more than 50 percent from one story to the next.	1631.2.9, Item 6
4. **Out-of-plane offsets** Discontinuities in a lateral force path, such as out-of-plane offsets of the vertical elements.	1628.7; 1631.2.9, Item 6; 2211.8
5. **Nonparallel systems** The vertical lateral load-resisting elements are not parallel to or symmetric about the major orthogonal axes of the lateral force-resisting system.	1631.1

1994 UNIFORM BUILDING CODE

TABLE 16-N—STRUCTURAL SYSTEMS

BASIC STRUCTURAL SYSTEM[1]	LATERAL-FORCE-RESISTING SYSTEM—DESCRIPTION	R_w[2]	H[3] × 304.8 for mm
1. Bearing wall system	1. Light-framed walls with shear panels		
	a. Wood structural panel walls for structures three stories or less	8	65
	b. All other light-framed walls	6	65
	2. Shear walls		
	a. Concrete	6	160
	b. Masonry	6	160
	3. Light steel-framed bearing walls with tension-only bracing	4	65
	4. Braced frames where bracing carries gravity loads		
	a. Steel	6	160
	b. Concrete[4]	4	—
	c. Heavy timber	4	65
2. Building frame system	1. Steel eccentrically braced frame (EBF)	10	240
	2. Light-framed walls with shear panels		
	a. Wood structural panel walls for structures three stories or less	9	65
	b. All other light-framed walls	7	65
	3. Shear walls		
	a. Concrete	8	240
	b. Masonry	8	160
	4. Ordinary braced frames		
	a. Steel	8	160
	b. Concrete[4]	8	—
	c. Heavy timber	8	65
	5. Special concentrically braced frames		
	a. Steel	9	240
3. Moment-resisting frame system	1. Special moment-resisting frames (SMRF)		
	a. Steel	12	N.L.
	b. Concrete	12	N.L.
	2. Masonry moment-resisting wall frame	9	160
	3. Concrete intermediate moment-resisting frames (IMRF)[5]	8	—
	4. Ordinary moment-resisting frames (OMRF)		
	a. Steel[6]	6	160
	b. Concrete[7,8]	5	—
4. Dual systems	1. Shear walls		
	a. Concrete with SMRF	12	N.L.
	b. Concrete with steel OMRF	6	160
	c. Concrete with concrete IMRF[5]	9	160
	d. Masonry with SMRF	8	160
	e. Masonry with steel OMRF	6	160
	f. Masonry with concrete IMRF[4]	7	—
	2. Steel EBF		
	a. With steel SMRF	12	N.L.
	b. With steel OMRF	6	160
	3. Ordinary braced frames		
	a. Steel with steel SMRF	10	N.L.
	b. Steel with steel OMRF	6	160
	c. Concrete with concrete SMRF[4]	9	—
	d. Concrete with concrete IMRF[4]	6	—
	4. Special concentrically braced frames		
	a. Steel with steel SMRF	11	N.L.
	b. Steel with steel OMRF	6	160
5. Undefined systems	See Sections 1627.8.3 and 1627.9.2	—	—

N.L.—No limit.

[1]Basic structural systems are defined in Section 1627.6.

[2]See Section 1628.3 for combination of structural system.

[3]*H*—Height limit applicable to Seismic Zones 3 and 4. See Section 1627.7.

[4]Prohibited in Seismic Zones 3 and 4.

[5]Prohibited in Seismic Zones 3 and 4, except as permitted in Section 1632.2.

[6]Ordinary moment-resisting frames in Seismic Zone 1 meeting the requirements of Section 2211.6 may use an R_w value of 12.

[7]Prohibited in Seismic Zones 2, 3 and 4.

[8]Prohibited in Seismic Zones 2A, 2B, 3 and 4. See Section 1631.2.7.

TABLE 16-O—HORIZONTAL FORCE FACTOR, C_p

ELEMENTS OF STRUCTURES, NONSTRUCTURAL COMPONENTS AND EQUIPMENT[1]	VALUE OF C_p	FOOTNOTE
1. **Elements of structures**		
1. Walls including the following:		
a. Unbraced (cantilevered) parapets	2.00	
b. Other exterior walls above the ground floor	0.75	2,3
c. All interior bearing and nonbearing walls and partitions	0.75	3
d. Masonry or concrete fences over 6 feet (1829 mm) high	0.75	
2. Penthouse (except when framed by an extension of the structural frame)	0.75	
3. Connections for prefabricated structural elements other than walls, with force applied at center of gravity	0.75	4
4. Diaphragms	—	5
2. **Nonstructural components**		
1. Exterior and interior ornamentations and appendages	2.00	
2. Chimneys, stacks, trussed towers and tanks on legs:		
a. Supported on or projecting as an unbraced cantilever above the roof more than one half their total height	2.00	
b. All others, including those supported below the roof with unbraced projection above the roof less than one half its	0.75	
height, or braced or guyed to the structural frame at or above their centers of mass	2.00	
3. Signs and billboards	0.75	10
4. Storage racks (include contents)	0.75	
5. Anchorage for permanent floor-supported cabinets and book stacks more than 5 feet (1524 mm) in height (include contents)	0.75	4,6,7,11
6. Anchorage for suspended ceilings and light fixtures	0.75	4,9
7. Access floor systems		
3. **Equipment**		
1. Tanks and vessels (include contents), including support systems and anchorage	0.75	
2. Electrical, mechanical and plumbing equipment and associated conduit, ductwork and piping, and machinery	0.75	8

[1]See Section 1630.2 for items supported at or below grade.
[2]See Section 1631.2.4 and Section 1630.2.
[3]Where flexible diaphragms, as defined in Section 1628.6, provide lateral support for walls and partitions, the value of C_p for anchorage shall be increased 50 percent for the center one half of the diaphragm span.
[4]Applies to Seismic Zones 2, 3 and 4 only.
[5]See Section 1631.2.9.
[6]Ceiling weight shall include all light fixtures and other equipment or partitions which are laterally supported by the ceiling. For purposes of determining the seismic force, a ceiling weight of not less than 4 pounds per square foot (19.5 kg/m^2) shall be used.
[7]Ceilings constructed of lath and plaster or gypsum board screw or nail attached to suspended members that support a ceiling at one level extending from wall to wall need not be analyzed provided the walls are not over 50 feet (15 240 mm) apart.
[8]Equipment includes, but is not limited to, boiler, chiller, heat exchanger, pump, air-handling unit, cooling tower, control panel, motor, switch gear, transformer and life-safety equipment. It includes major conduit, ducting and piping serving such machinery and equipment and fire sprinkler systems. See Section 1630.2 for additional requirements for determining C_p for nonrigid or flexibly mounted equipment.
[9]W_p for access floor systems shall be the dead load of the access floor system plus 25 percent of the floor live load plus a 10 psf (0.479 kN/m^2) partition load allowance.
[10]In lieu of the tabulated values, steel storage racks may be designed in accordance with Chapter 22, Division VI.
[11]Light fixtures and mechanical services installed in metal suspension systems for acoustical tile and lay-in panel ceilings shall be independently supported from the structure above as specified in U.B.C. Standard 25-2, Part III.

1994 UNIFORM BUILDING CODE

TABLE 16-P—R_W FACTORS FOR NONBUILDING STRUCTURES

	STRUCTURE TYPE	R_W
1.	Vessels, including tanks and pressurized spheres, on braced or unbraced legs.	3
2.	Cast-in-place concrete silos and chimneys having walls continuous to the foundation.	5
3.	Distributed mass cantilever structures such as stacks, chimneys, silos and skirt-supported vertical vessels.	4
4.	Trussed towers (freestanding or guyed), guyed stacks and chimneys.	4
5.	Inverted pendulum-type structures.	3
6.	Cooling towers.	5
7.	Bins and hoppers on braced or unbraced legs.	4
8.	Storage racks.	5
9.	Signs and billboards.	5
10.	Amusement structures and monuments.	3
11.	All other self-supporting structures not otherwise covered.	4

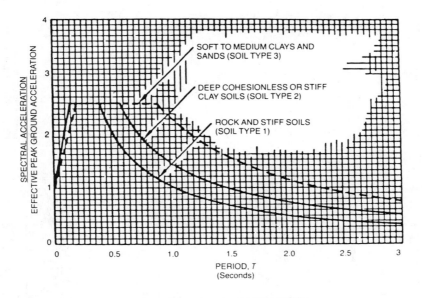

FIGURE 16-3—NORMALIZED RESPONSE SPECTRA SHAPES

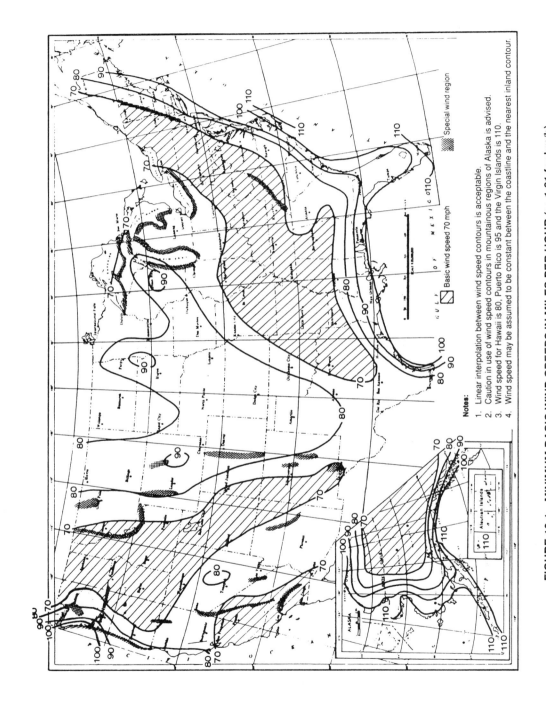

FIGURE 16-1—MINIMUM BASIC WIND SPEEDS IN MILES PER HOUR (× 1.61 for km/h)

Notes:

1. Linear interpolation between wind speed contours is acceptable.
2. Caution in use of wind speed contours in mountainous regions of Alaska is advised.
3. Wind speed for Hawaii is 80, Puerto Rico is 95 and the Virgin Islands is 110.
4. Wind speed may be assumed to be constant between the coastline and the nearest inland contour.

1994 UNIFORM BUILDING CODE

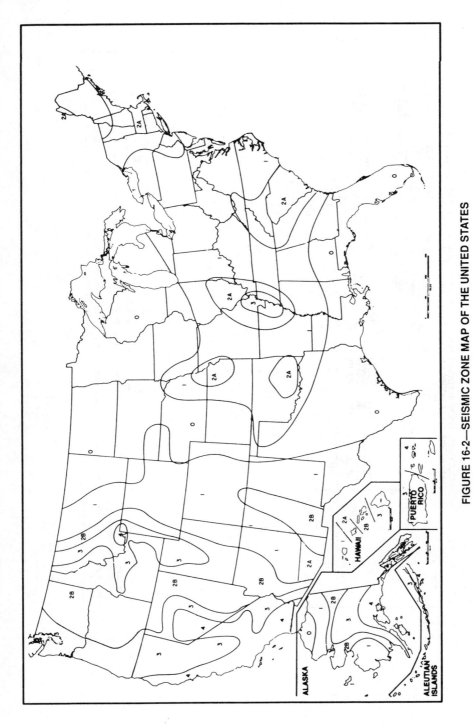

FIGURE 16-2—SEISMIC ZONE MAP OF THE UNITED STATES

For areas outside of the United States, see Appendix Chapter 16.

TABLE 23-I-J-1—ALLOWABLE SHEAR IN POUNDS PER FOOT FOR HORIZONTAL WOOD STRUCTURAL PANEL DIAPHRAGMS WITH FRAMING OF DOUGLAS FIR-LARCH OR SOUTHERN PINE[1]

PANEL GRADE	COMMON NAIL SIZE	MINIMUM NAIL PENETRATION IN FRAMING (inches) ×25.4 for mm	MINIMUM NOMINAL PANEL THICKNESS (inches) ×25.4 for mm	MINIMUM NOMINAL WIDTH OF FRAMING MEMBER (inches)	BLOCKED DIAPHRAGMS				UNBLOCKED DIAPHRAGMS	
					Nail spacing (in.) at diaphragm boundaries (all cases), at continuous panel edges parallel to load (Cases 3 and 4) and at all panel edges (Cases 5 and 6) ×25.4 for mm: 6 / 4 / 2½ / 2				Nails spaced 6″ (152 mm) max. at supported edges	
					Nail spacing (in.) at other panel edges (Cases 5 and 6) ×25.4 for mm: 6 / 6 / 4 / 3 × 0.0146 for N/mm				Case 1 (No unblocked edges or continuous joints parallel to load)	All other configurations (Cases 2, 3, 4, 5 and 6)
Structural I	6d	1¼	5/16	2 / 3	185 / 210	250 / 280	375 / 420	420 / 475	165 / 185	125 / 140
	8d	1½	3/8	2 / 3	270 / 300	360 / 400	530 / 600	600 / 675	240 / 265	180 / 200
	10d[3]	1⅝	15/32	2 / 3	320 / 360	425 / 480	640 / 720	730 / 820	285 / 320	215 / 240
C-D, C-C, Sheathing, and other grades covered in U.B.C. Standard 23-3 or 23-9	6d	1¼	5/16	2 / 3	170 / 190	225 / 250	335 / 380	380 / 430	150 / 170	110 / 125
			3/8	2 / 3	185 / 210	250 / 280	375 / 420	420 / 475	165 / 185	125 / 140
	8d	1½	3/8	2 / 3	240 / 270	320 / 360	480 / 540	545 / 610	215 / 240	160 / 180
			7/16	2 / 3	255 / 285	340 / 380	505 / 570	575 / 645	230 / 255	170 / 190
			15/32	2 / 3	270 / 300	360 / 400	530 / 600	600 / 675	240 / 265	180 / 200
	10d[3]	1⅝	15/32	2 / 3	290 / 325	385 / 430	575 / 650	655 / 735	255 / 290	190 / 215
			19/32	2 / 3	320 / 360	425 / 480	640 / 720	730 / 820	285 / 320	215 / 240

[1]These values are for short-time loads due to wind or earthquake and must be reduced 25 percent for normal loading. Space nails 12 inches (305 mm) on center along intermediate framing members.

Allowable shear values for nails in framing members of other species set forth in Table 23-III-FF of Division III shall be calculated for all other grades by multiplying the shear capacities for nails in Structural I by the following factors: 0.82 for species with specific gravity greater than or equal to 0.42 but less than 0.49, and 0.65 for species with a specific gravity less than 0.42.

1994 UNIFORM BUILDING CODE

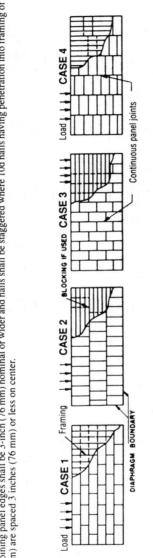
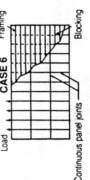
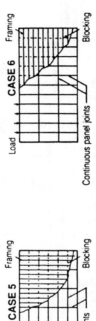

²Framing at adjoining panel edges shall be 3-inch (76 mm) nominal or wider and nails shall be staggered where nails are spaced 2 inches (51 mm) or $2^{1}/_{2}$ inches (64 mm) on center.

³Framing at adjoining panel edges shall be 3-inch (76 mm) nominal or wider and nails shall be staggered where 10d nails having penetration into framing of more than $1^{5}/_{8}$ inches (41 mm) are spaced 3 inches (76 mm) or less on center.

NOTE: Framing may be oriented in either direction for diaphragms, provided sheathing is properly designed for vertical loading.

TABLE 23-I-J-2—ALLOWABLE SHEAR IN POUNDS PER FOOT FOR HORIZONTAL PARTICLEBOARD DIAPHRAGMS WITH FRAMING OF DOUGLAS FIR-LARCH OR SOUTHERN PINE[1]

PANEL GRADE	COMMON NAIL SIZE	MINIMUM NAIL PENETRATION IN FRAMING (inches)	MINIMUM NOMINAL PANEL THICKNESS (inches) × 25.4 for mm	MINIMUM NOMINAL WIDTH OF FRAMING MEMBER (inches)	BLOCKED DIAPHRAGMS — Nail spacing (in.) at diaphragm boundaries (all cases), at continuous panel edges parallel to load (Cases 3 and 4) and at all panel edges (Cases 5 and 6) / Nail spacing (in.) at other plywood panel edges (Cases 5 and 6) × 25.4 for mm				UNBLOCKED DIAPHRAGMS — Nails spaced 6" (152 mm) max. at supported edges	
					6 / 6	4 / 6	2½ / 4	2 / 3	Case 1 (No unblocked edges or continuous joints parallel to load) × 0.0146 for N/mm	All other configurations (Cases 2, 3, 4, 5 and 6)
	6d	1¼	5/16	2	170	225	335	380	150	110
				3	190	250	380	430	170	125
	6d	1¼	3/8	2	185	250	375	420	165	125
				3	210	280	420	475	185	140
2-M-W	8d	1½	3/8	2	240	320	480	545	215	160
				3	270	360	540	610	240	180
	8d	1½	7/16	2	255	340	505	575	230	170
				3	285	380	570	645	255	190
	8d	1½	1/2	2	270	360	530	600	240	180
				3	300	400	600	675	265	200
	10d[3]	1 5/8	1/2	2	290	385	575	655	255	190
				3	325	430	650	735	290	215
	10d[3]	1 5/8	5/8	2	320	425	640	730	285	215
				3	360	480	720	820	320	240
2-M-3	10d[3]	1 5/8	3/4	2	320	425	640	730	285	215
				3	360	480	720	820	320	240

[1]These values are for short-time loads due to wind or earthquake and must be reduced 25 percent for normal loading. Space nails 12 inches (305 mm) on center along intermediate framing members.
Allowable shear values for nails in framing members of other species set forth in Table 23-III-J of Division III shall be calculated for all grades by multiplying the values for nails by the following factors: Group III, 0.82 and Group IV, 0.65.
[2]Framing at adjoining panel edges shall be 3-inch (76 mm) nominal or wider and nails shall be staggered where nails are spaced 2 inches (51 mm) or 2½ inches (64 mm) on center.
[3]Framing at adjoining panel edges shall be 3-inch (76 mm) nominal or wider and nails shall be staggered where 10d nails having penetration into framing of more than 1 5/8 inches (41 mm) are spaced 3 inches (76 mm) or less on center.

1994 UNIFORM BUILDING CODE

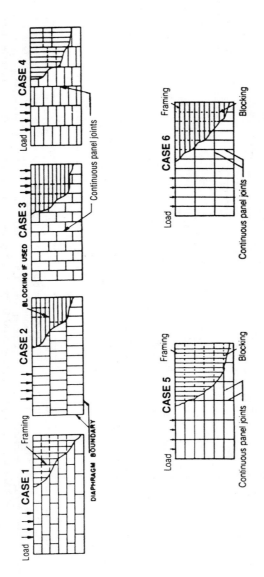

NOTE: Framing may be oriented in either direction for diaphragms, provided sheathing is properly designed for vertical loading.

TABLE 23-I-K-1—ALLOWABLE SHEAR FOR WIND OR SEISMIC FORCES IN POUNDS PER FOOT FOR WOOD STRUCTURAL PANEL SHEAR WALLS WITH FRAMING OF DOUGLAS FIR-LARCH OR SOUTHERN PINE[1,2]

PANEL GRADE	MINIMUM NOMINAL PANEL THICKNESS (inches) ×25.4 for mm	MINIMUM NAIL PENETRATION IN FRAMING (inches)	PANELS APPLIED DIRECTLY TO FRAMING					PANELS APPLIED OVER 1/2-INCH (13 mm) OR 5/8-INCH (16 mm) GYPSUM SHEATHING				
			Nail Size (Common or Galvanized Box)	Nail Spacing at Panel Edges (in.)				Nail Size (Common or Galvanized Box)	Nail Spacing at Panel Edges (in.)			
				6	4	3	2[3]		6	4	3	2[3]
				×25.4 for mm					×25.4 for mm			
				×0.0146 for N/mm					×0.0146 for N/mm			
Structural I	5/16	1 1/4	6d	200	300	390	510	8d	200	300	390	510
	3/8	1 1/2	8d	230[4]	360[4]	460[4]	610[4]	10d[5]	280	430	550	730
	7/16	1 1/2	8d	255[4]	395[4]	505[4]	670[4]	—	—	—	—	—
	15/32	1 1/2	8d	280	430	550	730	—	—	—	—	—
	15/32	1 5/8	10d[5]	340	510	665	870	—	—	—	—	—
C-D, C-C Sheathing, plywood panel siding and other grades covered in U.B.C. Standard 23-2 or 23-3	5/16	1 1/4	6d	180	270	350	450	8d	180	270	350	450
	3/8	1 1/4	6d	200	300	390	510	8d	200	300	390	510
	3/8	1 1/2	8d	220[4]	320[4]	410[4]	530[4]	10d[5]	260	380	490	640
	7/16	1 1/2	8d	240[4]	350[4]	450[4]	585[4]	—	—	—	—	—
	15/32	1 1/2	8d	260	380	490	640	—	—	—	—	—
	15/32	1 5/8	10d[5]	310	460	600	770	—	—	—	—	—
	19/32			340	510	665	870					
			Nail Size (Galvanized Casing)					Nail Size (Galvanized Casing)				
Plywood panel siding in grades covered in U.B.C. Standard 23-2	5/16	1 1/4	6d	140	210	275	360	8d	140	210	275	360
	3/8	1 1/2	8d	160	240	310	410	10d[5]	160	240	310	410

[1] All panel edges backed with 2-inch (51 mm) nominal or wider framing. Panels installed either horizontally or vertically. Space nails at 6 inches (152 mm) on center along intermediate framing members for 3/8-inch (9.5 mm) and 7/16-inch (11 mm) panels installed on studs spaced 24 inches (610 mm) on center and 12 inches (305 mm) on center for other conditions and panel thicknesses. These values are for short-time loads due to wind or earthquake and must be reduced 25 percent for normal loading. Allowable shear values for nails in framing members of other species set forth in Table 23-III-FF of Division III shall be calculated for all other grades by multiplying the shear capacities for nails in Structural I by the following factors: 0.82 for species with specific gravity greater than or equal to 0.42 but less than 0.49, and 0.65 for species with a specific gravity less than 0.42.

[2] Where panels are applied on both faces of a wall and nail spacing is less than 6 inches (152 mm) on center on either side, panel joints shall be offset to fall on different framing members or framing shall be 3-inch (76 mm) nominal or thicker and nails on each side shall be staggered.

[3] Framing at adjoining panel edges shall be 3-inch (76 mm) nominal or wider and nails shall be staggered where nails are spaced 2 inches (51 mm) on center.

[4] The values for 3/8-inch (9.5 mm) and 7/16-inch (11 mm) panels applied direct to framing may be increased to values shown for 15/32-inch (12 mm) panels, provided studs are spaced a maximum of 16 inches (406 mm) on center or panels are applied with long dimension across studs.

[5] Framing at adjoining panel edges shall be 3-inch (76 mm) nominal or wider and nails shall be staggered where 10d nails having penetration into framing of more than 1 5/8 inches (41 mm) are spaced 3 inches (76 mm) or less on center.

1994 UNIFORM BUILDING CODE

TABLE 23-I-K-2—ALLOWABLE SHEAR FOR WIND OR SEISMIC FORCES IN POUNDS PER FOOT FOR PARTICLEBOARD SHEAR WALLS WITH FRAMING OF DOUGLAS FIR-LARCH OR SOUTHERN PINE[1,2]

PANEL GRADE	MINIMUM NOMINAL PANEL THICKNESS (inches) ×25.4 for mm	MINIMUM NAIL PENETRATION IN FRAMING (inches) ×25.4 for mm	PANELS APPLIED DIRECTLY TO FRAMING — Nail Size (Common or Galvanized Box)	Nail Spacing at Panel Edges (inches) — 6 (×0.0146 for N/mm)	4	3	2[3]	PANEL APPLIED OVER 1/2-INCH (13 mm) GYPSUM SHEATHING — Nail Size (Common or Galvanized Box)	Nail Spacing at Plywood Panel Edges (inches) — 6 (×0.0146 for N/mm)	4	3	2[3]
2-M-W	5/16	1 1/4	6d	180	270	350	450	8d	180	270	350	450
	3/8		6d	200	300	390	510		200	300	390	510
	3/8	1 1/2	8d	220[4]	320[4]	410[4]	530[4]					
	7/16		8d	240[4]	350[4]	450[4]	585[4]	10d[5]	260	380	490	640
	1/2		8d	260	380	490	640					
	1/2	1 5/8	10d[5]	310	460	600	770		—	—	—	—
	5/8		10d[5]	340	510	665	870		—	—	—	—

[1] All panel edges backed with 2-inch (51 mm) nominal or wider framing. Plywood installed either horizontally or vertically. Space nails at 6 inches (152 mm) on center along intermediate framing members for 3/8-inch (9.5 mm) panel installed with the long dimension parallel to studs spaced 24 inches (610 mm) on center and 12 inches (305 mm) on center for other conditions and panel thicknesses. These values are for short-time loads due to wind or earthquake and must be reduced 25 percent for normal loading.
Allowable shear values for nails in framing members of other species set forth in Table 23-III-J of Division III shall be calculated for all grades by multiplying the values for common and galvanized box nails by the following factors: Group III, 0.82 and Group IV, 0.65.

[2] Where particleboard is applied on both faces of a wall and nail spacing is less than 6 inches (152 mm) on center on either side, panel joints shall be offset to fall on different framing members, or framing shall be 3-inch (76 mm) nominal or thicker and nails on each side shall be staggered.

[3] Framing at adjoining panel edges shall be 3-inch (76 mm) nominal or wider and nails shall be staggered where nails are spaced 2 inches (51 mm) on center.

[4] The allowable shear values may be increased to the values shown for 1/2-inch-thick (13 mm) sheathing with the same nailing, provided:
4.1 The studs are spaced a maximum of 16 inches (406 mm) on center, or
4.2 The panels are applied with the long dimension perpendicular to studs.

[5] Framing at adjoining panel edges shall be 3-inch (76 mm) nominal or wider and nails shall be staggered where 10d nails having penetration into framing of more than 1 5/8 inches (41 mm) are spaced 3 inches (76 mm) or less on center.

TABLE 23-I-G—SAFE LATERAL STRENGTH AND REQUIRED PENETRATION OF BOX AND COMMON WIRE NAILS DRIVEN PERPENDICULAR TO GRAIN OF WOOD

	STANDARD LENGTH (inches)		PENETRATION REQUIRED (inches)	LOADS (pounds)[1, 2, 3]	
				× 4.45 for N	
SIZE OF NAIL	× 25.4 for mm	WIRE GAGE	× 25.4 for mm	Douglas Fir Larch or Southern Pine	Other Species
Box Nails					
6d	2	$12^1/_2$	$1^1/_8$	51	
8d	$2^1/_2$	$11^1/_2$	$1^1/_4$	63	
10d	3	$10^1/_2$	$1^1/_2$	76	
12d	$3^1/_4$	$10^1/_2$	$1^1/_2$	76	See Division III
16d	$3^1/_2$	10	$1^1/_2$	82	
20d	4	9	$1^5/_8$	94	
30d	$4^1/_2$	9	$1^5/_8$	94	
40d	5	8	$1^3/_4$	108	
Common Nails					
6d	2	$11^1/_2$	$1^1/_4$	63	
8d	$2^1/_2$	$10^1/_4$	$1^1/_2$	78	
10d	3	9	$1^5/_8$	94	
12d	$3^1/_4$	9	$1^5/_8$	94	
16d	$3^1/_2$	8	$1^3/_4$	108	See Division III
20d	4	6	$2^1/_8$	139	
30d	$4^1/_2$	5	$2^1/_4$	155	
40d	5	4	$2^1/_2$	176	
50d	$5^1/_2$	3	$2^3/_4$	199	
60d	6	2	$2^7/_8$	223	

[1]See Division III for lateral strength values where metal side plates are used.
[2]See Division III for lateral strength values when considering wood diaphragm calculations.
[3]Tabulated values are on a normal load-duration basis and apply to joints made of seasoned lumber used in dry locations. See Division III for other service conditions.

TABLE 23-I-H—SAFE RESISTANCE TO WITHDRAWAL OF COMMON WIRE NAILS
Inserted Perpendicular to Grain of the Wood, in Pounds per Linear Inch
of Penetration into the Main Member

	SIZE OF NAIL									
	6d	8d	10d	12d	16d	20d	30d	40d	50d	60d
KIND OF WOOD	× 0.175 for N/mm									
1. Douglas fir, larch	29	34	38	38	42	49	53	58	63	67
2. Southern pine	35	41	46	46	50	59	64	70	76	81
3. Other species	See U.B.C. Standard 23-17									

TABLE 23-I-I—MAXIMUM DIAPHRAGM DIMENSION RATIOS

MATERIAL	HORIZONTAL DIAPHRAGMS	VERTICAL DIAPHRAGMS
	Maximum Span-Width Ratios	Maximum Height-Width Ratios
1. Diagonal sheathing, conventional	3:1	2:1
2. Diagonal sheathing, special	4:1	$3^1/_2$:1
3. Wood structural panels and particleboard, nailed all edges	4:1	$3^1/_2$:1
4. Wood structural panels and particleboard, blocking omitted at intermediate joints	4:1	2:1

DATA FOR
MASONRY STRUCTURES

Tables C.1 and C.2 present factors that are used for the evaluation of relative stiffness of masonry piers (short segments of masonry walls). This relates to the determination of load distribution for a series of interacting piers. Use of these data is illustrated in the design examples in Part III. Table data are reprinted from the *Masonry Design Manual* (Ref. 16) with permission of the publishers.

C.1 BENDING IN MASONRY WALLS

The following example illustrates the use of Fig. C.1. Other uses are illustrated in Chapter 10.

Investigate the wall and the required reinforcing for the following data:

Wall height = 16.7 ft

8-in. block: t = 7.625 in.

Use single row of reinforcing in center: d = 3.813 in., n = 40, allowable f_m = 250 × 1.33 = 333 psi

Grade 40 bars: f_s = 1.33 × 20,000 = 26,667 psi

Wind pressure = 20 psf

Find

$$M = \frac{wL^2}{8} = \frac{(20)(16.7)^2}{8} \times 12$$

$$= 8367 \text{ in.-lb.}$$

$$K = \frac{M}{bd^2} = \frac{8367}{(12)(3.813)^2} = 48$$

Enter the diagram (Fig. C.1) at the left with K = 48, proceed to the right to intersect f_m

TABLE C.1   RIGIDITY COEFFICIENTS FOR CANTILEVERED MASONRY WALLS

h/d	Rc	h/d	Rc	h/d	Rc	h/d	Rc	h/d	Rc	h/d	Rc
9.90	.0006	5.20	.0043	1.85	.0810	1.38	.1706	0.91	.4352	0.45	1.4582
9.80	.0007	5.10	.0046	1.84	.0821	1.37	.1737	0.90	.4452	0.44	1.5054
9.70	.0007	5.00	.0049	1.83	.0833	1.36	.1768	0.89	.4554	0.43	1.5547
9.60	.0007	4.90	.0052	1.82	.0845	1.35	.1800	0.88	.4659	0.42	1.6063
9.50	.0007	4.80	.0055	1.81	.0858	1.34	.1832	0.87	.4767	0.41	1.6604
9.40	.0007	4.70	.0058	1.80	.0870	1.33	.1866	0.86	.4899	0.40	1.7170
9.30	.0008	4.60	.0062	1.79	.0883	1.32	.1900	0.85	.4994	0.39	1.7765
9.20	.0008	4.50	.0066	1.78	.0896	1.31	.1935	0.84	.5112	0.38	1.8380
9.10	.0008	4.40	.0071	1.77	.0909	1.30	.1970	0.83	.5233	0.37	1.9098
9.00	.0008	4.30	.0076	1.76	.0923	1.29	.2007	0.82	.5359	0.36	1.9738
8.90	.0009	4.20	.0081	1.75	.0937	1.28	.2044	0.81	.5488	0.35	2.0467
8.80	.0009	4.10	.0087	1.74	.0951	1.27	.2083	0.80	.5621	0.34	2.1237
8.70	.0009	4.00	.0093	1.73	.0965	1.26	.2122	0.79	.5758	0.33	2.2051
8.60	.0010	3.90	.0100	1.72	.0980	1.25	.2162	0.78	.5899	0.32	2.2913
8.50	.0010	3.80	.0108	1.71	.0995	1.24	.2203	0.77	.6044	0.31	2.3828
8.40	.0010	3.70	.0117	1.70	.1010	1.23	.2245	0.76	.6194	0.30	2.4802
8.30	.0011	3.60	.0127	1.69	.1026	1.22	.2289	0.75	.6349	0.29	2.5838
8.20	.0012	3.50	.0137	1.68	.1041	1.21	.2333	0.74	.6509	0.28	2.6945
8.10	.0012	3.40	.0149	1.67	.1058	1.20	.2378	0.73	.6674	0.27	2.8130
8.00	.0012	3.30	.0163	1.66	.1074	1.19	.2425	0.72	.6844	0.26	2.9401
7.90	.0013	3.20	.0178	1.65	.1091	1.18	.2472	0.71	.7019	0.25	3.0769
7.80	.0013	3.10	.0195	1.64	.1108	1.17	.2521	0.70	.7200	0.24	3.2246
7.70	.0014	3.00	.0214	1.63	.1125	1.16	.2571	0.69	.7388	0.23	3.3845
7.60	.0014	2.90	.0235	1.62	.1143	1.15	.2622	0.68	.7581	0.22	3.5583
7.50	.0015	2.80	.0260	1.61	.1162	1.14	.2675	0.67	.7781	0.21	3.7479
7.40	.0015	2.70	.0288	1.60	.1180	1.13	.2729	0.66	.7987	0.20	3.9557
7.30	.0016	2.60	.0320	1.59	.1199	1.12	.2784	0.65	.8201	.195	4.0673
7.20	.0017	2.50	.0357	1.58	.1218	1.11	.2841	0.64	.8422	.190	4.1845
7.10	.0017	2.40	.0400	1.57	.1238	1.10	.2899	0.63	.8650	.185	4.3079
7.00	.0018	2.30	.0450	1.56	.1258	1.09	.2959	0.62	.8886	.180	4.4379
6.90	.0019	2.20	.0508	1.55	.1279	1.08	.3020	0.61	.9131	.175	4.5751
6.80	.0020	2.10	.0577	1.54	.1300	1.07	.3083	0.60	.9384	.170	4.7201
6.70	.0020	2.00	.0658	1.53	.1322	1.06	.3147	0.59	.9647	.165	4.8736
6.60	.0021	1.99	.0667	1.52	.1344	1.05	.3213	0.58	.9919	.160	5.0364
6.50	.0022	1.98	.0676	1.51	.1366	1.04	.3281	0.57	1.0201	.155	5.2095
6.40	.0023	1.97	.0685	1.50	.1389	1.03	.3351	0.56	1.0493	.150	5.3937
6.30	.0025	1.96	.0694	1.49	.1412	1.02	.3422	0.55	1.0797	.145	5.5904
6.20	.0026	1.95	.0704	1.48	.1436	1.01	.3496	0.54	1.1112	.140	5.8008
6.10	.0027	1.94	.0714	1.47	.1461	1.00	.3571	0.53	1.1439	.135	6.0261
6.00	.0028	1.93	.0724	1.46	.1486	0.99	.3649	0.52	1.1779	.130	6.2696
5.90	.0030	1.92	.0734	1.45	.1511	0.98	.3729	0.51	1.2132	.125	6.5306
5.80	.0031	1.91	.0744	1.44	.1537	0.97	.3811	0.50	1.2500	.120	6.8136
5.70	.0033	1.90	.0754	1.43	.1564	0.96	.3895	0.49	1.2883	.115	7.1208
5.60	.0035	1.89	.0765	1.42	.1591	0.95	.3981	0.48	1.3281	.110	7.4555
5.50	.0037	1.88	.0776	1.41	.1619	0.94	.4070	0.47	1.3696	.105	7.8215
5.40	.0039	1.87	.0787	1.40	.1647	0.93	.4162	0.46	1.4130	.100	8.2237
5.30	.0041	1.86	.0798	1.39	.1676	0.92	.4255				

TABLE C.2   RIGIDITY COEFFICIENTS FOR FIXED MASONRY WALLS

h/d	R_f	h/d	R_f	h/d	R_f	h/d	R_f	h/d	R_f	h/d	R_f
9.90	.0025	5.20	.0160	1.85	.2104	1.38	.3694	0.91	.7177	0.45	1.736
9.80	.0026	5.10	.0169	1.84	.2128	1.37	.3742	0.90	.7291	0.44	1.779
9.70	.0027	5.00	.0179	1.83	.2152	1.36	.3790	0.89	.7407	0.43	1.825
9.60	.0027	4.90	.0189	1.82	.2176	1.35	.3840	0.88	.7527	0.42	1.874
9.50	.0028	4.80	.0200	1.81	.2201	1.34	.3890	0.87	.7649	0.41	1.924
9.40	.0029	4.70	.0212	1.80	.2226	1.33	.3942	0.86	.7773	0.40	1.978
9.30	.0030	4.60	.0225	1.79	.2251	1.32	.3994	0.85	.7901	0.39	2.034
9.20	.0031	4.50	.0239	1.78	.2277	1.31	.4047	0.84	.8031	0.38	2.092
9.10	.0032	4.40	.0254	1.77	.2303	1.30	.4100	0.83	.8165	0.37	2.154
9.00	.0033	4.30	.0271	1.76	.2330	1.29	.4155	0.82	.8302	0.36	2.219
8.90	.0034	4.20	.0288	1.75	.2356	1.28	.4211	0.81	.8442	0.35	2.287
8.80	.0035	4.10	.0308	1.74	.2384	1.27	.4267	0.80	.8585	0.34	2.360
8.70	.0037	4.00	.0329	1.73	.2411	1.26	.4324	0.79	0.873	0.33	2.437
8.60	.0038	3.90	.0352	1.72	.2439	1.25	.4384	0.78	0.888	0.32	2.518
8.50	.0039	3.80	.0377	1.71	.2468	1.24	.4443	0.77	0.904	0.31	2.605
8.40	.0040	3.70	.0405	1.70	.2497	1.23	.4504	0.76	0.920	0.30	2.697
8.30	.0042	3.60	.0435	1.69	.2526	1.22	.4566	0.75	0.936	0.29	2.795
8.20	.0043	3.50	.0468	1.68	.2556	1.21	.4628	0.74	0.952	0.28	2.900
8.10	.0045	3.40	.0505	1.67	.2586	1.20	.4692	0.73	0.969	0.27	3.013
8.00	.0047	3.30	.0545	1.66	.2617	1.19	.4757	0.72	0.987	0.26	3.135
7.90	.0048	3.20	.0590	1.65	.2648	1.18	.4823	0.71	1.005	0.25	3.265
7.80	.0050	3.10	.0640	1.64	.2679	1.17	.4891	0.70	1.023	0.24	3.407
7.70	.0052	3.00	.0694	1.63	.2711	1.16	.4959	0.69	1.042	0.23	3.560
7.60	.0054	2.90	.0756	1.62	.2744	1.15	.5029	0.68	1.062	0.22	3.728
7.50	.0056	2.80	.0824	1.61	.2777	1.14	.5100	0.67	1.082	0.21	3.911
7.40	.0058	2.70	.0900	1.60	.2811	1.13	.5173	0.66	1.103	0.20	4.112
7.30	.0061	2.60	.0985	1.59	.2844	1.12	.5247	0.65	1.124	.195	4.220
7.20	.0063	2.50	.1081	1.58	.2879	1.11	.5322	0.64	1.146	.190	4.334
7.10	.0065	2.40	.1189	1.57	.2914	1.10	.5398	0.63	1.168	.185	4.454
7.00	.0069	2.30	.1311	1.56	.2949	1.09	.5476	0.62	1.191	.180	4.580
6.90	.0072	2.20	.1449	1.55	.2985	1.08	.5556	0.61	1.216	.175	4.714
6.80	.0075	2.10	.1607	1.54	.3022	1.07	.5637	0.60	1.240	.170	4.855
6.70	.0078	2.00	.1786	1.53	.3059	1.06	.5719	0.59	1.266	.165	5.005
6.60	.0081	1.99	.1805	1.52	.3097	1.05	.5804	0.58	1.292	.160	5.164
6.50	.0085	1.98	.1824	1.51	.3136	1.04	.5889	0.57	1.319	.155	5.334
6.40	.0089	1.97	.1844	1.50	.3175	1.03	.5977	0.56	1.347	.150	5.514
6.30	.0093	1.96	.1864	1.49	.3214	1.02	.6066	0.55	1.376	.145	5.707
6.20	.0097	1.95	.1885	1.48	.3245	1.01	.6157	0.54	1.407	.140	5.914
6.10	.0102	1.94	.1905	1.47	.3295	1.00	.6250	0.53	1.438	.135	6.136
6.00	.0107	1.93	.1926	1.46	.3337	0.99	.6344	0.52	1.470	.130	6.374
5.90	.0112	1.92	.1947	1.45	.3379	0.98	.6441	0.51	1.504	.125	6.632
5.80	.0118	1.91	.1969	1.44	.3422	0.97	.6540	0.50	1.539	.120	6.911
5.70	.0124	1.90	.1991	1.43	.3465	0.96	.6641	0.49	1.575	.115	7.215
5.60	.0130	1.89	.2013	1.42	.3510	0.95	.6743	0.48	1.612	.110	7.545
5.50	.0137	1.88	.2035	1.41	.3555	0.94	.6848	0.47	1.651	.105	7.908
5.40	.0144	1.87	.2058	1.40	.3600	0.93	.6955	0.46	1.692	.100	8.306
5.30	.0152	1.86	.2081	1.39	.3647	0.92	.7065				

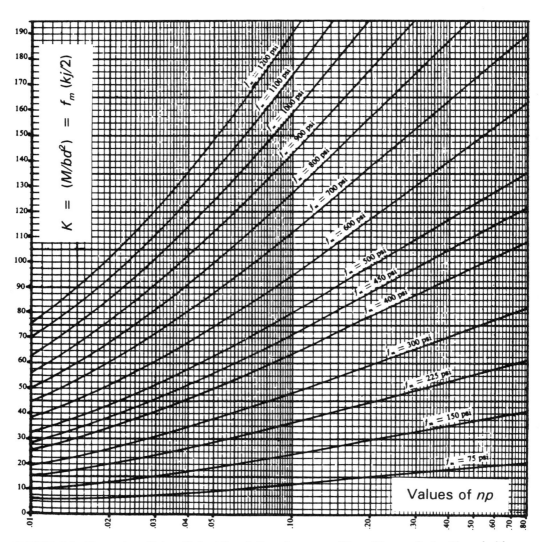

FIGURE C.1 Flexural coefficient K chart for reinforced masonry. (From *Masonry Design Manual* with permission of the publishers; Masonry Institute of America.)

= 333 psi, read at the bottom $np = 0.073$. Then

$$p = \frac{0.073}{n} = \frac{0.073}{40} = 0.001825$$

$$A_s = pbd = 0.001825 \times 12 \times 3.813$$
$$= 0.0835 \text{ in.}^2/\text{ft}$$

Try No. 5 at 40 in.:

$$A_s = \frac{12}{40} \times 0.31 = 0.093 \text{ in.}^2/\text{ft}$$

Check f_s using approximate $j = 0.90$:

$$f_s = \frac{M}{A_s jd} = \frac{8367}{(0.093)(0.9)(3.813)}$$
$$= 26{,}217 \text{ psi}$$

The reinforcing is adequate unless a combined stress must be investigated.

WEIGHT OF BUILDING CONSTRUCTION

This appendix contains material that will assist in the determination of the weight of various parts of the building as is required in the process of finding the lateral inertial force. Table D.1 contains data for the determination of the weight of specific materials where a unique assemblage must be evaluated. Table D.2 presents the average weights of commonly used elements of construction. For simpler and faster approximations of weights in preliminary calculations, Table D.3 presents average weights of ordinary total assemblages of typical construction.

TABLE D.1   WEIGHT OF MATERIALS

Material	Weight (lb/ft^3)	Specific Gravity[a]
Soil		
Top soil		1/21
Dry, loose	75	1.61
Moist, packed	100	
Sand, gravel		1.53
Dry, loose	90–100	1.85
Moist, packed	110–120	
Clay		1.05
Dry		1.69
Wet		
Wood		
Cedar	22	0.35
Fir, northern	35	0.56
Fir, white	25	0.40
Hemlock	30	0.48
Oak, red	40	0.65
Pine, white	25	0.40
Pine, yellow, long-leaf	45	0.73
Poplar	30	0.48
Redwood	25	0.40
Spruce	25	0.40
Metal		
Aluminum	165	2.66
Copper	556	8.97
Lead	710	11.45
Steel	490	7.90
Masonry		
Brick	90	1.45
Finished, pressed	65	1.05
Common, soft	85	1.37
Concrete block	140	2.26
Normal weight, voids empty		
Lightweight, voids empty		
Lightweight, voids 25% filled		
Lightweight, voids 100% filled		
Concrete		
Normal, gravel	145	2.34
(with average reinforcing)	150	2.42
Lightweight		
Structural	110	1.77
Masonry units	90	1.45
Insulating	30–40	0.56
Miscellaneous		
Asphalt paving	100	1.61
Glass	160	2.58

TABLE D.1 *(Continued)*

Material	Weight (lb/ft^3)	Specific Gravity[a]
Plaster		
Cement, stucco	120	1.94
Gypsum	40	0.65
Water	62	1.00

[a] Average weight divided by weight of water at 62 lb/ft^3.

TABLE D.2 WEIGHT OF ELEMENTS OF BUILDING CONSTRUCTION

	lb/ft^2	kN/m^2
Roofs		
3-ply ready roofing (roll, composition)	1	0.05
3-ply felt and gravel	5.5	0.26
5-ply felt and gravel	6.5	0.31
Shingles		
Wood	2	0.10
Asphalt	2–3	0.10–0.15
Clay tile	9–12	0.43–0.58
Concrete tile	8–12	0.38–0.58
Slate, $\frac{1}{4}$ in.	10	0.48
Fiber glass	2–3	0.10–0.15
Aluminum	1	0.05
Steel	2	0.10
Insulation		
Fiber glass batts	0.5	0.025
Rigid foam plastic	1.5	0.075
Foamed concrete, mineral aggregate	2.5/in.	0.0047/mm
Wood rafters		
2 × 6 at 24 in.	1.0	0.05
2 × 8 at 24 in.	1.4	0.07
2 × 10 at 24 in.	1.7	0.08
2 × 12 at 24 in.	2.1	0.10
Steel deck, painted		
22 ga	1.6	0.08
20 ga	2.0	0.10
18 ga	2.6	0.13
Skylight		
Glass with steel frame	6–10	0.29–0.48
Plastic with aluminum frame	3–6	0.15–0.29
Plywood or softwood board sheathing	3.0/in.	0.0057/mm
Ceilings		
Suspended steel channels	1	0.05
Lath		
Steel mesh	0.5	0.025
Gypsum board, $\frac{1}{2}$ in.	2	0.10
Fiber tile	1	0.05
Drywall, gypsum board, $\frac{1}{2}$ in.	2.5	0.12

TABLE D.2 (*Continued*)

	lb/ft^2	kN/m^2
Plaster		
Gypsum, acoustic	5	0.24
Cement	8.5	0.41
Suspended lighting and air distribution	3	0.15
Systems, average		
Floors		
Hardwood, $\frac{1}{2}$ in.	2.5	0.12
Vinyl tile, $\frac{1}{8}$ in.	1.5	0.07
Asphalt mastic	12/in.	0.023/mm
Ceramic tile		
$\frac{3}{4}$ in.	10	0.48
Thin set	5	0.24
Fiberboard underlay, $\frac{5}{8}$ in.	3	0.15
Carpet and pad, average	3	0.15
Timber deck	2.5/in.	0.0047/mm
Steel deck, stone concrete fill, average	35–40	1.68–1.92
Concrete deck, stone aggregate	12.5/in.	0.024/mm
Wood joists		
2 × 8 at 16 in.	2.1	0.10
2 × 10 at 16 in.	2.6	0.13
2 × 12 at 16 in.	3.2	0.16
Lightweight concrete fill	8.0/in.	0.015/mm
Walls		
2 × 4 studs at 16 in., average	2	0.10
Steel studs at 16 in., average	4	0.20
Lath, plaster; see Ceilings		
Gypsum drywall, $\frac{5}{8}$ in. single	2.5	0.12
Stucco, $\frac{7}{8}$ in., or wire and paper or felt	10	0.48
Windows, average, glazing + frame		
Small pane, single glazing, wood or metal frame	5	0.24
Large pane, single glazing, wood or metal frame	8	0.38
Increase for double glazing	2–3	0.10–0.15
Curtain walls, manufactured units	10–15	0.48–0.72
Brick veneer		
4 in., mortar joints	40	1.92
$\frac{1}{2}$ in., mastic	10	0.48
Concrete block		
Lightweight, unreinforced		
4 in.	20	0.96
6 in.	25	1.20
8 in.	30	1.44
Heavy, reinforced, grouted		
6 in.	45	2.15
8 in.	60	2.87
12 in.	85	4.07

TABLE D.3 AVERAGE VALUES OF TOTAL DEAD WEIGHT FOR TYPICAL BUILDING CONSTRUCTION

Description of Construction	Weight[a] (lb/ft^2)
Roofs	
1. Light wood frame, shingles, no ceiling	5
2. Light wood frame, 3-ply felt + gravel, gypsum drywall ceiling	12
3. Medium span wood or steel trusses, plywood deck, 3-ply felt + gravel, no ceiling	15
4. Steel open-web joists, metal deck, lightweight concrete insulating fill, 3-ply felt + gravel, suspended drywall ceiling	18–25
5. Glue-laminated wood girders, timber purlins, plywood deck, clay or concrete tile shingles, no ceiling	25–30
6. Steel beams, otherwise same as item 5	22–30
7. Precast concrete plank, 3-ply felt + gravel, 2-in. concrete fill, no ceiling	22–30
6-in.-thick units	50
8-in.-thick units	65
10-in.-thick units	80
8. Precast concrete double-tees, 2-in. concrete fill, 3-ply felt + gravel, no ceiling	80–100
9. Cast-in-place concrete slab and beams, insulating concrete fill, 3-ply felt + gravel, suspended drywall ceiling	100–125
Floors	
10. Wood joist, plywood deck, carpet, drywall ceiling	10
11. Steel- or wood-trussed joists, lplywood deck, 2-in. concrete fill, carpet, drywall ceiling	30
12. Steel beam, steel deck, concrete fill, suspended lay-in the ceiling	60–70
13. Precast concrete plank, 2-in. concrete fill, carpet, suspended drywall ceiling	
6-in-thick units	50
8-in.-thick units	65
10-in.-thick units	80
14. Poured-in-place concrete slab and beams, 2-in. concrete fill, suspended lay-in tile ceiling	100–125
Walls	
15. Exterior, wood studs, stucco, drywall	15–20
16. Exterior, wood studs, wood siding, drywall	10–15
17. Exterior, brick veneer, wood studs, plywood, drywall (real bricks)	60
18. Exterior, fake brick (thin tile), plywood, wood studs, drywall	25
19. Exterior, reinforced concrete block, 8-in. units, 25% voids filled, average reinforcing, insulation, furred-out drywall	60
20. Steel frame, aluminum + glass windows (curtain wall), 25% windows—small size	25
21. Windows, wood or metal frame, small panes of glass (5–10 ft^2/window)	5
22. Interior, wood stud, drywall both sides	8
23. Interior, wood stud, cement plaster both sides	20

[a]Average weight per square foot of surface.

STUDY AIDS

In this section we provide the reader with some means to measure his or her comprehension and skill development with regard to the book presentations. When the study of a chapter is completed, the reader should use the materials in this section to find out what has been accomplished. The Glossary and Index can be used to find definitions of the words and terms. Answers to the general questions are given at the end of this section.

WORDS AND TERMS

For each chapter indicated, review the meaning and significance of the following words and terms.

Chapter 1

Basic wind speed (velocity)
Clean-off effect

Design wind pressure
Drift
Exposure condition
Gust
Harmonic effects (of wind)
Negative pressure (of wind)
Normal force method (for wind load)
Overturn
Projected area method (for wind load)
Torsion on building
Uplift
Wind stagnation pressure

Chapter 2

Base
Base isolation
Base shear
Coupling
Decoupling
Epicenter
Fundamental period (T)
Importance factor (I)

Irregularity
Lateral bracing system
Mass
Regularity
Response factor (R_W)
Seismic separation
Site coefficient (S)
Spectrum response
Zone factor (Z) for probability of seismic activity

Chapter 3

Aspect ratio
Eccentric bracing
Lateral resistive systems: Box (shear wall), braced frame (trussed, rigid frame (moment-resistive), self-stabilizing
Load sharing
Positive connection (anchorage, etc.)
Propagation of loads
Seismic separation
Site–structure interaction
Unreinforced masonry

Chapter 4

Battering
Box system
Dual bracing system
Multimassed building
Plan irregularity
Setback
Soft story
Tag-along element
Three-sided building
Vertical irregularity
Vulnerable element
Weak story

Chapter 5

Braced frame
Eccentric bracing
K-bracing
Overturn
Peripheral (perimeter) bracing

Rigid frame (moment-resistive)
Shear wall
V-bracing
X-bracing

Chapter 6

Collector
Diaphragm: horizontal, vertical
Diaphragm chord
Horizontal anchor
Separation joint
Tiedown (hold-down)

Chapter 8

Battered pile
Cracked section
Equivalent eccentricity (of compressive force)
Equivalent fluid pressure
Kern limit
Passive soil pressure
Pole
Pole structure: pole-platform, pole-frame
Pressure wedge method
Shear key

Appendix A

Equivalent static effect
Harmonic motion: amplitude, cycle, period, frequency, damping, resonance
Kinematics: displacement, velocity, acceleration
Kinetics: force, mass, work, energy, momentum
Motion: translation, rotation, of a rigid body, of a deformable body

GENERAL QUESTIONS

Chapter 1

1. What general relation exists between the magnitude of wind velocity (speed)

and the resulting pressure effect on a stationary object?

2. Why are wind pressures greater on the upper portions of tall buildings?

3. What effect is caused on the building surface by negative wind pressure?

4. Why is building dead load usually considered as a positive factor in design for wind resistance?

5. How do nonstructural elements of the building construction sometimes make unintended contributions to the bracing of the building against wind?

Chapter 2

1. Why are the horizontal components of ground motion generally most destructive to buildings?

2. What basic dynamic properties of a building critically affect the building's response to an earthquake in terms of the magnitude of the force exerted on the building?

3. What generally qualifies a building as "regular" for seismic response, according to the *UBC*?

4. Why is the period of the site of concern for the seismic response of a building?

Chapter 3

1. When wind force acts on a building, various elements of the building construction typically perform functions in the development of the building's response. Briefly describe the usual role of the following elements in this regard.

   a. The building exterior walls facing the wind

   b. The building exterior walls parallel to the wind direction

c. Roofs and upper-level floors

d. The building foundations

2. How does wind or seismic force usually generate a torsional (twisting, rotational) effect on a building?

3. What basic structural mechanisms (actions, forms of response) are developed by the following bracing systems in the resisting of lateral loads?

   a. Box system (shear walls horizontal diaphragm)

   b. Braced (trussed) frame

   c. Rigid (moment-resistive) frame

4. Ordinary light wood frame construction (2 × 4 studs, etc.), as developed over a long period of use, may have some inherent deficiencies with regard to the resistance of seismic forces. Describe some of these potential weaknesses.

5. Why are lateral (horizontal) effects, rather than vertical effects, of wind and earthquakes generally more critical for building design?

6. What form of masonry construction is least desirable for resistance to seismic forces?

7. What inherent capacity for stability possessed by cast-in-place concrete frame construction is lost when precasting is employed?

Chapter 4

1. Architectural design decisions greatly affect the determination of the character of response of a building to lateral loads. Describe some design features of buildings that tend to complicate or make difficult the design for lateral load response.

   a. For wind

   b. For seismic effects

2. What characterizes the structural na-

ture of the fourth side in a so-called *three-sided building*?

3. What makes a so-called *soft story* soft?

4. What is the difference between a soft story and a weak story?

Chapter 5

1. Of what basic structural elements are box systems usually formed?

2. Why is a braced (trussed) frame considered to be a relatively stiff lateral bracing system?

3. What actions in the structure are added to the usual truss actions when knee-bracing or K-bracing is used?

4. Why is the term *rigid* inappropriate in describing the so-called rigid frame bracing system for lateral loads?

Chapter 6

1. What basic function does a chord perform for a horizontal diaphragm?

2. Other than base anchorage, what resists overturn in a shear wall?

3. What is the significance of nail spacing at the edges of plywood panels in a horizontal diaphragm?

4. What does a seismic separation joint separate? Why?

5. What is generally required of a structural connection to qualify it as positive?

6. When used in conjunction with a horizontal diaphragm, what does a collector usually collect?

Chapter 7

1. How do the following affect the choice for the lateral resistive structure?

a. General nature of the type of lateral resistive system

b. General building form

c. Specific materials of the general building construction

d. Building code requirements

2. How is it possible that an unsymmetrical building may have a torsion-free response to lateral loads?

3. Why must the structural responses of nonstructural elements of the building construction be considered in design of the structure?

Chapter 8

1. What two soil stress mechanisms usually combine to resist the horizontal movement of foundations?

2. How does surcharge affect the development of resistance to lateral movement by a foundation?

3. With regard to vertical soil pressure, what is significant about a loading condition on a footing that results in the resultant force being outside the kern limit for the footing?

4. What two basic properties of the pressure wedge permit the static analysis for soil pressure on the cracked section?

5. How is vertical uplift force resisted by deep foundation elements?

Appendix A

1. What generally distinguishes the difference between the fields of kinematics and kinetics?

2. How do the means of measurement differentiate motion of translation from motion of rotation?

3. How are work and energy related?

4. What is the usual reason for using the equivalent static force method for the

investigation of dynamic actions on structures?

ANSWERS TO THE GENERAL QUESTIONS

Chapter 1

1. Magnitude of pressure is proportional to the square of wind velocity.
2. Wind velocity increases with distance above the ground, as ground surface drag effect decreases and effects of sheltering by trees, hills, and other buildings is less significant.
3. Outward, or suction, force.
4. It provides a natural anchor against effects of sliding, uplift, and overturning.
5. By their own inherent stiffness and their attachment to the lateral force-resisting system.

Chapter 2

1. Two reasons: (1) the horizontal ground accelerations are usually larger than the vertical; (2) buildings are not ordinarily designed for horizontal force but rather, for vertical gravity force.
2. Mass (weight), fundamental period of harmonic vibration, any presence of damping sources for harmonic motion, potential for resonance in the building structure itself and for any resonant building–site interaction, potential for progressive deterioration of structural resistance (lack of positive connections, etc.).
3. The absence of significant discontinuities, abrupt changes, or other irregularities in the lateral-force-resisting structure or the building mass.

4. Since the building and site are coupled, the potential exists for destructive interaction.

Chapter 3

1. a. Walls receive wind pressure directly, span between other parts of the structure (usually roof and floors), and transfer the lateral forces to those supporting elements (i.e., to the edge of the roof, etc.).
   b. Walls may act as shear walls, if developed to do so. If not intended as shear walls, they may be structurally isolated to prevent damage to them.
   c. See answer a. Normally function as horizontal diaphragms.
   d. Anchor the building against sliding and overturn and transfer the lateral force to the soil.
2. When the resultant of the horizontal force does not coincide with the center (centroid, etc.) of the lateral bracing system.
3. a. Resistance to deformations of shear and bending in the plane of the surface elements; walls as shear walls; roofs and floors as horizontal diaphragms.
   b. Connected columns, beams, and diagonal members form vertical, planar, trussed bents.
   c. Connected columns and beams form vertical, planar bents, developing rigid frame actions (shear and bending in columns and beams; moments transferred through connections).
4. Anchorage of elements (walls to foundations, roof to walls, etc.) may not be adequate. Structural framing may not be strong enough or attached sufficiently to develop diaphragm ac-

tions. Framing may not have the necessary continuity from piece to piece to function as chords, ties, or collectors. Wall, roof, and floor surfacing may not provide diaphragm actions.

5. Buildings are routinely (normally) developed in response to gravity (vertical) loads.

6. Unreinforced masonry made with low-quality mortar, soft, weak masonry units, and little or no steel reinforcement.

7. Capacity for rigid frame actions, due to the natural member-to-member continuity caused by the continuous casting and the extension of reinforcement through the cast joints.

Chapter 4

1. a. Really bad aerodynamic form in general; long overhangs; high parapets; deep insets or other open spots that cup the wind; highly flexible (thin, soft, etc.) surface elements that can vibrate in the wind; flexible elements coupled with stiff ones and not adequately attached.

   b. Unsymmetrical plans; abrupt changes in form (multimassed buildings, setbacks, etc.); use of rigid but nonstructural forms of construction (plastered ceilings, masonry veneers, etc.); creation of forms that develop soft stories; use of heavy materials in upper levels (heavy roof tiles, concrete roof structures, etc.).

2. Inability to develop lateral resistance in the plane of the wall.

3. Relative lateral stiffness significantly less than that of other stories that are above or below.

4. "Soft" refers to excessive difference in stiffness (resistance to lateral deflection); "weak" refers to excessive difference in strength, meaning resistance to force (static conditions) or energy capacity (dynamic conditions).

Chapter 5

1. Vertical, planar walls and horizontal, planar roof and floor decks.

2. Principal deformation of the frame is due to lengthening and shortening of the frame members; normally very small dimensions. Trussed frames are thus in most cases much stiffer than rigid frames, in which deformations are mostly due to bending in the members. Rigid frames can be made stiff only by using very short or very stiff members.

3. Essentially, rigid frame actions involving the bending of members.

4. See the answer to Question 2.

Chapter 6

1. Development of tension and compression for resistance to the bending created by the spanning action of the diaphragm.

2. The dead weight of the wall and of any construction it supports or is attached to.

3. It indicates potential for transfer of diaphragm shear between the plywood panels.

4. Separate units of the building. When their interaction or coupled response is potentially harmful.

5. Lack of potential for loosening or fracture due to dynamic actions. Most critically, the repeated actions of seismic vibrations or wind flutter and rocking.

6. Shear in the diaphragm, for transfer to vertical bracing.

Chapter 7

1. These are enumerated in Sec. 7.1.
2. Torsion due to lateral loads involves the relationship between the building mass and the centroid of the lateral resistive structure. It is theoretically possible (but not easy to achieve) to have a structure that relates to the unsymmetrical building mass to eliminate torsion.
3. If attached securely to the structure, the nonstructural elements (walls especially) can significantly modify the dynamic response of the structure. Even if not securely attached, they can be a major source of damping or resonance.

Chapter 8

1. Soil friction and passive lateral pressure.
2. Added weight on footings may develop more soil friction (especially in sandy soils). Passive pressure is usually proportional to the depth below the ground surface; thus higher lateral pressures are possible with deeper footings.

3. A "cracked section" is produced, with no vertical contact pressure on some part of the footing-to-soil interface surface.
4. Volume of the wedge is equal to the total vertical load on the soil. The centroid of the wedge is coincident with the resultant load (at its actual or equivalent eccentricity).
5. By skin friction on the deep foundation elements (mostly on piles) and the dead weight of the elements (mostly significant for concrete piles or piers).

Appendix A

1. Kinematics involves only motion with time (variables of geometry and time). Kinetics adds considerations for mass and force (involving work and energy).
2. Translation is measured in length change (inches, etc.); rotation is measured in angular change.
3. Work requires and uses up energy; energy can be measured as capacity for doing work.
4. Basically, it is usually much easier; at least for hand computations. It is also easier to visualize combined effects; gravity + wind, and so on.

BIBLIOGRAPHY

The following list contains materials that have been used as references in the development of various portions of the text. Also included are some widely used publications that serve as general references for seismic design and for general design of building structures, although no direct use of materials from them has been made in this book. The numbering system is random and merely serves to simplify referencing by text notation.

EARTHQUAKES AND BUILDINGS

1. *Uniform Building Code*, Vol. 2: *Structural Engineering Provisions*, 1994 ed., International Conference of Building Officials, 5360 South Workmanmill Road, Whittier, CA 90601 (called simply the *UBC*).
2. *Minimum Design Loads for Buildings and Other Structures* (ANSI/ASCE 7-88), American Society of Civil Engineers (ASCE), New York, 1990 (revision of ANSI A58.1-1982).
3. *Recommended Lateral Force Requirements and Commentary*, Seismology Committee, Structural Engineers Association of California (SEACA), 1990 (known simply as the *SEACA Bluebook*).
4. C. Arnold and R. Reitherman, *Building Configuration and Seismic Design*, Wiley, New York, 1982.
5. Henry J. Lagorio, *Earthquakes: An Architect's Guide to Nonstructural Seismic Hazards*, Wiley, New York, 1990.
6. Ellis L. Krinitzsky, James P. Gould, and Peter H. Edinger, *Fundamentals of Earthquake Resistant Construction*, Wiley, New York, 1993.
7. *Introduction to Earthquake Retrofitting: Tools and Techniques*, 1994, Building Education Center, 812 Page Street, Berkeley, CA 94710.
8. *Strengthening Wood Frame Houses for Earthquake Safety*, Bay Area Regional Earthquake Preparedness Project, Office of

Emergency Services, State of California, Oakland, CA.

BUILDING CONSTRUCTION AND STRUCTURES

9. Charles G. Ramsey and Harold R. Sleeper, *Architectural Graphic Standards*, 9th ed., Wiley, New York, 1994.

10. Edward Allen, *Fundamentals of Building Construction: Materials and Methods*, 2nd ed., Wiley, New York, 1990.

11. *Manual of Steel Construction*, 8th ed., American Institute of Steel Construction (AISC), Chicago, 1980 (called simply the *AISC Manual*).

12. *National Design Specification for Wood Construction*, National Forest Products Association, Washington DC, 1991.

13. *Timber Construction Manual*, 3rd ed., American Institute of Timber Construction, Wiley, New York, 1985.

14. *Building Code Requirements for Reinforced Concrete*, ACI 318-89, American Concrete Institute, Detroit, MI, 1989 (called simply the *ACI Code*).

15. *CRSI Handbook*, Concrete Reinforcing Steel Institute, Schaumburg, IL, 1992.

16. *Masonry Design Manual*, 3rd ed., 1979. Masonry Institute of America, 2550 Beverly Boulevard, Los Angeles, CA 90057.

17. R. R. Schneider and W. L. Dickey, *Reinforced Masonry Design*, 3rd ed., Prentice Hall, Englewood Cliffs, NJ, 1987.

18. *Steel Deck Institute Design Manual for Composite Decks, Form Decks, and Roof Decks*, Steel Deck Institute, P.O. Box 3812, St. Louis, MO 63122.

19. Jack C. McCormac, *Structural Analysis*, 4th ed., Harper & Row, New York, 1984.

20. Stan W. Crawley and Robert M. Dillon, *Steel Buildings: Analysis and Design*, 4th ed., Wiley, New York, 1993.

21. Donald E. Breyer, *Design of Wood Structures*, 3rd ed., McGraw-Hill, New York, 1993.

22. Phil M. Ferguson, *Reinforced Concrete Fundamentals*, 5th ed., Wiley, New York, 1988.

23. James Ambrose, *Simplified Design of Building Foundations*, 2nd ed., Wiley, New York, 1988.

INDEX

349

INDEX FOR DESIGN AND INVESTIGATION EXAMPLES

355